123539

Elke Hahn-Deinstrop

Applied Thin-Layer Chromatography

1807–2007 Knowledge for Generations

Each generation has its unique needs and aspirations. When Charles Wiley first opened his small printing shop in lower Manhattan in 1807, it was a generation of boundless potential searching for an identity. And we were there, helping to define a new American literary tradition. Over half a century later, in the midst of the Second Industrial Revolution, it was a generation focused on building the future. Once again, we were there, supplying the critical scientific, technical, and engineering knowledge that helped frame the world. Throughout the 20th Century, and into the new millennium, nations began to reach out beyond their own borders and a new international community was born. Wiley was there, expanding its operations around the world to enable a global exchange of ideas, opinions, and know-how.

For 200 years, Wiley has been an integral part of each generation's journey, enabling the flow of information and understanding necessary to meet their needs and fulfill their aspirations. Today, bold new technologies are changing the way we live and learn. Wiley will be there, providing you the must-have knowledge you need to imagine new worlds, new possibilities, and new opportunities.

Generations come and go, but you can always count on Wiley to provide you the knowledge you need, when and where you need it!

William J. Pesce
President and Chief Executive Officer

Peter Booth Wiley
Chairman of the Board

Elke Hahn-Deinstrop

Applied Thin-Layer Chromatography

Best Practice and Avoidance of Mistakes

Second, Revised and Enlarged Edition

Translated by R. G. Leach

WILEY-VCH Verlag GmbH & Co. KGaA

Elke Hahn-Deinstrop
Kleingeschaidter Str. 23
90542 Eckental
Germany
elke.hahn_deinstrop@arcor.de

1st Edition 2000
2nd Edition 2007
1st Reprint 2007

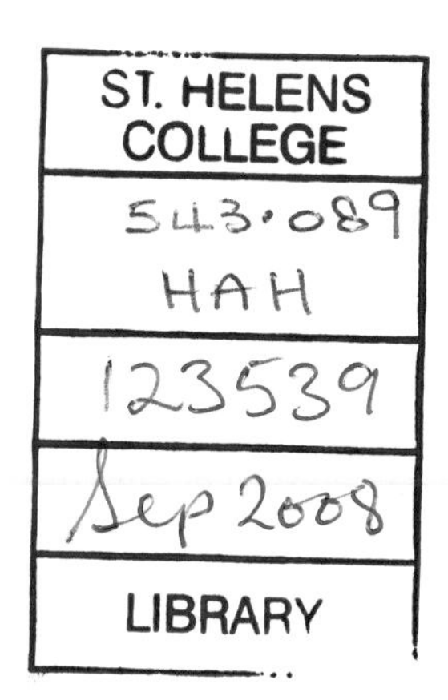

Library of Congress Card No.: Applied for
British Library Cataloging-in-Publication Data
A catalogue record for this book is available from the British Library

Bibliographic information published by the Deutsche Nationalbibliothek
The Deutsche Nationalbibliothek lists this publication in the Deutsche Nationalbibliografie; detailed bibliographic data are available in the Internet at http://dnb.d-nb.de.

Typesetting Mitterweger & Partner, Plankstadt
Printing betz-druck GmbH, Darmstadt
Bookbinding Litges & Dopf Buchbinderei GmbH, Heppenheim

Printed in the Federal Republic of Germany
Printed on acid-free paper

ISBN: 978-3-527-31553-6

To the Man by my Side

Preface to the Second English Edition

It has been almost seven years since the publication of the first English Edition of my book on TLC. The following improvements in technology over the years have made it necessary for me to update the first edition: new precoated layers for both existing and new fields of applications, a new generation of equipment for safe operations and reproducible results, new devices such as the Diode Array Detector and Bioluminescence Analyzer, new methods of interface between TLC and analysis methods, especially the use of digital cameras for the documentation of thin layer chromatograms. For the reader's benefit, I have updated my description of available products on the market.

I had a wealth of assistance and support, including many telephone exchanges within Germany to Hamburg, Berlin, Stuttgart, Darmstadt and Offenburg as well as the exchanging of many files via e-mail to and from Muttenz (Switzerland) and Houston (Texas, USA).

During the last few years I held a series of lectures on chromatography and partook in TLC workshops at high schools and universities within Germany. In attending those events, it was reinforced to me how important sound and comprehensive knowledge of TLC is, in particular for recognizing and avoiding errors.

If this book can contribute to confer my 40 years long enthusiasm for thin layer chromatography to the reader, then the energy expended was worthwhile and I take leave in my retirement.

Eckental, September 2006

Elke Hahn-Deinstrop

Preface to the First English Edition

Shortly after the announcements in the scientific press in early 1998 of the publication of the German edition of my book on TLC, I received many enquiries from outside Germany asking when the English version would be available.

The decision by Wiley-VCH to publish an English edition in 1999 was the start of many hectic months for me. To produce the present book, not only had the text, references and market overview to be updated, but also two more Sections describing new equipment had to be included. Documents 4–13 were revised and Tables 21a and 26 were added. The Sections on video documentation were also extensively revised to take account of technical advances in this area.

Numerous discussions by telephone and fax have helped the translator, Mr. R. G. Leach, to import a flavor of my personal writing style into the English edition. The main aim is to prevent fatigue and to inspire the reader to read on. Also, as a small "extra", my ideas for two new cartoons have been excellently translated into actual drawings by Norbert Barth.

I dedicate the second of these new cartoons to Dr. Angelika Koch, with whom I published several papers last year on the subject of the ancient remedy frankincense (olibanum), and who, in the course many conversations, gave me the strength to complete all the work for this book.

I hope that all my friends and colleagues, in nearby Europe and also in distant Japan, China, India and Australia, and all other TLC users worldwide will derive pleasure from reading my book and will have great success in their work with TLC!

Eckental, September 1999 *Elke Hahn-Deinstrop*

Preface to the First German Edition

During twenty-five years of practical experience with thin-layer chromatography (TLC), I have learned to appreciate the advantages of this method of analysis, especially its power, flexibility and cost-effectiveness. The aim of this book is to pass on to a new generation of analysts any useful knowledge and practical tips that I have accumulated during this time. It includes some descriptions, illustrated by cartoons, of amusing incidents in the everyday laboratory life of a second-year apprentice and a trainee pharmacist. I have already found these anecdotes to be useful teaching aids. If the cartoons seem to suggest that a new university graduate is likely to be less knowledgeable about TLC than a young girl trained in an industrial laboratory, this is quite deliberate. However, I hope that established practitioners will also be able to pick up some tips that may be useful in their everyday work.

Formalism is nowadays unavoidable even in the field of TLC, and this book consequently contains a great many descriptions of practical procedures. The author has nevertheless taken great care to describe these accurately and reliably and also to give copious hints on how to avoid mistakes.

The theme of TLC as an art form is also discussed in a short section in the Appendix.

A writer starts the day with a blank sheet of paper and is happy with any prose or poetry successfully written on it. In TLC we start with a white plate. If the chromatograms we eventually obtain fulfill their intended purpose, we have had a successful day.

Eckental, October 1997 *Elke Hahn-Deinstrop*

Contents

List of Tables

1 Introduction

Thin-layer chromatography (TLC) is a very old method of analysis that has been well proven in practice. For more than thirty years, it has occupied a prominent position, especially in qualitative investigations. With the development of modern precoated layers and the introduction of partially or completely automated equipment for the various stages of operation of TLC, not only are highly accurate quantitative determinations now possible, but also the requirement that the work should comply with the GMP/GLP guidelines can be fulfilled.

Following the widespread use of high-performance liquid chromatography (HPLC), the importance of TLC, mainly measured by the work rate of the method, has been forced into the background. This is reflected in the unfavorable treatment of TLC as taught in universities, higher technological teaching establishments, technical colleges and industry. In addition to this, the restructuring of the chemical industry begun some years ago and the consequent job losses have led to considerable loss of specialist know-how in the use of TLC.

For these reasons, it is hoped that the present book will point towards good practical methods of performing TLC. Special attention is paid to possible sources of error. Theoretical aspects are not placed in the foreground, but emphasis is rather placed on the current state of the technology and the scope of modern TLC. The arrangement of the book strictly follows the individual operating steps of TLC, so that the user will be able to locate these various steps with ease.

This book is mainly intended for the younger scientific generation. For teachers it tries to encourage a form of teaching close to practical "real-life" TLC analysis, and the many practical tips also offer invaluable support for the less experienced users in industrial and official laboratories. Last but not least, it can be used by the analyst in a pharmaceutical laboratory as a work of reference.

1.1 What Does TLC Mean?

Chromatography means a method of analysis in which a mobile phase passes over a stationary phase in such a way that a mixture of substances is separated into its components. The term "thin-layer chromatography", introduced by E. Stahl in 1956, means a chromatographic separation process in which the stationary phase consists of a thin layer applied to a solid substrate or "support" [1,2]. For some years, TLC has also been referred to as planar chromatography. However, apart from the fact that paper

Applied Thin-Layer Chromatography: Best Practice and Avoidance of Mistakes, 2nd Edition
Edited by Elke Hahn-Deinstrop

ISBN: 978-3-527-31553-6

chromatography, which is also a planar method, is now hardly used, I do not think that this term will ever be widely accepted because the abbreviation PC could easily be confused with the abbreviation for personal computer.

The term TLC is now used to describe the method in all its forms, including manual or semiautomatic operation on conventional, high-performance or modified layers.

1.2 When Is TLC Used?

An essential precondition is that the substances or mixtures of substances to be analyzed should be soluble in a solvent or mixture of solvents.

TLC is used if

- the substances are nonvolatile or of low volatility
- the substances are strongly polar, of medium polarity, nonpolar or ionic
- a large number of samples must be analyzed simultaneously, cost-effectively, and within a limited period of time
- the samples to be analyzed would damage or destroy the columns of LC (liquid chromatography) or GC (gas chromatography)
- the solvents used would attack the sorbents in LC column packings
- the substances in the material being analyzed cannot be detected by the methods of LC or GC or only with great difficulty
- after the chromatography, all the components of the sample have to be detectable (remain at the start or migrate with the front)
- the components of a mixture of substances after separation have to be detected individually or have to be subjected to various detection methods one after the other (e.g. in drug screening)
- no source of electricity is available

1.3 Where Is TLC Used?

Pharmaceuticals and Drugs
Identification, purity testing and determination of the concentration of active ingredients, auxiliary substances and preservatives in drugs and drug preparations, process control in synthetic manufacturing processes.

Clinical Chemistry, Forensic Chemistry and Biochemistry
Determination of active substances and their metabolites in biological matrices, diagnosis of metabolic disorders such as PKU (phenylketonuria), cystinuria and maple syrup disease in babies.

Cosmetology
Dye raw materials and end products, preservatives, surfactants, fatty acids, constituents of perfumes.

Food Analysis
Determination of pesticides and fungicides in drinking water, residues in vegetables, salads and meat, vitamins in soft drinks and margarine, banned additives in Germany (e.g. sandalwood extract in fish and meat products), compliance with limit values (e.g. polycyclic compounds in drinking water, aflatoxins in milk and milk products).

Environmental Analysis
Groundwater analysis, determination of pollutants from abandoned armaments in soils and surface waters, decomposition products from azo dyes used in textiles.

Analysis of Inorganic Substances
Determination of inorganic ions (metals).

Other Areas
Electrolytic technology (*meta*-nitrobenzoic acid in nickel plating baths).

A graphical representation of the distribution of TLC publications among the most important fields of application during the years 1993 and 1994 is given in Fig. 1 [3]. However, this diagram does not give any indication of the distribution of the actual use of the technique. Reliable information on this subject is difficult to obtain. Information on quantities of materials used for TLC must mainly come from the manufacturers, but they are unwilling to release this on grounds of industrial secrecy. Our own research in northern and southern Germany has revealed that 40 % of the precoated layers go to universities and other higher educational establishments for use in the areas of pharmacy, medicine and biology, while a further ca. 40 % are used in the pharmaceutical industry, including use by pharmacists, and the remainder is divided between official investigative organizations (e.g. food monitoring, police and customs) and private institutions. This leads us to conclude that the majority of TLC users work in the area of pharmaceutical investigations. Recent polls confirm this distribution.

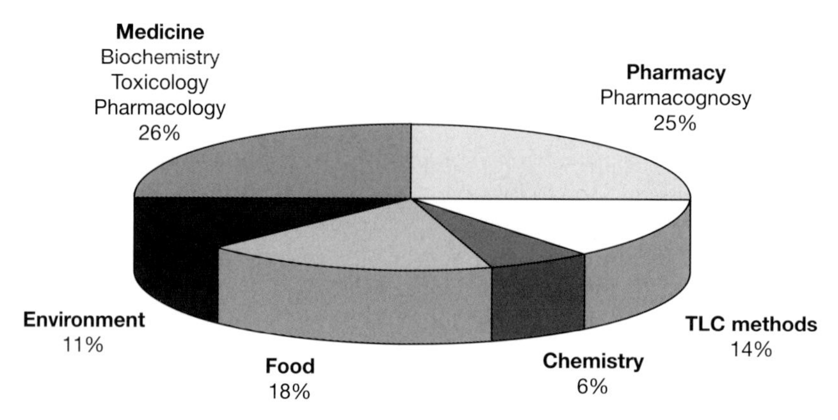

Figure 1. Fields of application of thin-layer chromatography (TLC/HPTLC) over the years 1993–1994

1.4 How is the Result of a TLC Represented?

Please do not expect a profound treatment of chromatographic parameters at this point. As beginners in TLC you should not be frightened off at the very beginning of this book. Any reader interested in the theory of TLC should read books devoted to this subject, the two by Geiss being especially recommended [4, 5].

The subject of TLC has its own special parameters and concepts, the most important of these for practical purposes being described below.

1.4.1 Retardation Factor

The position of a substance zone (spot) in a thin-layer chromatogram can be described with the aid of the retardation factor $\boldsymbol{R_f}$. This is defined as the quotient obtained by dividing the distance between the substance zone and the starting line by the distance between the solvent front and the starting line (see Fig. 3):

$$R_f = \frac{Z_S}{Z_F - Z_0}$$

where

R_f = retardation factor
Z_S = distance of the substance zone from the starting line [mm]
Z_F = distance of the solvent front from the solvent liquid level [mm]
Z_0 = distance between the solvent liquid level and the starting line [mm]

From this formula, one obtains an "observed" R_f value, which describes the position of a spot in the chromatogram in a simple numerical way. It gives no information about the chromatographic process used or under what other "boundary conditions" this result was obtained. This calculated R_f is always ≤ 1. As it has been found to be inconvenient in routine laboratory work always to write a zero and a decimal point, the R_f value is multiplied by 100, referred to as the **h**$\mathbf{R_f}$ value,[1] quoted as a whole number, and used for the qualitative description of thin-layer chromatograms.

In the calculation of hRf values as described in the literature, the distance Z_S is measured from the starting line to the mid-point of the substance zone. In general, this is correct and is also accurate enough for small spots. However, in purity tests on pharmaceutical materials, amounts of substance up to and even exceeding 1000 µg/spot are used, and this can lead to hRf value ranges up to ca. 18. If, in addition, limit-value amounts of at least 0.1 % of the same substance are applied and chromatographed on the same plate, these ideally lie exactly in the calculated central point of the main spot. However, this does not always happen. They are more likely to deviate from this position and be distributed over the whole hRf value range. Here, the term **"hRf value range"** means the imaginary hRf value range from the beginning to the end of a substance spot. In Fig. 2a–c, the chromatogram of purity tests of three active substances are given in which the position of the small amount of substance is respectively at the top end, approximately in the center, and at the bottom end of the hRf range.

▶ **Figure 2:** see Photograph Section.

Practical Tip for calculation of the hRf values:

- In purity tests, always quote hRf values as a range extending from the beginning to the end of a substance spot.

Figure 3 gives a graphical representation of the parameters and terms used in this book to describe a thin-layer chromatogram. Explanations of other terms are given in Section 1.4.3.

Because of the often poor reproducibility, especially when TLC plates prepared in-house are used and the conditions necessary for a good chromatographic result are in consequence not complied with, the so-called R_{St} value, based on a standard substance, was formerly often also given. This is defined as

$$R_{St} = \frac{Z_S}{Z_{St}}$$

where

Z_S = distance from the substance zone to the starting line [mm]
Z_{St} = distance of the standard substance from the starting line [mm]

[1] Because of the formatting difficulties associated with subscripts in the computer age, the term **hR**f value has become established and is used throughout this book.

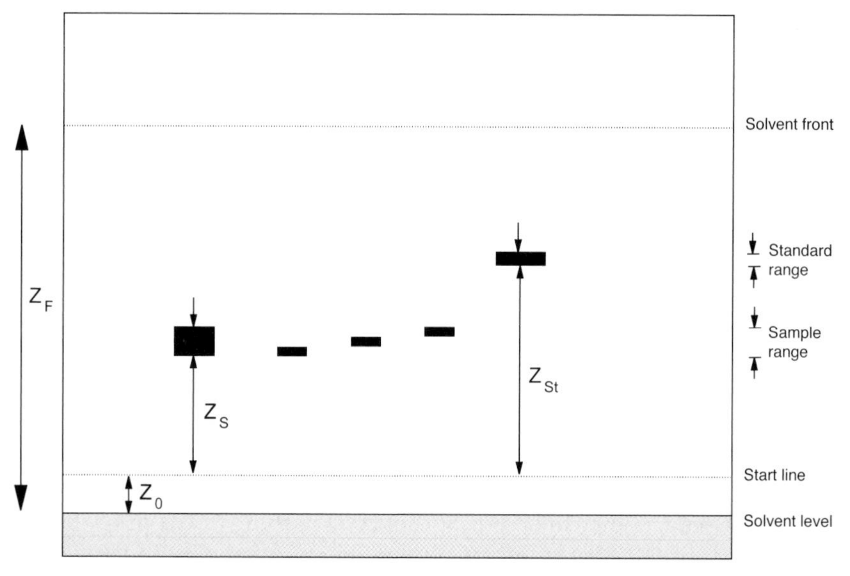

Figure 3. Terms used to describe a thin-layer chromatogram

According to Geiss [4, p. 65], it is not a good principle to quote R_{St} values as they are practically worthless and only give the appearance of certainty. In pharmacopoeias also, the still common linking of samples to standard substances with known R_f values has been shown to be of doubtful value as routine laboratory practice. Therefore, only the hRf value is used to evaluate results in this book.

1.4.2 Flow Constant

The flow constant or velocity constant (κ) is a measure of the migration rate of the solvent front. It is an important parameter for TLC users and can be used to calculate, for example, development times with different separation distances, provided that the sorbent, solvent system, chamber type and temperature remain constant. The flow constant is given by the following equation:

$$\kappa = \frac{Z_F^2}{t}$$

where

κ = flow constant [mm^2/s]
Z_F = distance between the solvent front and the solvent level [mm]
t = development time [s]

The following example illustrates the usefulness of the flow constant in laboratory work. In a TLC, if the development time for a migration distance of 10 cm was 30 min and the Z_0 distance is 5 mm, the κ value is 6.125 mm^2/s.

Question: How much time is required to develop a 15-cm migration distance if the sorbent, solvent system, Z_0 and laboratory temperature remain constant?

Answer: 65.4 min.

This means that more than twice the development time is required for a migration distance which is only 5 cm longer!

It should be mentioned here that the flow constant is influenced by other effects also, e.g. the surface tension and viscosity of the solvent system. In general, the greater the viscosity and the smaller the surface tension of the solvent system, the smaller is the migration rate of the front.

1.4.3 Other TLC Parameters

In the TLC literature, different terms are often used for the same characteristic values and parameters. As this can lead to confusion, especially for beginners, the most commonly used terms are listed below, those used in this book being in bold type.

- **Solvent system** — Developing solvent, mobile phase, eluent (only used in OPLC)
- **Migration distance** — Run distance, run height, developing distance
- **Developing time** — Run time
- **Derivatization reagent** — Detection reagent

Other terms commonly used in TLC are:

- **Fluorescence quenching.** If a TLC plate has a layer which contains a fluorescence indicator, UV-active substances cause the fluorescence to be totally or partially extinguished and can be seen as dark spots on a bright background (see also Section 2.2.3 "Additives").
- **Separation efficiency** describes the spread of the spots caused by chromatographic effects in the chosen system.
- **System suitability.** This is an expression used in the German Pharmacopoeia (Deutsche Arzneibuch, DAB), and describes a method of testing a system whereby two or more substances have to be separated from each other on a TLC plate prepared in-house, in order to establish whether samples under investigation can in fact be analyzed using the system.
- **Selectivity** describes the varying strengths of the interactions between the sample substances to be separated and the stationary phase (ΔhRf) in the chosen TLC system.

1.5 What Kinds of Reference Substances Are Used in TLC?

Because of their great importance to TLC, the various types of reference substances are described in the following Section. These are often known as "standards" and must only be used if they are of suitable quality for the intended application. These levels of quality are of especial importance in the field of pharmacy. All the relevant requirements must therefore be controlled in an SOP (standard operating procedure, see Chapter 9 "GMP/GLP-Conforming Operations in TLC").

1. Pharmacopoeia Substance (PS)

This term indicates a commercially available substance that meets the requirements of the relevant pharmacopoeia. For example, the American pharmacopoeia is indicated by the suffix **USP**, the British by **BP**, and the European by **CRS**. The possible use of a **PS** is specified by the relevant institutions, and is terminated by a change in the LOT number in the suppliers' catalogs. The Commission of the USP lists so-called "official distributors", of which the company LGC Promochem is a member (see Section 12.5 "Market Overview"). Care must be taken when ordering a substance listed in a pharmacopoeia to use the precise term for the substance. Although it is extremely rare, it does happen that the related compounds (rel. c.) of a substance have different names in the DAB and CRS lists.

It is especially confusing if, for example, the "rel. c. A" of a substance (e.g. ranitidine HCl) in the USP list appears as "rel. c. B" in the BP list, "rel. c. A" in the BP list bears the name "rel. c. C" in the USP list, and the "rel. c. B" in the USP list does not appear at all in the BP list.

2. Primary Reference Substance (PRS)

This term denotes a substance referred to as a Class 1 Standard by suppliers (analysis certificate with, e.g., at least two assays performed by different methods) or defined by the user's own tests without reference to other substances.

3. Secondary Reference Substance (SRS)

A tested and accepted batch of a substance which, after comparison with a **PS** or **PRS**, is declared as a "house standard". Can also be termed a **working standard**.

4. Related Compound (rel. c.)

This is usually a substance obtained from a supplier, but may be a substance produced by the user for purity testing, which is not a **PS** or a **PRS** and does not require information about its concentration. Such substances are in most cases decomposition products or intermediates in the synthesis process, and can be linked to a particular active substance.

5. Identity Substance (IS)

This is usually a commercially obtained substance with high chromatographic purity, although the content of the desired substance need not be known exactly if it is only to be used for identification purposes, e.g. for determining the hRf value and/or the documentation of the color or the fluorescence.

☞ **Practical hints** for ordering reference substances:

- In view of the existence of different descriptions of a substance in the lists in pharmacopoeias, it can be useful to make a note of the structural formula or to ask for help from an experienced person. Both names should then be written on a data base label indicating the relevant pharmacopoeias. Using the substance glibenclamid as an example:

  In the DAB literature, this is named

  5-Chlor-2-methoxy-N[2–4-sulfamoyl-phenyl]benzamid CRS,

  and in LGC Promochem's list "Reference Substances from the European Pharmacopoeia" it can be found under

  4-[2-(Chloro-2-methoxybenzamido)ethyl]-benzene-sulphonamide.

- A large number of reference substances that are constituents of plants can be obtained from the companies Phytolab and Carl Roth (see Section 12.5). When placing an order, the question of an analysis certificate should be always be raised.

1.6 The Literature on TLC

The first edition of the laboratory handbook "Dünnschicht-Chromatographie" edited by E. Stahl was published by Springer in 1962. Many subsequent publications can now be regarded as obsolete because of the developments in methodology and technique. I hope that the personal comments on the publications listed below will help the user in his or her choice.

1.6.1 General Literature

1.6.1.1 Books and Information Sheets in German

Wintermeyer, U.: **Die Wurzeln der Chromatographie**, GIT Verlag, Darmstadt 1989.
This book is recommended for anyone who would like to know how chromatography began. It goes up to the mid-1970s and also describes the development of CHROMart (the artistic use of chromatography).

Stahl. E. (ed): **Dünnschichtchromatographie,** Ein Laboratoriumshandbuch, 2nd edn, Springer-Verlag, Berlin 1967.
This book is popularly known as "thick Stahl" to distinguish it from the earlier thinner edition. It is a standard work, providing a good basic knowledge and including a large number of practical examples which are still valid, although modern stationary phases are not included.

Geiss, F.: **Die Parameter der Dünnschichtchromatographie**, eine moderne Einführung in Grundlagen und Praxis, Vieweg Verlag, Braunschweig 1972.
This is the most frequently cited work when the theory of TLC is under discussion: it is **the** treat for formula freaks.
English edition: **Fundamentals of Thin Layer Chromatography** (Planar Chromatography), Huethig Verlag, Heidelberg 1987, ISBN 3-7785-0854-7.

Kaiser, R.E. (ed): **Einführung in die Hochleistungs-Dünnschicht-Chromatographie HPDC**, published by the Institut für Chromatographie, Bad Dürkheim 1976.
The now usual English abbreviation HPTLC for high-performance thin-layer chromatography becomes a mixture of English and German for this book title (high-performance Dünnschicht-Chromatographie). It describes the fundamental principles of quantitative TLC, and (almost) all the great names among the founders of TLC have contributed.

Pachaly, P.: **DC-Atlas, Dünnschicht-Chromatographie in der Apotheke**, Wissenschaftliche Verlagsgesellschaft, Stuttgart, 4th Supplement 1999.
A supplement in loose-leaf form (following the bound volume published in 1982), which has the great advantage, for non-pharmacists, of giving comprehensibility to the rather unusual style of the German Pharmacopoeia (DAB); an extremely impressive model of the chromatographic system in the form of a shopping street; alphabetical sequence of monographs; clear and well-organized description of the process of research; color plates of the chromatograms.

Wagner, H.; S. Bladt, E.M. Zgainski: **Drogenanalyse**, Springer-Verlag, Berlin/Heidelberg 1983, ISBN 3-540-11867-5.
The bible for all plant analysts arranged according to drug classes, although the search for chromatographic parameters and detection reagents can sometimes be laborious. The color plates of the chromatograms are very good.
English edition: **Plant Drug Analysis**, Thin Layer Chromatography Atlas, 2nd edn 1996, ISBN 3-540-596-76-8.

Jork, Funk, Fischer, Wimmer: **Dünnschicht-Chromatographie, Reagenzien und Nachweismethoden**, Vol 1a, 1989 and Vol 1b, 1993, VCH Verlagsgesellschaft, Weinheim.
The absolute must in every TLC laboratory! Many formulae of reagents and reaction equations, descriptions of the practical performance of reactions with tested examples, and some scanned chromatograms and color plates of chromatograms.
English editions**: Thin-Layer Chromatography, Reagents and Detection Methods**, Vol. 1a, 1990, ISBN 3-527-27834-6 and Vol 1b, 1994, ISBN 3-527-28205-X.

Frey, H.-P., K. Zieloff: **Qualitative und quantitative Dünnschichtchromatographie**, Grundlagen und Praxis, VCH Verlagsgesellschaft, Weinheim 1993, ISBN 3-527-28373-0.
This book answers the principally theoretical questions about TLC which are not discussed in the present practically oriented volume.

Kraus, Lj., A. Koch, S. Hoffstetter-Kuhn: **Dünnschichtchromatographie**, Springer Laboratory Manual, Berlin 1996.
Building on his "Kleines Praktikumsbuch der Dünnschicht-Chromatographie", whose first edition appeared in 1985 and 4th edition (published by DESAGA, Heidelberg) in 1992, Kraus (who died in 1994) has attempted to give a comprehensive account of the theory and practice of TLC. The examples given show that this book is mainly intended for pharmaceutical biologists. However, the theoretical principles are also quite suitable for other TLC students.

1.6.1.2 Books in English

Most of the publications listed in Section 1.6.1.1 have also appeared in English, usually after a short delay, but nevertheless describe the most recent state of the technique at their publication date. The following books published after and including 1994 have not appeared in German.

Pachaly, P. (ed): **Simple Thin-layer Chromatographic Identification of Active Ingredients in Essential Drugs**, Gesundheitshilfe Dritte Welt, German Pharma Health Fund e.V., ECV, Aulendorf 1994.
Brochure for the identification and purity testing of pharmaceutical agents in the Third World.

Sherma, J., B. Fried (ed): **Handbook of Thin-Layer Chromatography**, Chromatographic Science Series, Vol. 71, Marcel Dekker, New York 1996, ISBN 0-8247-9454-0.
In this textbook, theoretical aspects as well as a large number of practical examples, arranged by substance classes, are described. This second edition contains literature references up to and including 1995.

Sethi, P.D.: HPTLC, **High Performance Thin-Layer Chromatography, Quantitative Analysis of Pharmaceutical Formulations**, CBS Publisher & Distributors, New Dehli (India) 1996, ISBN 81-239-0439-8.
After a short introduction to TLC and quantitative HPTLC, ca. 200 examples of pharmaceutical materials are listed in order of substance class. Detailed examination has shown that, at least for some substances, the analytical methods are a little slipshod. In spite of this, the book has considerable value for its information in the field of pharmacy.

Fried, B., J. Sherma: **Thin-Layer Chromatography**, 4th edn, revised and expanded, Chromatographic Science Series, Vol. 81, Marcel Dekker, New York 1999, ISBN 0-8247-0222-0.
A good overview of the modern state of TLC is provided in two parts (general practical training in TLC and examples of various substance groups). The comprehensive list of references, which refers to publications (unfortunately only those that appeared in English) up to and including 1997, is especially valuable. The chapter on documentation in TLC is rather meager, and there are neither color plates of chromatograms nor advice on working according to the GMP/GLP guidelines. Also, new developments in sorbents, (e.g. the HPTLC LiChrospher Si 60 F_{254s} coated plates, which contain spherical silica gel) are not included in the book.

Wall, P.E.: **Thin-Layer Chromatography – A modern Practical Approach**, The Royal Society of Chemistry (RSC), Cambridge 2005, ISBN 0-85404-535-X.
The book derives from Wall's own long-term experience in TLC up to the year 2000. Chapters about the new precoated layers (Lux®, UTLC, ProteoChrom®) as well as documentation are missing. The print is in black and white only.

Kowalska, T., J. Sherma (ed): **Preparative Layer Chromatography**, CRC Press, February 2006, ISBN 0-8493-4039-X.
A good description of all important areas of preparative layer chromatography, theory and a wide range of applications (e. g. the use of PLC for isolation and identification of unknown compounds from the frankincense resin (Olibanum), strategies for finding marker substances).

1.6.1.3 Book in Another Language

Chinese

Xie Peishan (ed): *Chinese Pharmacopoea TLC ATLAS of Traditional Chinese Herb Drugs,* June 1993, ISBN 7-5359-1023-81/R.192.
A very elaborately produced book with exceptional color plates of the chromatograms. English translations of four monographs are provided, and these give a very good description of the derivation of the chromatographic parameters, with comprehensive data. Unfortunately, at least 95 % of the experiments were performed with TLC plates prepared in-house, so that full evaluation of these would be problematical. The English edition is in preparation.

1.6.2 Journals

No journal in the German language exists which deals exclusively with TLC. Many authors are of the opinion that only articles in English can provide the necessary wide distribution and thus attract enough attention to their research results.

1.6.2.1 German Language Journals Containing Articles on TLC (Selection)

Chrom View, Eigenverlag Merck KGaA, Darmstadt

Deutsche Apotheker Zeitung, (DAZ), Deutscher Apotheker Verlag, 70191 Stuttgart

Die pharmazeutische Industrie (pharmind), Editio Cantor Verlag, 88322 Aulendorf

Die Pharmazie, up to 1991 VEB Volk und Gesundheit, Berlin, from 1991 Govi-Verlag, 65760 Eschborn/Taunus

GIT Fachzeitschrift für das Laboratorium und *GIT SPEZIAL Chromatographie*, GIT-Verlag, Darmstadt

LABO, Kennziffer-Fachzeitschrift für Labortechnik, Verlag Hoppenstedt GmbH, Darmstadt

Laborpraxis, Vogel-Verlag, Würzburg

Naturwissenschaften im Unterricht Chemie (NiU-Chemie), E. Friedrich Verlag, 30917 Seelze

Pharmazeutische Zeitung, (PZ), Govi-Verlag Pharmazeutischer Verlag GmbH, 95760 Eschborn/ Taunus

Pharmazie in unserer Zeit, Wiley-VCH, Weinheim

Planta Medica, (Journal of Medicinal Plant Research), Georg Thieme Verlag, 70469 Stuttgart

Scientia Pharmaceutica, (Sci. Pharm.), Österreichische Apotheker-Verlagsgesellschaft mbH, Wien

1.6.2.2 English Language Journals on TLC

I am aware of only one journal that deals exclusively with TLC:
JPC – Journal of Planar Chromatography – Modern TLC, Published by the Research Institute for Medicinal Plants, P.O. Box 11, H-2011 Budakalász, in association with Springer Verlag, Hungary.

1.6.2.3 English Language Journals Containing General Articles on Chromatography (Selection)

Analytical and Bioanalytical Chemistry, Springer, Heidelberg, D

Chromatographia, Friedr. Vieweg und Sohn, Weisbaden, D

Journal of Agriculture and Food Chemistry, American Chemical Society, Washington DC, USA

Journal of Chromatography, Elsevier, Amsterdam, NL

Journal of Liquid Chromatography & Related Technologies, Taylor & Francis, London, UK

Journal of Separation Science, Wiley-VCH, Weinheim, D

LC-GC North America and *LC-GC Europe,* Advanstar, Iselin NJ, USA and Chester, UK, respectively

Pharmaceutical Biology, Taylor & Francis, London, UK

Phytochemical Analysis, John Wiley & Sons, Chichester, UK

Phytochemistry, Elsevier, Oxford, UK

1.6.3 Abstracts

CA Selects, Paper and Thin-Layer Chromatography, Chemical Abstracts Service, P.O. Box 3012, Columbus, Ohio 43210, USA

CBS – CAMAG Literaturdienst Dünnschicht-Chromatographie (CAMAG Bibliography Service) appears biannually in the form of yellow pages in the house journal of the company CAMAG, and gives an excellent review of articles on TLC published worldwide. It is arranged according to substance classes.

1.6.4 Pharmacopoeias

The first reference to TLC in a pharmacopoeia was as "Dünnschichtchromatografische Prüfung" ("thin-layer chromatographic testing") in the DAB7-DDR 1964 for the monograph "Oleum Menthae piperitae" (peppermint oil) [6]. It was first described as a stand-alone analytical method in the European Pharmacopoeia of 1974 (Ph. Eur. 1), where TLC was specified for the identification of 23 drugs. One year later, in the second issue of the DAB7, TLC identification was specified for a further 18 drugs.

TLC is now established in almost all the pharmacopoeias of the world, although only for identification and purity testing. The improvements achieved over 30 years in sorbents, solvent systems and detection reagents and the use of partially and fully automated equipment for sample application, development and densitometric determinations on substances are beginning to appear in the monographs of all pharmacopoeias, although rather slowly. Germany has proposed to the European Pharmacopoeia Commission a supplement to the Chapter "TLC" to include horizontal development and quantitative methods. The Commission accepted this recommendation and published it in Pharmeuropa [7] in 1994 as a draft for comment. A commentary to the draft monograph followed in 1996 [8]. A new section on the theme of TLC in the European Pharmacopoeia in the 1997 edition was an important step forward [9]. Precoated layers were first described under the heading "Reagents". Suitability tests were prescribed for the determination of the separation efficiency and selectivity, also for the effectiveness of fluorescence indicators in the precoated layers used. In the revised chapter on TLC, the quantitative determination of substances by means of a scanner in the UV-VIS and a counter for radioactive substances is now included under Method 2.2.27. In the DAB, the monograph on "Oil-Free Soya Lecithin" includes the world's first quantitative determination by means of a TLC scanner.

All modifications recommended in The European Pharmacopoeia 4th edition (English) 2002 with respect to the work method in TLC were also recommended on a European level but with a time delay. For the first time the documentation of chromatograms is required with photographs and/or data files.

Specialists demand a worldwide agreement on the analytical methods of TLC in the pharmacopeias. The latest editions are, however, far from it. Those previously mentioned improvements in Pharm. Eur. (Method 2.2.27) still are not present in 2006 in USP 29 with NF 24 (Method 621).

2 Precoated Layers

The methods of analysis using TLC plates prepared in-house which are described in the pharmacopoeias are not discussed in this book. Rather, in this chapter, we argue for the use of precoated layers, and advise on the choice of stationary phases in compliance with the recommendations of the pharmacopoeias.

2.1 Precoated Layers – Why?

In the first edition of Stahl's laboratory handbook "Dünnschicht-Chromatographie" published in 1962, equipment for coating glass plates is illustrated, and detailed procedures for the preparation of TLC plates are described. Even today, TLC sorbents for a manual coating operation are produced in an only slightly modified form, especially where labor costs are small and time is a minor consideration. The layers obtained often show good selectivity, but do not achieve the separation efficiency obtained by precoated layers and seldom lead to reproducible results.

If TLC is used as an analytical method in quality control, the reproducibility, i.e. lack of scatter (precision of both system and method), and also the accuracy of the analytical results must be determined. The GMP/GLP guidelines also require validation of the method, i.e. testing for linearity, selectivity, robustness and limits of detection and determination (see also Section 9.1 "Validation of TLC Methods"). For method development, high demands are placed on the stationary phases of the chromatographic system:

- Consistent chromatographic behavior with respect to chromatographic parameters (retardation properties, selectivity, separation efficiency).
- Surface homogeneity combined with a constant layer thickness optimized for the intended use.
- Good adhesion and good abrasion resistance of the layer.
- Traceability by use of coded precoated plates or by marking the article and batch number in the right hand upper corner of the TLC plate.

When precoated layers supplied by a manufacturer are accompanied by a DIN/ISO 9001 certificate, it can be assumed that constant quality of "hardware" for the TLC is being maintained. By standardizing the most important operations in the TLC over several years, very reproducible results are obtained [10].

Applied Thin-Layer Chromatography: Best Practice and Avoidance of Mistakes, 2nd Edition
Edited by Elke Hahn-Deinstrop

ISBN: 978-3-527-31553-6

Today, in-house production of layers for TLC is not therefore a sensible option in industrial, pharmaceutical or official laboratories. To save time and to obtain good results, commercially available precoated layers must be used.

2.2 What Are Precoated Layers Composed Of?

2.2.1 Sorbents

It was already clear in the early 1930s to the first German group of researchers, Kuhn and Brockmann, that in experiments performed to solve various problems by the adsorption method of Tswett the use of various sorbents would be necessary. They tested various substances, including aluminum oxides, aluminum silicates, calcium carbonate, kaolin, kieselguhr, magnesium oxide, powdered sugar, silica gels, starch and talc [11].

The introduction of finely divided porous silica gel produced by first grinding and then fractionating precipitated silicic acid (described in simple terms) enabled TLC to make its breakthrough. Silica gel 60 is today the most commonly used sorbent. In the nomenclature used for these materials, the name of the sorbent is followed by a number representing the mean pore diameter of the sorption agent measured in the now obsolete Ångstrom units (10 Å = 1 nm). As well as pure silica gel 60, caffeine-impregnated silica gel and mixtures with kieselguhr are used.

The separation efficiency obtained in TLC is essentially determined by the mean particle size and the size distribution of the sorption agent used in the preparation of the layer. As can be seen from Fig. 4, the mean particle size of silica gel of a quality suitable for HPTLC is 5 μm, that of TLC quality ca. 11 μm and that of PSC quality (for preparative work) over 20 μm. In a very recent development, spherical particles with

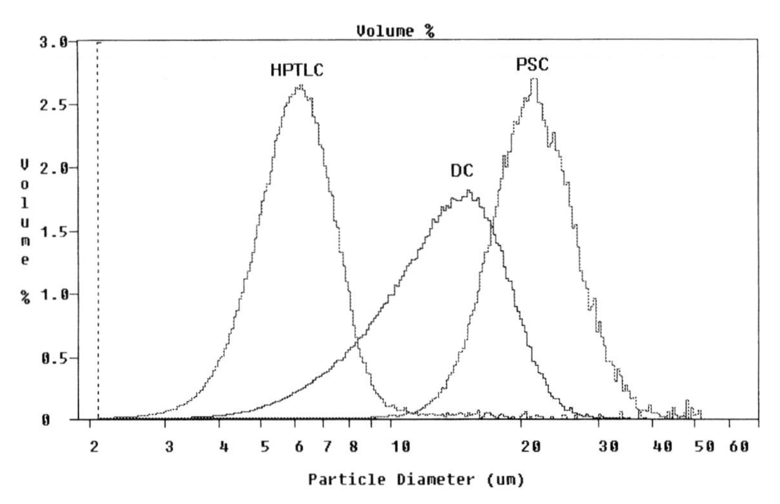

Figure 4. Typical particle size distribution Curves of HPTLC, TLC and PSC silica gel 60 material. Measurement method: Coulter counter (Merck, Darmstadt).

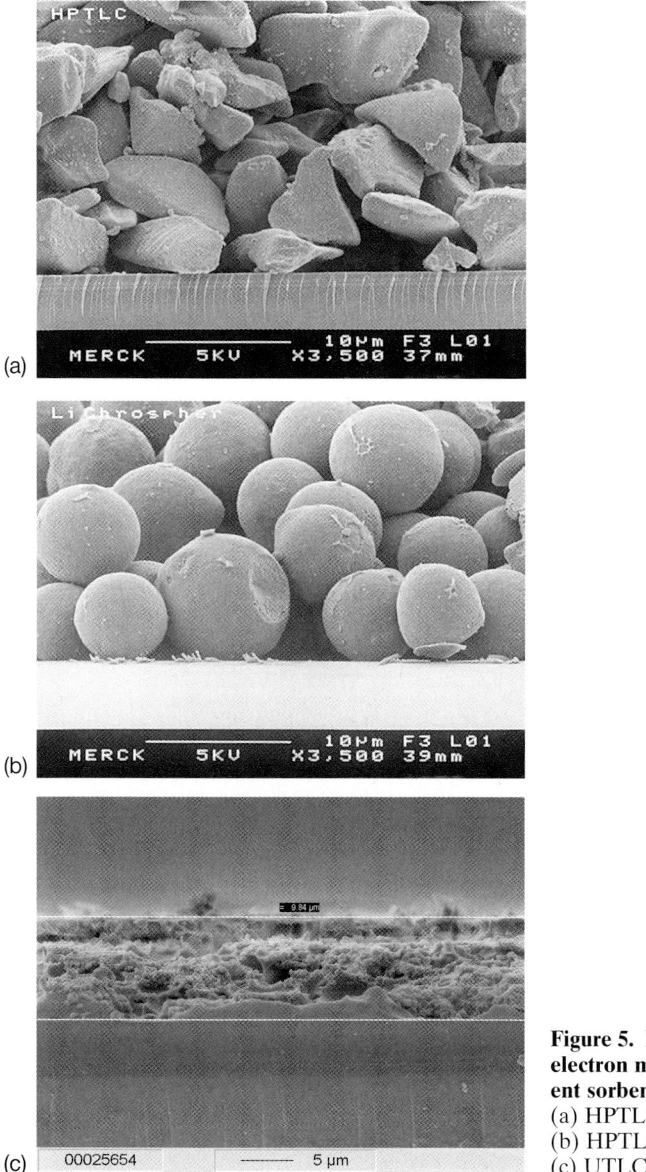

Figure 5. Photographs taken by scanning electron microscopy showing two different sorbent materials
(a) HPTLC silica gel 60
(b) HPTLC LiChrospher® Si 60
(c) UTLC

a mean particle size of ca. 7 µm are used to prepare the precoated plates known as "HPTLC-LiChrospher® Si $60F_{254s}$" (20 × 10 cm, Merck Article No. 15445). The difference between this and the older material which consisted of irregularly shaped particles is clearly shown in photographs taken by scanning electron microscopy (Fig. 5a–c). Chromatographs produced with this material in many cases show considerably increased separation efficiency (see Section 2.6 "Effect of the Stationary Phase when Mobile Phases are Identical").

In the early days of TLC, the use of cellulose, polyamide and florisil (ca. 15 % MgO + 85 % SiO_2) as sorption agents was described, but these are of minor importance in modern TLC.

On the other hand, the "modern" sorbents are used more and more widely. The letter-number combinations such as "RP-8", "RP-18" or "NH_2" indicate chemically modified layers all of which are based on a silica gel matrix. "RP" indicates "reversed phase" and the number represents the length of the hydrocarbon chain used for the modification. The material "RP-2" is seen to be an exception, as this is a dimethylsilyl

Table 1: Types of sorbents and supports for precoated layers

Sorbent material	Support
Aluminum oxide 60, 150	Aluminum foil, glass plate, plastic film
Cellulose (unmodified)	Aluminum foil, glass plate, plastic film
Cellulose (acetylated)	Glass plate, plastic sheet
PEI-Cellulose	Glass plate, plastic sheet
Silica gel 40	Glass plate
Silica gel 60	Aluminum foil, glass plate, plastic film
Kieselguhr	Aluminum foil, glass plate
LiChrospher® Si 60	Glass plate
Si 50000	Glass plate
Si 60 RAMAN	Aluminum foil
Silica gel, modified	
CHIR (chiral)	Glass plate
CN (cyano)	Glass plate
DIOL	Glass plate
NH_2 (amino)	Aluminum foil, glass plate
Silica gel 60 caffeine-impregnated	Glass plate
Silical G ammonium sulfate-impregnated	Glass plate
Silica gel 60 silanized (RP-2), RP-8	Glass plate
RP-18	Aluminum foil, glass plate
Mixed layers	
Aluminum oxide/acetylated cellulose	Glass plate
Cellulose/silica	Glass plate
Cellulose 300 DEAE/cellulose 300 HR	Glass plate
Silica gel 60/ kieselguhr	Aluminum foil, glass plate
Two-zone layers[a)]	
Si 50000 (conc.), silica gel 60 (sep.)	Aluminum foil, glass plate
Si 50000 (conc.), RP-18 (separ.)	Glass plate
Kieselguhr (conc.), silica gel 60 (sep.)	Glass plate
Silica gel 60 (1st sep.), RP-18 (2nd sep.)	Glass plate
Special layers	
IONEX (ion exchange resin)	Plastic sheet
Polyamide 6	Plastic sheet
Polyamide 11	Aluminum foil, glass plate

This table does not claim to be fully comprehensive
[a)] conc. = concentrating zone
sep. = separation zone

modification. "NH_2" indicates a modification incorporating an end amino group, which gives a weakly basic ion exchange function. "CN" and "DIOL" are hydrophilic neutral modifications with end cyano and diol groups respectively, and "CHIR" indicate that this is a chiral layer used for ligand exchange chromatography of optically active substances [12].

The most important types of sorption agent used in proprietary products for TLC are listed in Table 1 in alphabetical order with the corresponding supports. In order to be able to find one's way in the manufacturers' catalogs and obtain the most suitable product for one's own application, it is important to become familiar with the meanings of the letters and numbers used. As an example, the letters and numbers used by Merck are listed in Table 2 [12].

Table 2: Meanings of code letters and numbers in product designations

CHIR	Chiral layer for separating enantiomers
CN	Hydrophilic layer with cyano modification
DIOL	Hydrophilic layer with diol modification
F	Containing fluorescence indicator
$F_{254+366}$	Excitation wavelengths of fluorescence indicator
F_{254s}	Acid-stable fluorescence indicator
G	Gypsum
H	Containing no foreign binders
NH_2	Hydrophilic layer with amino modification
P	For preparative work
R	Specially purified
RP	Reversed phase
RP-8, RP-18	Reversed phase with C-8 or C-18 hydrocarbon chain
Silanized, RP-2	Reversed phase with dimethylsilyl modification
W	Water-tolerant, wettable layer
40, 60 etc.	Mean pore size in Å

2.2.2 Supports for Stationary Phases

Supports used for the stationary phases include aluminum foil, terephthalate films, thin glass plates and braided PTFE or glass fiber. Without the use of these inert substrates, it would be quite impossible to handle the finely divided sorbents.

For a precoated layer subjected to a number of operations performed one after the other such as prewashing, activation, spraying or dipping with subsequent heating, the more robust glass plate has been found to be a very good support. For quantitative work it is indispensable. Aluminum foils are preferable to all others materials where qualitative analyses are to be performed at minimum cost. Compared with glass plates they have less weight, require less storage space and can easily be cut with shears into pieces of any desired dimensions. However, aluminum foils will not tolerate the use of some solvent systems that contain strong acids or treatment in an iodine chamber for several hours.

Terephthalate films are becoming less frequently used. Their translucency is regarded as an advantage by some users, but their considerably poorer thermal stability is a disadvantage.

Braided PTFE and glass fiber products are no longer obtainable commercially in Europe.

The thickness of the sorbent layer for analytical purposes lies between 50 and 100 μm (extra thin films for automated multiple development, AMD, are described in Section 11.1), 200 μm for HPTLC plates and 250 μm for the normal TLC plates. The layer thickness on aluminum foil is 200 μm, while layers that can be used for preparative work can be up to 2 mm thick.

The maximum size of the glass plates used as supports for TLC precoated layers is 20 × 20 cm, and for HPTLC layers 20 × 10 cm. The smallest usable glass plates have a 5 × 5 cm format. These are produced as scored precoated 10 × 10 cm plates for HPTLC, which can easily be broken up into four 5 × 5 cm plates. The 20 × 20 cm "Multiformat" plates (Merck Article No. 1.05620) can be broken up as necessary into the formats 20 × 15 cm, 20 × 10 cm, 20 × 5 cm, 15 × 10 cm, 10 × 10 cm and 5 × 10 cm or can be used in their unbroken original size. Although only used for identity testing for teaching purposes, the 2.5 × 7.5 cm glass plate is thought of as having a microscope slide format. These are available as TLC materials with and without fluorescence indicator. The glass plates of dimensions 40 × 20 cm formerly often used in preparative TLC work are now no longer obtainable. The only precoated plates without binder are the new UTLC-plates (see Section 2.3).

Packages of aluminum foil or plastic film for use as supports are available in the 20 × 20 cm format, and other formats can easily be produced by cutting with shears. Aluminum foil (AF) can also be obtained as pocket-sized packages in the 5 × 7.5 cm and 5 × 10 cm formats. The 500 cm long and 20 cm wide rolls of TLC materials in the Merck catalog are unusual. As the price of the rolls is identical to that of a 25-item package, the question arises *where* plates with a width exceeding 20 cm are used, especially as there are currently no developing chambers suitable for dimensions in excess of 20 × 20 cm.

2.2.3 Additives

So that the sorbent will adhere to the substrate and the individual particles will be bonded together, small amounts of binder are added to the stationary phase. Depending on the manufacturer and the sorbent material, inorganic, organic or sometimes polymeric substances are used. For the most part, these do not interfere with the chromatography.

Another additive, a so-called "fluorescence indicator", e.g. manganese-activated zinc silicate, can be mixed with the sorbent. Its particle size does not exceed that of the sorbent. On irradiating with short-wave UV light (λ = 254 nm), a yellow-green emission is produced. The "acid-stable" fluorescence indicators of the modified silica gel layers (tungstates) emit a pale blue color on irradiation. The advantage of the indicators is that all UV-active substances, e.g. those with conjugated π-electron systems such as all aromatic compounds, become visible as dark zones on a bright emitting background when the chromatogram is observed in short-wave UV light. This method of detection does not normally cause decomposition. It should also be mentioned that fluorescence indicators also exist which, when excited in long-wave UV light (λ = 366 nm), emit a blue, yellow-green or red radiation

For special separations, an impregnation of the layer is often necessary, e.g. of unsaturated fatty acids with silver nitrate or of antibiotics of the tetracycline type with EDTA. Whereas the user must perform these impregnations himself, a caffeine-impregnated HPTLC precoated layer is available for the determination of polycyclic aromatic hydrocarbons.

2.3 What Types of Precoated Layers Are There?

All TLC plates are white!

This airy comment by a practical pharmacist is true in principle, but is nevertheless a bold generalization in view of the many possibilities presented by the ca. 23 different types of sorbent and ca. 140 different commercial grades of precoated layers currently marketed by Merck alone. As other manufacturers such as Macherey-Nagel, Whatman or Analtech offer other products with their own names and product numbers it is essential to study the catalogs of the manufacturers of precoated layers in great detail. TLC products are primarily classified according to sorption agents and secondarily according to particle size, any fluorescence indicator present, supports and other characteristic properties. For documentation in compliance with GMP/GLP, the item number of the manufacturer is a characteristic that is important for the user, there being no overlapping of item numbers worldwide.

Table 3 shows the most important commercially available precoated layers and some typical examples of their use. In the choice of the "hardware" for TLC it cannot be assumed that nominally "identical" sorbents from different manufacturers will lead to equal separations [13]. It follows that trade names and item numbers used in publications or testing procedures should not be regarded simply as surreptitious advertising, but are absolutely essential if reproducible results are to be obtained. This is especially true for validated methods. An example is given in Fig. 6, which shows separations of peppermint oils on TLC plates whose sorbent specifications in the catalogs of different manufacturers are identical. When the development heights are equal, the variation in separation efficiency is clearly visible, and different hRf values are obtained for the same substances (see Table 4).

▶ **Figure 6:** see Photograph Section.

Table 3a: Important commercially available precoated layers and examples of typical applications[a)]

Sorbent material	Chromatographic principle	Typical applications
Aluminum oxide	Adsorption chromatography due to polar interactions	Alkaloids, steroids, terpenes, aliphatic, aromatic and basic compounds
Cellulose		
Unmodified cellulose	Partition chromatography due to polar interactions	Amino acids and other carboxylic acids as well as carbohydrates
Acetylated cellulose	Depending on acetyl content transition from normal phase to reversed phase chromatography	Anthraquinones, antioxidants, polycyclic aromatics, carboxylic acids, nitrophenols, sweeteners
Cellulose ion exchangers	Anion exchange	Amino acids, peptides, enzymes, nucleic acids constituents (nucleotides, nucleosides) etc.
Mixed layers Cellulose DEAE/cellulose HR	Ion exchange	Mono- and oligonucleotides in nucleic acid hydrolyzates
Ionex ion exchangers	Cation and anion exchange	Amino acids, nucleic acid hydrolyzates, amino sugars, antibiotics, inorganic phosphates, cations; racemate separation in peptide synthesis
Kieselguhr	Commonly impregnated for reversed phase separations	Aflatoxins, herbicides, tetracyclines
Polyamide	Partition chromatography due to polar interactions (e.g. hydrogen bonds)	Phenolic and polyphenolic natural substances
Silica		
Unmodified silica gel		
Standard and nano silica gel, also with concentrating zone	Normal phase chromatography	Most frequent application of all TLC layers
High purity silica gel 60		Aflatoxins
Silica gel G, impregnated with ammonium sulfate		Surfactants, lipids (neonatal respiratory syndrome)
Silica gel 60, impregnated with caffeine for PAH determination	Charge transfer complexes	Polycyclic aromatic hydrocarbons (PAH) acc. to German drinking water specification (TVO)
Chemically modified layers: CHIRalplate	Enantiomer separation based on ligand exchange chromatography	Chiral amino acids, α-hydroxy-carboxylic acids and other compounds which can form chelate complexes with Cu(II) ions

Table 3a: Continued

Sorbent material	Chromatographic principle	Typical applications
Cyano-modified layer CN	Normal phase and reversed phase chromatography	Pesticides, phenols, preservatives, steroids
DIOL-modified layer		Steroids, hormones
Amino-modified layer NH_2	Anion exchange, normal phase and reversed phase chromatography	Nucleotides, pesticides, phenols, purine derivates, steroids, vitamins, sulfonic acids, carboxylic acids, xanthines
RP layers:		
RP-2, RP-8, RP-18		Nonpolar substances (lipids, aromatics)
Silica gel 60 silanized		Polar substances (basic and acidic pharmaceutical active ingredients)
RP-18 W/UV_{254}, wettable	Normal phase and reversed phase chromatography	Aminophenols, barbiturates, preservatives, nucleobases, PAH, steroids, tetracyclines, phthalates
Spherical silica gel		
LiChrospher® Si 60	Normal phase chromatography	Pesticides, phytopharmaceuticals
Mixed layers		
Aluminum oxide/acetylated cellulose	Normal phase and reversed phase chromatography	Polycyclic aromatic hydrocarbons (PAH)
Cellulose/silica gel	Normal phase chromatography	Preservatives
Kieselguhr/silica gel	Normal phase chromatography, reduced Adsorption capacity compared to silica gel	Carbohydrates, antioxidants, steroids, photographic developer substances

[a] Following Macherey-Nagel's and Merck's catalog

Table 3b: New precoated layers

Name of the plate	Particularities	Typical applications
Adamant® (Macherey-Nagel) **Lux®** (Merck)	Both: Increased amount of fluorescence indicator	Universal
UTLC (Merck)	Ultra thin monolithic silica gel	Steroids, azepams, amino acids, phthalates and phenols
ProteoChrom® (Merck)	a) HPTLC silica gel 60 F_{254s}, 20 × 10 cm glass plate b) HPTLC cellulose, 10 × 10 cm aluminum sheet	Amino acids, peptides (from protein digest) Amino acids, peptides, 2D-TLC
HPTLC Premium Purity Plate (Merck)	Wrapped in a special plastic-coated foil	All pharmacopoeia applications

Table 4: Peppermint oil

	DAB 10	**Pachaly TLC-Atlas**	**Alternative I**	**Alternative II**
Sample solution	0,1 g/ml toluene	10 µg/ml toluene	10 µl/ml toluene	10 µl/ml toluene
Sorbent	Prescribed: silica gel G F_{254} used: TLC silica gel 60 F_{254} (Merck 1.05715)	TLC silica gel 60 F_{254} 20 × 20 cm (Merck 1.05715)	TLC-silica gel 60 F_{254} 20 × 20 cm (Merck 1.05715)	Durasil-25 UV 254 nm 20 × 20 cm (Machery-Nagel 812008)
Solvent system	Toluene + ethyl acetate (95 + 5)	Toluene + ethyl acetate (93 + 7)	Toluene + ethyl acetate (93 + 7)	Toluene + ethyl acetate (93 +7)
Applied sample volume	20 µl 20 × 3 mm	15 µl 15 × 3 mm	10 µl 10 mm bandwise	10 µl 10 mm bandwise
Chamber saturation	Yes	Yes	Yes	Yes
Migration distance	15 cm	15 cm	13 cm	13 cm
Development time	35 min	39 min	27 min	25 min
Derivatization reagent	Anisaldehyde-sulfuric acid	Anisaldehyde-sulfuric acid	Vanillin-sulfuric acid	Vanillin-sulfuric acid
Color of substances after derivatization:				
	Acc. to literature	Acc. to literature		
Menthol	Bluish violet	Bluish violet	Bluish violet	Bluish violet
Menthone	Grayish blue	Yellowish green	Green	Green
Menthyl acetate	Bluish violet	Bluish violet	Blue	Blue
Menthofuran	Brownish yellow	Orange + violet overlap	Brownish orange	Orange
hRf-values:		Acc. to literature		
Menthol	11–15	20	18–22	28–32
Menthone	37–40	46	46–49	56–60
Menthyl acetate	40–45	51	50–54	61–64
Menthofuran	64–68	66	66–70	70–73

Since the publishing of the 1st edition of this book, some new precoated layers have been introduced. To simplify the visual evaluation for the user, Macherey-Nagel and Merck have developed precoated silica gel 60 layers with an increased amount of fluorescence indicator that go by the trade names of Adamant®- and Lux®-plates. Furthermore these plates contain a higher amount of binder which improves the abrasion resistance. Figure 7 shows the comparison at UV-light 254 nm between a normal TLC- and a Lux®-plate.

▶ **Figure 7:** see Photograph Section.

Following the common trend towards miniaturization also in analytical techniques, Merck has developed an UTLC-plate (**UTLC: ultra thin-layer chromatography**). Its precoated layer consists of a monolithic, porous silica gel. This non-particular ready-to-use plate contains no binder. With a layer thickness of only 10 μm these plates show very fast separations in combination with an extremely low consumption of solvents. Only very small sample volumes have to be applied on the layer, e.g. 20 nl solution of dyes (see Fig. 8).

▶ **Figure 8:** see Photograph Section.

To improve analytical tools within the life sciences, Merck has introduced a new product line with two new ready-to-use plates called "ProteoChrom":

ProteoChrom® HPTLC silica gel glass plate
ProteoChrom® HPTLC cellulose aluminium sheet

Figures 9 a and b show the separation of peptides obtained by tryptic digest of Cytochrome C. For more information see Table 3b.

▶ **Figures 9 a and b:** see Photograph Section.

The new **HPTLC Premium Purity Plate** (Merck) is especially designed for demanding pharmacopoeia applications. It is carefully wrapped in a special plastic coated aluminum foil to prevent any plastic deposits from the wrapping material that could appear as an unknown extra zone when using medium-polar solvent systems such as toluene/ethylacetate (95/5). These plates have no GLP-laser code.

Figure 10 shows three commercial labels for precoated layers with explanatory notes. The concepts explained in this Chapter are again clearly represented for the user, so that taking a plate out of the "wrong" box should be a thing of the past!

Figure 10. Examples of labels used to identify precoated layers.
(a) Silica gel 60 with normal separation efficiency
(b) Water-tolerant high-performance plate
(c) Silica gel high-performance plate suitable for AMD (Automated Multiple Development).

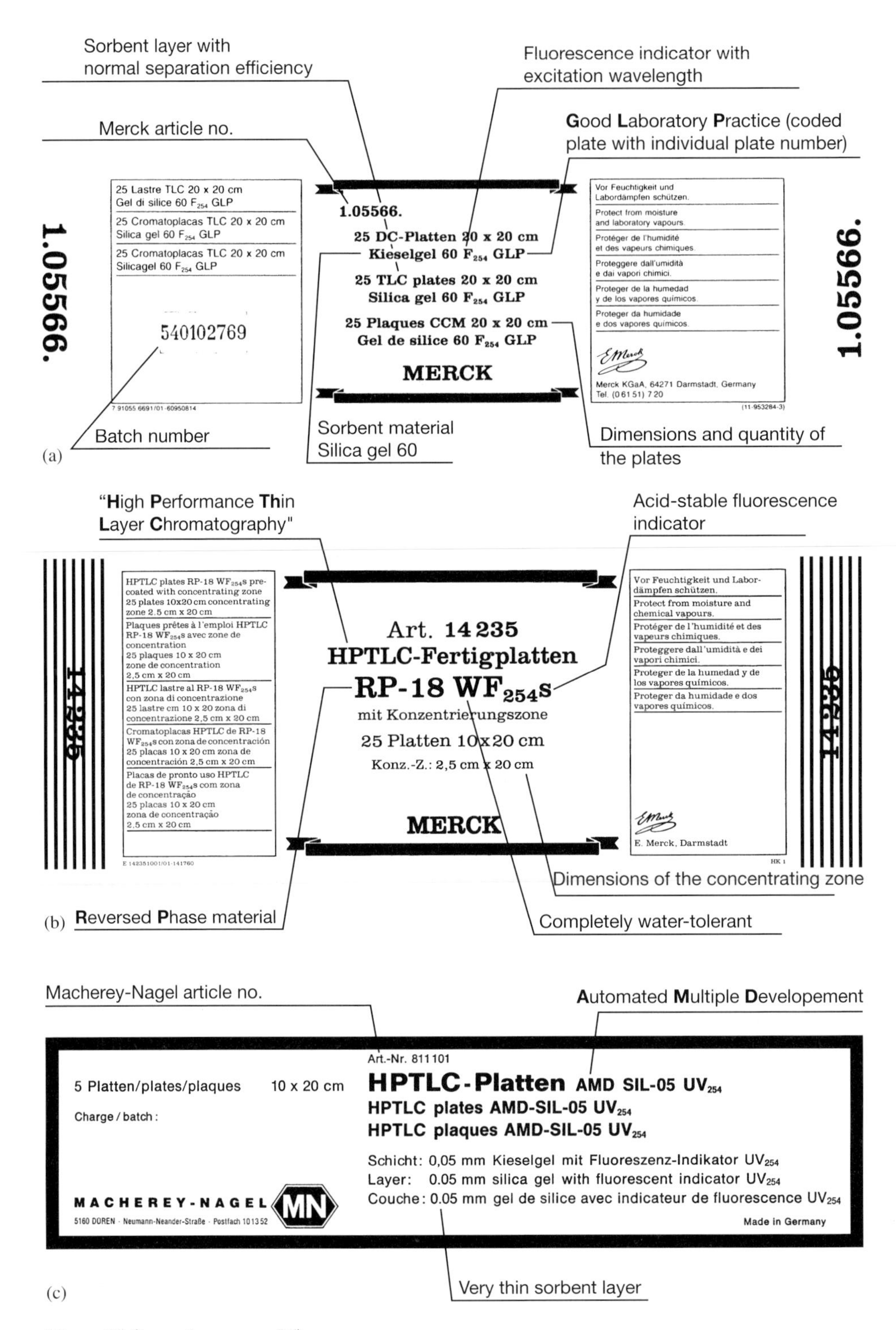

Figure 10 (legend see page 25)

2.4 What Are the Uses of Precoated Layers?

TLC investigations are mainly concerned with the determination of

- Identity
- Purity
- Assay

or a combination of these parameters. The TLC can be performed not merely one-dimensionally but, very simply and at no great cost, two- or multi-dimensionally, e.g., with two different solvent systems in two different directions or by multiple development with different solvent systems in one direction.

Moreover, TLC can as a rule be very elegantly combined on-line with both spectroscopic methods and with other analytical and preparative separation methods. In this way, its ability to separate complex mixtures and its reliability in identifying individual substances are greatly improved. Separation processes can, for example, be based on the following well-known combinations:

LC	→	UV/VIS
HPLC	→	HPTLC
HPLC	→	HPTLC/AMD
TLC	→	Electrophoresis

Successful on-line coupling of TLC with spectroscopic processes include:

TLC/HPTLC	→	UV/VIS
TLC/HPTLC	→	Fluorescence
TLC/HPTLC	→	FTIR
TLC/HPTLC	→	Raman
TLC/HPTLC	→	SERS
TLC/HPTLC	→	MS

2.5 Criteria for the Selection of Stationary Phases in TLC

The choice of the stationary phase for a given separation problem is the most difficult decision in TLC, especially, of course, for beginners. The most important aspect of this is to find a suitable selectivity. That is to say, the chemical composition of the stationary phase, and in particular that of its surface, must be suitable for the task. To obtain satisfactory separation efficiency, the mean particle size, the particle size distribution and the morphology of the particle must be considered. The support for the stationary phase should be as well suited for the intended application as the additives in the layer. Some attention should be paid to the cost-effectiveness of the analysis. For example, rather than preparing a large number of chromatograms individually as and when required, it may well be advisable to secure better reproducibility by using proprietary precoated solvent systems from a commercial manufacturer who provides a certificate guaranteeing the quality of his product.

2.5.1 How Can the Choice of the Stationary Phase Be Made?

The first step when tackling any new problem should be to carry out a literature search. Here, not only can articles on TLC be helpful, but sometimes publications from other areas of chromatography, e.g. HPLC, can provide exactly the information needed to solve the problem. The following example is from my own experience. An HPLC publication on the characterization of some extracts from medicinal plants used in cosmetics yielded the information that some amino acids in *Althaeae officinalis* (marshmallow) root showed a certain "fingerprint". I therefore proposed a possible method of identifying this extract, whose identification would otherwise be very difficult. A quantitative determination of alanine from a dry extract of marshmallow root in an instant tea was also possible after optimization of this HPLC system by means of TLC (see also Sections 7.2.1 and 7.2.2.1).

The results of the literature search serve as a basis for preliminary investigations. Without such information one usually starts with aluminum foil silica gel 60, from which suitable pieces can quickly be cut out. After this, if necessary, other layers can can be tested to optimize the resolution and the hRf values. For example, according to the literature, plates precoated with silica gel 60 were often used for the analysis of sandalwood, but the chromatograms obtained were not very good. Many other sorbents were therefore tried, but without success. Only the use of HPTLC precoated plates of the type NH_2 F_{254} with the solvent system *n*-butanol + formic acid + water (19 + 2 + 5) gave good results in the identification of sandalwood, a dye additive prohibited in Germany [75].

2.5.2 How Can the Recommendations for Stationary Phases Found in Pharmacopoeias Be Applied to Precoated Layers?

In the preparation of monographs, members of Commissions have not wished to mention any proprietary products so as not to appear to show any preference, and descriptions are mainly of plates prepared in-house from loose sorbent materials. Consequently, uncertainties nowadays frequently arise when trying to follow recommendations in pharmacopoeias. For example, in many monographs in the German pharmacopoeia (DAB) it is stated that the test parameters "identity" and "purity" can be tested for by TLC (Method V.6.20.2) using a layer of silica gel G *R*, G F_{254} *R*, H F_{254} *R* etc. Method V.6.20.2 prescribes the use of either precoated or homemade plates, but although the silica gels mentioned are reagents described in the DAB, there is no information about the corresponding precoated plates. Plates precoated with the silica gel sorbents G*R* and G F_{254} *R* are not manufactured commercially (at least in Europe), as the prescribed proportion of gypsum (13 %) leads to very soft layers, so that the plates are unsuitable for transportation and present severe handling problems. In the case of coatings of silica gel H *R* and H F_{254} *R* ("with no foreign binders"), these problems are even more severe, so that these materials also cannot be obtained as precoated layers.

The question of which is the corresponding precoated layer therefore arises in routine laboratory work. General recommendations on the subject can only be made with some reservations, as a complete rearrangement of the chromatographic system, including not only the precoated layer but also the solvent system and other parameters, is not always readily feasible. Many examples from pharmacopoeia monographs are given in 11 Tables in this book (Tables 4, 6, 9–12, 14, 15, and 17–19) in which a complete change of all the conditions has sometimes led to optimum results.

☞ **Practical Tips** for the choice of precoated layers based on pharmacopoeias:

- In the case of all sorbents for which DAB recommends the silica gels G *R*, $G_{254}R$, H *R* and H F_{254} *R* and which should have a mean grain size of ca. 15 μm, preliminary investigations can be made using TLC silica gel 60 with or without fluorescence indicator (FI).
- Chromatographic silica gel *R* (DAB) is a very fine grade with a grain size in the range 3–10 μm. For this, the HPTLC silica gel 60 precoated layers with or without FI are recommended.
- If the chromatographic silica gel cyanopropylsilylated *R* is recommended, the HPTLC plates CN F_{254s} can be used.
- For the chromatographic silica gel octadecylsilylated *R* (3–10 μm), the HPTLC plates RP-18, RP-18 W, RP-18 F_{254s}, RP-18 WF_{254s} and RP-18 with concentration zones give good results.
- If the chromatographic silica gel octylsilylated *R* is recommended, HPTLC precoated plates RP-8 F_{254s} can be used.
- Silica gel H and H F_{254}, silanized *R*, can be replaced by HPTLC precoated plates RP-2 F_{254s}.
- Silica gel anion exchanger *R* is a very fine silica gel (3–10 μm), whose surface is chemically changed by modification with quaternary ammonium groups. In this case, good chromatograms can be obtained on HPTLC plates NH_2 and NH_2 F_{254s}.
- In the case of, e.g., propyl-4-hydroxybenzoate (and all other PHB esters according to Ph Eur), a suitable octadecylsilanized silica gel that contains a fluorescence indicator with very intensive fluorescence excitation at 254 nm is:
  DC-RP-18 F_{254s}: Merck 1.15389 20 × 20 cm
  Merck 1.15423 10 × 20 cm

2.6 Effect of the Stationary Phase When Mobile Phases Are Identical

An example of the difference between nominally identical precoated layers from different manufacturers has already been graphically described in Section 2.3. The discussions in this Section are concerned with precoated layers with different types of sorbent, all produced by one manufacturer.

In the past, a large number of TLC investigations were carried out on TLC silica gel 60 precoated plates, finer materials not being on the market at that time. As this then became available as HPTLC silica gel 60 precoated plates, this new precoated layer was rarely checked to find out whether it showed any advantages. Apart from the higher price, a change would also have meant another validation of the method. The new material was particularly recommended by the manufacturer for quantitative determinations, with the suggestion that only 20 % of the amount of substance formerly used was the optimum when chromatography was perfomed using this layer.

Tables 9 (carbamazepine), 17 (greater celandine) and 18 (sugar) show the results of comparative tests using precoated layers of TLC and HPTLC silica gel 60. Whereas in the identity test of the sugar using the HPTLC layer a drastic reduction in the time of the analysis from 210 min to 92 min could be achieved (2×7 cm instead of 2×15 cm migration distance, with intermediate drying and cooling of the plate), the time difference with carbamazepine was only 5 min (TLC: 10 cm, 21 min; HPTLC: 7.5 cm, 16 min). The hRf values of the sugar with HPTLC layers are somewhat lower than with the TLC material, and with carbamazepine the reverse is true. Comparative tests with the 5×5 cm horizontal chamber (described in detail in Section 4.2.1.2) were performed with various samples of greater celandine. Whereas with the TLC layer the solvent had reached the top end of the plate after 3.5 min, the time required with an HPTLC layer of the same migration distance was 6 min. Here too the hRf values for the HPTLC layer were higher than with the TLC material. The chromatograms of the components of the flavonoid-containing drugs also show considerable differences. On HPTLC silica gel 60 precoated plates, the substances appear as compact, sharply separated zones in Fig. 11b. Figure 11a shows the same chromatogram on TLC silica gel 60 material, both photographed under 254-nm UV light before derivatization with the Neu's reagent. This example confirms the improvement in the separation efficiency by changing from TLC to HPTLC precoated layers.

▶ **Figure 11:** see Photograph Section.

The newly introduced HPTLC-LiChrospher® Si 60 F_{254s} precoated plates in many cases show a still better separation efficiency than the HPTLC silica gel 60 precoated plates. For example, Fig. 12a (HPTLC silica gel 60) and Fig. 12b (HPTLC-LiChrospher Si 60_{254s}) how the scanned chromatograms of eight pesticides, all in the solvent system petroleum ether 40–60 °C + acetone (70 + 30 ml, without saturation of the chamber) [13a]. Also in the analysis of frankincense (olibanum), better separations were obtained on plates precoated with spherical silica gel than on those precoated with irregularly shaped particles [13b–d].

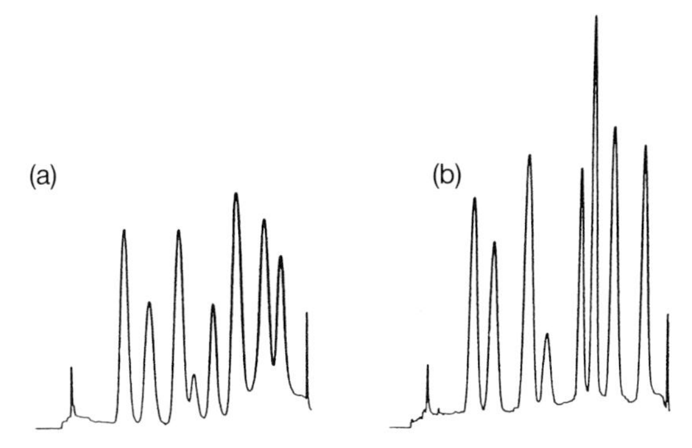

Figure 12. Comparison of separation efficiencies of HPTLC silica gel 60 F_{254} and HPTLC LiChrospher® Si 60 F_{254s} precoated plates of one manufacturer
(a) HPTLC silica gel 60 F_{254} GLP precoated plate (Merck Article No. 1.05613)
(b) HPTLC LiChrospher® Si 60 F_{254s} precoated plate (Merck Article No. 1.15445)

Detection: Direct optical evaluation under 254-nm UV light using the TLC-Scanner II (CAMAG)
The scans show the following pesticides from *left* to *right*:
Hexazin, Metoxuron, Monuron, Aldicarb, Azinphos methyl, Prometryn, Pyridat, Trifluralin
Sample volume: 50 nl, normal chamber without chamber saturation, solvent system: petroleum ether (40–60 °C) + acetonitrile (70 + 30 v/v), migration distance: 7 cm.

In this case, as with all other recommendations from the literature, the following advice applies:

Only your own work can confirm that you have obtained the results you were hoping for!

2.7 Advice on the Ordering and Storage of Precoated Layers

To have a clear picture of the storage of TLC precoated layers for routine and developmental work, and to comply with the GMP/GLP guidelines, the following method of working has been found to be satisfactory:

Using a card system (either actual cards or a modern PC data base), the various precoated layers are listed using the item number and name of the manufacturer as key words. These are recorded in the heading of the card, and the date received and batch number of the package are listed underneath. The withdrawal of a package from the store (for large businesses) should be noted and also the date of opening a package in smaller TLC laboratories to avoid breaking into several packages containing the same type of sorbent. Figure 13 shows a record from the file "precoat.crd" as an example.

In addition, the relevant pages can be copied from the current catalogs of the manufacturer and affixed to the inside of the storage cabinet door or displayed in a store-

1.005613 Merck				
HPTLC-silica gel 60 F_{254} GLP, 20 x 10 cm				
Plates received	*Batch No.*	*Date from store*	*Date opened*	*From*
15.03.96	06345678	26.04.	02.05.	HD
ditto	ditto	18.05.	18.05.	Ha
15.05.96	06432103	29.05.	30.05.	Heg
ditto	ditto			
04.06.96	06432103			
ditto	ditto			

Figure 13. Example of data stored in a card index system for HPTLC precoated plates

room for the TLC plates. The item numbers of the stored plates are marked with felt tipped pens, which should be readily available. The colors can have the following meanings:

Red: routine material
Yellow: often used material
Green: seldom used material
Blue: very seldom used material

Special samples are marked as such, and instructions on how they might used are included. The number of stored packages should also be marked on the copies, so that reordering can be done at the correct time. An average storage time of 3 months should ensure that TLC plates will be available even at peak periods.

In today's industry, most scientists have their own PC with access to an intranet. Inventory is kept online in a datasheet (e.g. excel file) or any other electronic database, which delivers all details on quantities and qualities of available TLC-materials as well as their location.

Precoated plates are the reactive and fragile basic materials used for TLC work. Because of their enormously large surface area, they take up water (atmospheric moisture) etc. from the laboratory air, and this can affect the chromatographic work for which they are later used. Therefore, not only packages that have been broken into but also unbroken packages should be stored in special TLC plate storage cabinets or desiccators.

2.8 Problems in the Naming and Arrangement of Precoated Layers

It should be possible to prevent selection of the wrong box of plates by recording the identifying letters and numbers as well as the item numbers of the manufacturer. Nevertheless, it often happens that even long-established users of TLC ask what is the item number of the new "aluminum oxide layer of Merck with the light blue background". In this case, this means the TLC-alusheet RP-18 F_{254s} with the item number 1.05559. Here, several things have been confused:

- Support with sorbent.
- In the case of the support, aluminum oxide has been confused with metallic aluminum.
- As the stationary phase, modified silica gel RP-18 with acid-stable fluorescence indicator was intended.

3 Before the TLC Development Process

In this Chapter we describe methods of handling precoated layers and also how to apply and dry samples before development.

3.1 Handling of Precoated Layers

TLC precoated plates are fragile, and must be properly handled from the opening of the package to the documentation stage, which can consist of photographs taken with a camera, video shots produced with a video system or densitometric measurements recorded on a printed document. In the intervening period a number of operations are performed in which the layer material is repeatedly manually handled. Aluminum foil, plastic film and glass plates are nowadays the most common sorbent supports. Being used for a wide range of purposes, they are also handled in many different ways.

3.1.1 Film and Foil

Plastic films are seldom used on account of their poor thermal stability, and therefore in this Section we only discuss the use of aluminum foil as a support for TLC and HPTLC materials. For preliminary investigations, ordinary scissors can be used to cut up prepared products into smaller sizes suitable for the intended application. Here, the direction in which the layer of sorbent was applied is important: abrasion marks produced in the production process will be visible on the back of the aluminum foil, and the eventual chromatographic process should be performed in the same direction.

When cutting the foil, the angle of the scissors is extremely important. Good cut edges are obtained if the scissors are inclined slightly from left to right during cutting, as shown in Fig. 14b. If, after this, the cut edges are gently wiped with a spatula to remove loose layer material, chromatograms with constant hRf values over the whole width of the layer are obtained. The scissors should not be inclined to the left as shown in Fig. 14a, as this usually leads to delamination of the layer at one of the two cut edges. As a result, a capillary gap forms between the chromatographic layer and the foil, in which the solvent moves forward much more rapidly than it does in the middle of the chromatogram. This means that solvent also flows from the cut edge towards the middle of the chromatogram causing spot deformation and skewed and distorted chromatogram lanes [14].

Applied Thin-Layer Chromatography: Best Practice and Avoidance of Mistakes, 2nd Edition
Edited by Elke Hahn-Deinstrop

ISBN: 978-3-527-31553-6

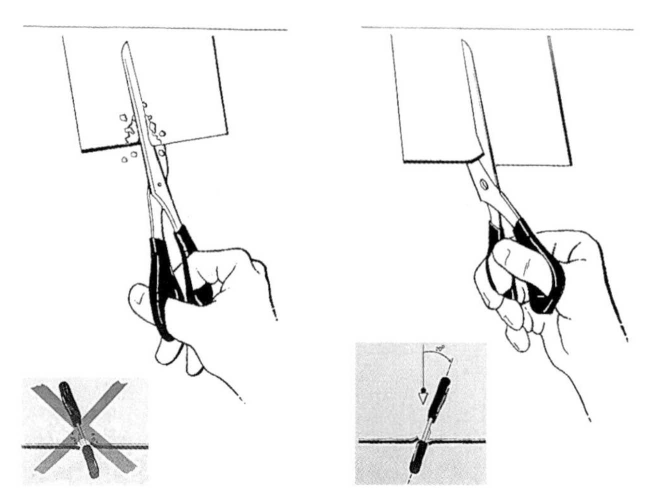

Figure 14. Method of holding scissors when cutting a TLC sorbent-coated aluminum sheet
(a) Unsuitable angle
(b) Suitable angle

As an example of these effects, Fig. 15a shows the chromatograms of samples of coffee and tea produced in a teaching experiment in which a piece of badly cut RP-18 foil was used. Figure 15b shows the same samples on a correctly cut piece of foil.

Aluminum sheets are not prewashed, as they become deformed when the subsequent necessary activation process is performed in a drying oven. Later operations can then become difficult or even impossible.

▶ **Figure 15:** see Photograph Section.

3.1.2 Glass Plates

Glass plates with precoated layers of sorbents are wrapped in plastic film, placed in styrofoam packaging material and packed in a cardboard carton by the manufacturer. These packages, protected in this way, are taken from the store or the TLC storage cabinet. Although the cardboard carton and styrofoam packaging are easy to remove (Fig. 16a,b), many fingernails have been broken opening the packages.

☞ **Practical Tip:** When opening the cardboard carton, use scissors or a screwdriver as a tool.

The plastic film can then be opened using a laboratory scalpel in two ways: either the plates are left in the styrofoam container and the top film is cut twice diagonally, enabling the plates to be removed (Fig. 16c,d), or the plastic-wrapped package is removed completely from the styrofoam packaging and cut with the scalpel along three

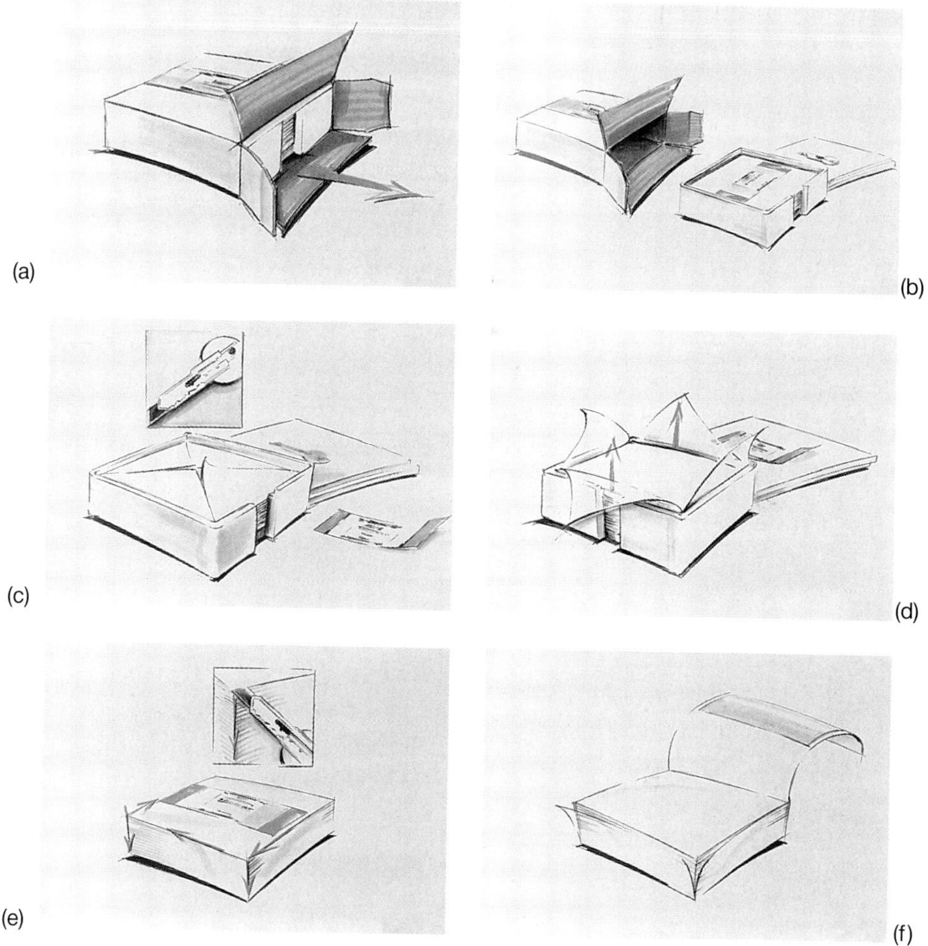

Figure 16. Ways of opening commercially available TLC packages
(a) and (b) Removal of cardboard carton and styrofoam lid
(c) and (d) Diagonally scored protective foil
(e) and (f) Protective foil cut on three sides.

top edges; the package is then replaced in the bottom part of the styrofoam packaging and the plates are taken out (Fig. 16e,f). If you prefer the first alternative, please note:

Practical Tip: Before you pick up the scalpel, pull the label off the plastic film and stick it to the outside of the styrofoam lid, one side of which should still be attached to the bottom part of the styrofoam by adhesive film.

It is now common practice in many laboratories to dispose of the outer packaging after opening a new TLC carton. In this case, it is essential to write the item number on the bottom part of the styrofoam packaging to enable the package to be identified with certainty. A mix-up of precoated plates can only be avoided by taking this precaution at the very beginning.

☞ **Practical Tips** for dealing with TLC plates:

- Never have more than one carton of the same product available in the workplace of any one laboratory or department.
- In the case of rarely used precoated layers do not remove the packaging, as this carries the manufacturers labels and gives better protection of the plates. Also, store the carton in the TLC store and not in the laboratory.
- After taking a plate out of the carton, the item number should immediately be written on the top right hand edge, as it is seldom possible to find out what kind of plate it is later on. The only exceptions to this are Merck's silica gel GLP plates, which carry the manufacturer's item number, batch number and individual plate number.
- When storing opened packages of precoated layers, care should be taken to place the plates with the sorbent underneath in the bottom part of the styrofoam and covered with plastic film. Components of the laboratory atmosphere otherwise soon become deposited on the active layers, and constituents of the glue on the back of the label can diffuse into the layer.

The following points are very important: plates must only be touched with the fingertips (of both hands if possible) on the outer cut edges, and transportation should mainly be in TLC drying racks. These are so constructed that the plates can be pushed into slots in a horizontal position. In the case of tiltable racks, e.g. for activation in the drying oven, the plates are vertical. So far, the only commercially available plate racks have been for plates 20 cm wide. Therefore, a home-constructed rack has been useful for the transportation and subsequent drying in a desiccator of small (10 cm) plates (Fig. 17). The company Baron has recently included a rack for 10 × 10 cm plates in its delivery program.

☞ **Practical Tip:** Do not leave your fingerprints on the plates! These can become clearly visible after derivatization by some reagents, e.g., ninhydrin or vanillin-sulfuric acid reagent. With certain solvent systems, the "constituents of perspiration" also spoil the chromatogram by appearing as unknown extra zones.

Preliminary tests are often performed on 5 × 10 cm plates, and these are commercially available coated with silica gel 60 F_{254}. To obtain plates of this size precoated with other sorbents it is best to break down larger plates. In the early days only simple diamond glass cutters looking like pens could be used to score glass plates, and a "clean" break was seldom produced. However, thanks to T. Omori a TLC plate-cutting tool has now been available for some years (Fig. 18a), and this can be used to cut glass plates into strips as narrow as 1 cm without difficulty; this equipment is supplied by DESAGA. The TLC plate is placed, with its coated layer facing downwards, on a supporting plate with raised 1-cm divisions along the side. It is located by placing it against the second guide rail, which is at right angles to the first. The carbide cutter, which is in a holder, can be caused to score the glass by pressing it with the index finger. It is first moved across the plate, its movement being controlled by a guide rail (Fig. 18c). Keeping it steady, the cutting head is moved back to its starting position. After removing the TLC

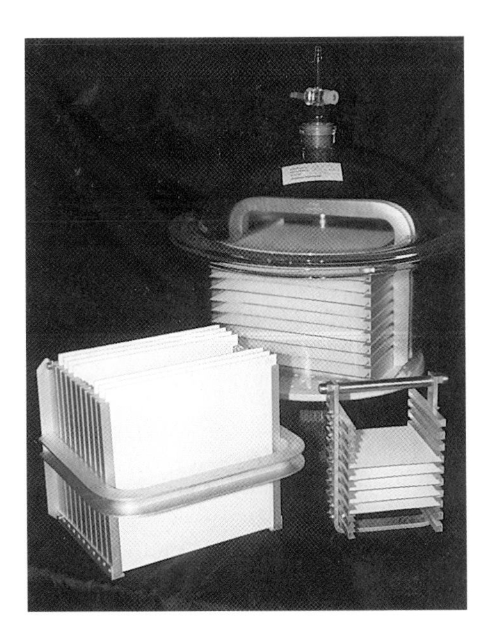

Figure 17. Drying rack for the transport and storage of TLC plates
Left front 20 × 20 cm TLC plates stacked vertically in a tilted drying rack (ready for "activation"), *right front* drying rack for plates with 10-cm sides (in-house construction), *back* storage of TLC plates in a desiccator over silica gel

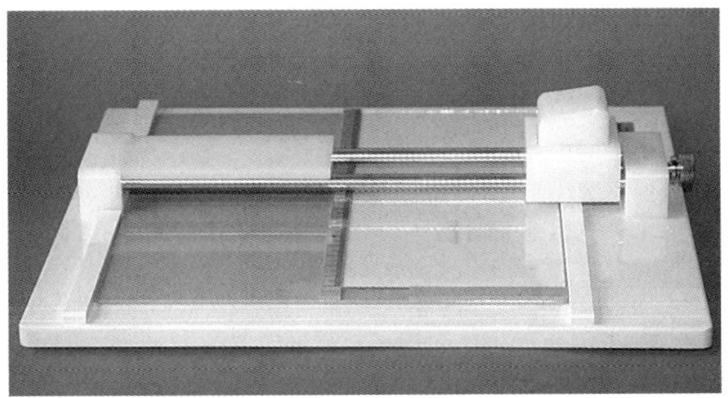

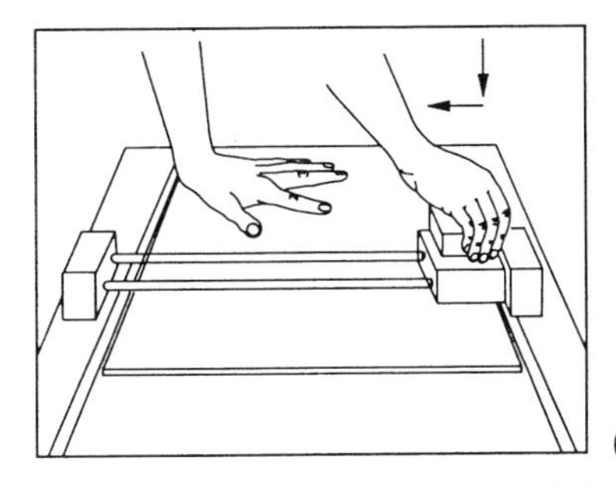

(b)

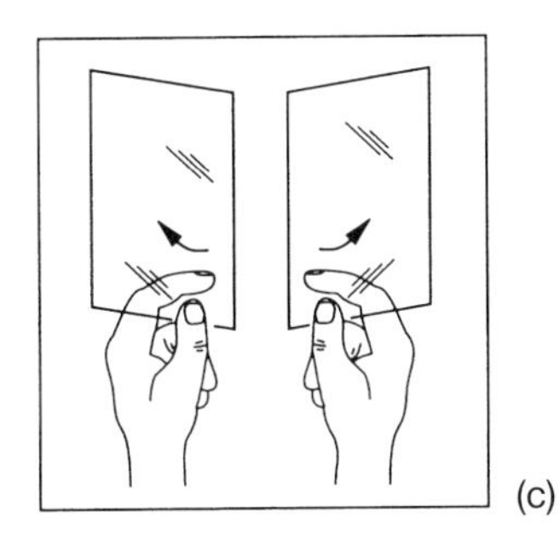

(c)

Figure 18. The TLC plate cutter of T. Omori
(a) A 20 × 10 cm plate, with its layer underneath, ready to be divided
(b) Scoring
(c) Snapping the TLC plate into two pieces

plate from the support plate, it is placed on a paper tissue and then broken into two pieces along the scored line by bending it with the hands (Fig. 18c).

Other examples of the use of TLC plate-cutting equipment are given below:

- An "unsuccessful" application zone is cut off and the samples are applied again.
- If the plate is cut into several pieces after development, different methods of derivatization can be performed.
- If it is desired to use a photo-copier to produce transparencies for documentation without first producing a paper copy, it is preferable to cut up the derivatized plate so that only the desired sections of the chromatogram are photocopied for a subsequent presentation.

3.2 Prewashing

Glass plates are usually coated with surface-active sorbents, which pick up not only water but also "dirt" from the surrounding atmosphere. This should be removed just as completely as the soluble binder components that can form dirty zones with certain solvent systems (mainly polar).

This purification step is known as "prewashing". It produces a uniformly "clean" plate background for visual assessment, and a more favorable base line and an improved signal/noise ratio are observed when measuring with a TLC scanner or other densitometric evaluation system [15]. This operation is therefore essential for quantitative evaluation. It is also to be recommended for stability testing and in trace analysis. In routine analysis also, e.g. raw materials testing, prewashed plates should always be used, as a dirty front would otherwise appear at the height of the substance zones under investigation (Fig. 19a,b).

▶ **Figure 19:** see Photograph Section.

Prewashing is performed either by dipping (once or twice, using dipping times in the range 1–7 min) or by blank chromatography of the plate [16]. If dipping is performed over shorter time periods a uniform layer is certainly obtained, but the desired cleaning effect is often not as good. Blank chromatography requires rather more time. Here, a good cleaning effect is obtained as the "dirt" becomes concentrated at the top edge of the plate. Methanol is usually used as the prewashing agent. If the cleaning power is insufficient for the solvent system which is to be used later in the chromatographic process, it is a good idea to perform a development using this solvent system, to which neither acids nor bases should normally be added. In the blank chromatography, the plate should be marked, e.g. at the right hand side of the solvent front, to ensure that the prewashing and the subsequent development are performed in the same direction.

As a drying rack for the subsequent necessary drying is designed for 10 plates, twin developing chambers with vertical slots in the transverse walls are used. Five plates are placed in each of these for a "washing process" for 20 × 20 cm plates. These are each provided with 75 ml washing liquid whose composition should be tested for each application and prescribed in the testing procedure. If coded GLP plates are not used, the plates are marked with a marker pen in the top right hand corner. It has been found to be advisable to mark with the item number of the manufacturer. Thus, plates with different sorbents can later be kept in one desiccator and used without fear of confusion. Smaller plate sizes, e.g. 10 × 20 cm, are placed in 10 × 20 cm developing chambers (containing a maximum of two plates). TLC and HPTLC plates are prewashed up to the top edge. However, if this is not possible for reasons of time, care must be taken that the prewashing distance should be 1 cm longer than the migration distance to be used in the subsequent chromatogram. The prewashing front is marked with a marker pen on the right hand edge directly after taking the plate out of the developing chamber. Almost all sorbents can be subjected to this cleaning process, depending on the application. However, it is not necessary for "high-purity" silica gel plates already cleaned by the manufacturer provided that these have been stored in their original packaging. Finally, the prewashed plates must be dried before use such that no solvent residues remain on the precoated layers.

Errors due to prewashing always result if the prewashing front formed in blank chromatography is not marked and the plate is rotated through 90° when the sample is applied, after which it is developed. The cartoon below shows an example of this. Prewashing agents with a poor elution strength such that the "dirt" is not removed from the plate or which give a short migration distance in the blank chromatography can lead to poor results. Thus, for example, a TLC plate to be used for stability testing of tablets was prewashed with ethyl acetate and methanol (1 + 1 ml) and then developed with the solvent system toluene + 2-propanol + conc. ammonia solution (70 + 25 + 1 ml). The dirt zone can be clearly seen in Fig. 20.

▶ **Figure 20:** see Photograph Section.

If, in order to save time, it is decided to prewash using the dipping technique, its effectiveness should be verified experimentally.

☞ **Practical Tip** for prewashing: if the signal/noise ratio is too low (see Section 5.2 "Direct Optical Evaluation Using Instruments"), the blank chromatography method should be given preference.

3.3 Activation

An example of activation is the drying of silica gel plates for 30 min at 120 °C. This causes the physically adsorbed water from the moisture present in the atmosphere to be expelled from the surface of the silica gel. If a much higher temperature is used, this could cause some release of chemisorbed water and hence an irreversible change in the chromatographic properties of the silica gel.

In adsorption chromatography, high activity leads to high retardation properties of the stationary phase and hence to shorter migration distances of the sample substances than those obtained with less active sorbents. This means that, to achieve reproducible retardation values, a well-defined level of activity of the layer is necessary.

To dry the marked plates, they are removed from the prewashing chamber and slid, in the direction of flow and with the layer at the top, along the slots of the drying rack until they hit the end stops. They are then left for a few minutes in this position in the fume hood before tilting the drying rack until the plates are standing on the end stops in a vertical position. The drying rack and plates are then placed in the drying oven preheated to 120 °C. On completion of the drying process, the hot rack is taken out of the drying oven using insulating gloves or a cloth, tilted back into its original horizontal position and, keeping it in this position, placed in a vacuum desiccator containing self-indicating silica gel. To equalize the pressure, the desiccator tap must be opened briefly. The plates are kept in the desiccator until they are used [17]. To avoid undesired contamination, the use of any type of grease (e.g. ground-glass joint lubricant) is avoided.

The activation process described above is mainly suitable for silica gel and aluminum oxide layers. For layers of other sorbents, information about suitable activation conditions can be requested from the manufacturers. For example, the company Merck, in the information sheets provided with the package, recommend activation of amino-modified HPTLC plates for 10 min at 120 °C before use.

A process of further activation, e.g. 3–4 h at 150 °C as described by Stahl in 1962 [17], leads to very active layers in which there is considerable danger of decomposition of the substances to be separated (formation of artifacts). For very sensitive sample substances, the use of layers with lipophilic modification, e.g. RP material, is generally recommended.

Temperature and time are possible **sources of error in activation**. Too short a drying time and/or too low a temperature can result in incomplete removal of the "prewashing agent", leaving the layer in a poorly defined condition and causing a shift in the hRf values. In contrast, too long an activation period and/or too high a temperature can lead to some removal of chemically bonded water and hence to changes in the chromatographic properties of the precoated layer. Moreover, in such circumstances, cracking of the coating or decomposition in the case of chemically modified sorbents can occur.

3.4 Conditioning

If special precautions are not taken during the application of the samples, the high activity of the precoated layers produced by the activation process can deteriorate, as an equilibrium, which is dependent on the relative humidity of the laboratory atmosphere, is established between the laboratory atmosphere and the sorbent within a few minutes. However, preconditioning of the layer can be used to influence the subsequent chromatographic process. This can be achieved, for example, by exposing the plate to a gas to modify the condition of the layer. Changes can be produced by a controlled moisture content. Saturated aqueous solutions of the salts listed in Table 5 produce definite relative humidities in closed vessels [18]. The solutions, which also contain suspended salt and sediment, are placed either in one trough of a twin-trough chamber or in the inner compartment of a horizontal chamber. A single plate can be conditioned overnight in the controlled humidity. Several plates can be conditioned in a drying rack above such a saturated solution in a desiccator. Figure 21a–c gives an example of the effect of preconditioning on the chromatography. Depending on the relative humidity to which the layer is exposed, marked changes can be observed in the hRf value and the separation behavior of some *meta*-oligophenylenes [19].

A very impressive example of a good separation obtained by controlling the humidity was described by Janssen et al. in their paper “Identifizierung von Zimtstoffen mittels DC und HPTLC” (“Identification of components of cinnamon by TLC and HPTLC”) [20]. The substances eugenol, cinnamaldehyde, coumarin, linalool etc. are applied to TLC and HPTLC plates next to the cinnamon bark preparation, and the plates are then placed in a chamber at various relative humidities (9 %, 24 %, 40 %, 50 % and 64 %) for 30 min. The best separations were obtained by using the layer

Table 5: Production of constant humidity in closed vessels

Saturated salt solution containing a large quantity of undissolved salt		Relative humidity over the solution (20 °C) [%]
di-Sodium hydrogen phosphate	$Na_2HPO_4 \times 12\ H_2O$	95
Sodium carbonate	$Na_2CO_3 \times 10\ H_2O$	92
Zinc sulfate	$ZnSO_4 \times 7\ H_2O$	90
Potassium chloride	KCl	86
Ammonium sulfate	$(NH_4)_2SO_4$	80
Sodium chloride	$NaCl$	76
Sodium chlorate	$NaClO_3$	75
Sodium nitrite	$NaNO_2$	65
Ammonium nitrate	NH_4NO_3	63
Calcium nitrate	$Ca(NO_3)_2 \times 4\ H_2O$	55
Sodium dichromate	$Na_2Cr_2O_7 \times 2\ H_2O$	52
Potassium carbonate	K_2CO_3	45
Zinc nitrate	$Zn(NO_3)_2 \times 6\ H_2O$	42
Chromium trioxide	CrO_3	35
Calcium chloride	$CaCl_2 \times 6\ H_2O$	32
Potassium acetate	$K(OOCCH_3)$	20
Lithium chloride	$LiCl \times H_2O$	15

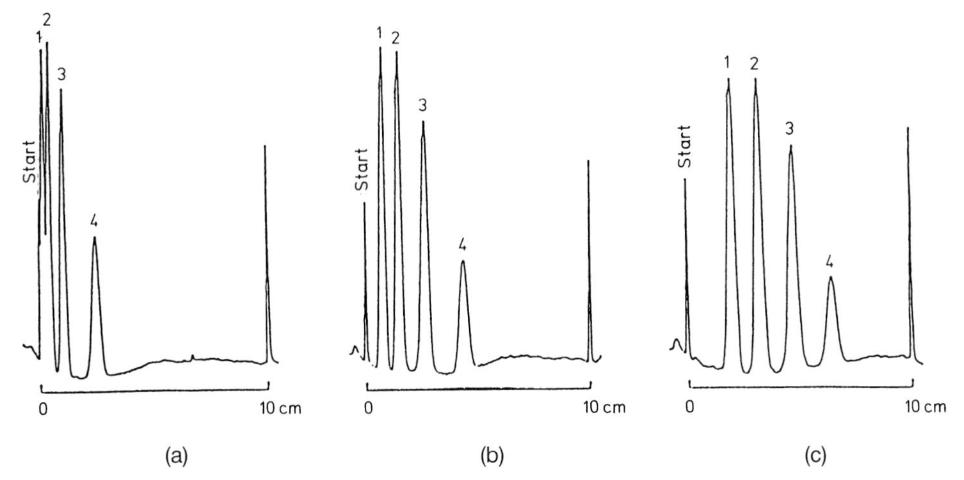

Figure 21. Influence of the relative humidity on the retardation and resolution, using the example of *meta*-oligophenylenes

Sorbent:	HPTLC silica gel 60 F_{254} (Merck 1.05642)
Solvent:	Cyclohexane
Chamber:	Vario KS chamber (CAMAG)
Preconditioning:	(a) 20 % relative humidity (b) 50 % relative humidity (c) 80 % relative humidity
Test substances:	1. *meta*-quinquephenyl 2. *meta*-quaterphenyl 3. *meta*-terphenyl 4. *meta*-biphenyl
Detection:	Direct optical evaluation under 254-nm UV light using the TLC-Scanner II (CAMAG)

activity obtained at 40 % relative humidity, as only here were the four above-mentioned substances clearly separated from each other.

Using similar techniques, preconditioning can be performed by treating plates with molecules of organic compounds (e.g. solvents) or with acids and bases, sometimes after the sample has been applied, and this often improves the chromatographic result.

☞ **Practical Tips** for conditioning TLC plates:

- When applying samples to a plate, place a clean glass plate above the application zone so that the high activity of a precoated layer will be retained for as long as possible until it is developed.

- When conditioning precoated layers with saturated aqueous solutions, ensure that enough undissolved salt is present. If this is not the case, reproducible conditioning will be impossible.

3.5 Impregnation

The solution to some separating problems can sometimes be to impregnate the sorption layer with inorganic or organic substances. This procedure requires not only time but also some experience with TLC. The impregnation is performed either by dipping, spraying, or developing with the impregnating solution. For example, to establish a homogeneous pH throughout the layer, it can be exposed to, e.g., ammonia or acid vapors in a conditioning or developing chamber.

Impregnation by dipping is used for separating unsaturated compounds (impregnation with silver nitrate), impregnation by spraying for separating antibiotics (EDTA impregnation) and impregnation by development for separating alkaloids (pH control).

3.5.1 Impregnation by Dipping

A TLC plate precoated with silica gel (0.25-mm or 0.5-mm layer thickness) is dipped for 15–20 min in a 20 % aqueous solution of silver nitrate. The plate is then dried, first in the air with exclusion of light, and is then activated at 80 °C in a drying oven. Using this procedure, each plate with a 0.5-mm layer takes up ca. 3.4 g silver nitrate. As this layer contains ca. 8–10 g silica gel, the silver nitrate content after the impregnation is ca. 40 % [12]. Kraus describes the slightly modified impregnation of 5 × 5 cm TLC silica gel 60 F_{254} plates: the plates are dipped for 3 min in a 3 % silver nitrate solution (methanol/water [93 + 7 ml]), and are then dried and activated as described above. If plates impregnated in this way are not immediately used for chromatography, they should be stored wrapped in black plastic film in a desiccator [21].

Separation on silver nitrate layers is based on complex formation by the Ag^+ ions with the π-electrons of one or more double or triple bonds of the substances being separated and depends on the strength of the bond [22]. As an example, Kraus describes the separation of some C10 alcohols which cannot be separated by either partition or adsorption chromatography without impregnation [21, experiment 13.9.9].

3.5.2 Impregnation by Spraying

The monograph "Doxycycline" in the DAB 96, 2nd addendum 1993, specifies a layer sprayed with a solution of sodium edetate solution for the identification of this substance. However, the problems begin with the choice of the layer, as DAB states: "the test is performed with the aid of TLC (V.6.20.2) using a layer of silica gel H *R*." The DAB, as always, specifies TLC plates produced in-house, as silica gel H is a loose sorbent without additions of extraneous binders (see Table 2), where *R* represents a silica gel with a mean particle size of 15 μm.

☞ **Practical Tip:** Forget the DAB recommendations and use "normal" TLC plates precoated with silica gel 60 without fluorescence indicator. If the only plates available contain this indicator, it does not matter for this example. This method, of course, does not imply treatment of only *one* plate, and the impregnation process is not always successful every time. If you impregnate three plates, you will be on the safe side as a beginner.

The substance name "sodium edetate R" in the DAB refers to the reagent *ethylenedinitrilotetraacetic acid disodium salt dihydrate*, also designated "Titriplex III" by the company Merck and available commercially as Article No. 108418.

Procedure: 80 ml of a 10 % aqueous solution of Titriplex III (m/v) is adjusted to pH 8.0 by addition of a 40 % aqueous solution of sodium hydroxide, using a pH-meter. This solution is sprayed uniformly on the plates, which have been previously activated for ca. 15 min at 120 °C. Approximately 15 ml of solution is sufficient for a 20 × 20 cm plate. The plates are then dried for ca. 30 min in a horizontal position and then placed in a drying oven at 120 °C for a further 30 min. They are then allowed to cool in a desiccator, in which they are stored until used.

What is the effect of impregnating with Titriplex III?
The complexing agent combines with the metallic ions present in the sorption layer which would otherwise interact with the substance under investigation. A TLC identification of doxycycline would be impossible.

Figure 22a shows the chromatogram on a sorbent layer which is not impregnated, and Fig. 22b shows the chromatogram on an impregnated TLC plate precoated with silica gel 60. However, a poorly performed impregnation leads to unusable TLCs, as shown in Fig. 22c.

▶ **Figure 22:** see Photographic Section.

☞ **Practical Tip** for impregnating by spraying: Application of the aqueous solution of Titriplex by spraying uniformly first gives a transparent effect (see also Section 6.3.1 "Spraying of TLC Plates"), after which the drying time at room temperature determines whether or not cracking of the layer will occur when further drying is performed in a drying oven. Of course, the prescribed temperature of the drying oven must also be used here.

3.5.3 Impregnation by Predevelopment

It is not always possible to obtain good separations of antibiotics of the tetracycline type on TLC plates impregnated by spraying. Oka describes the alternative method of predevelopment with saturated aqueous Na_2EDTA solution followed by activation for 2 h at 130 °C [22a]. Eight tetracycline antibiotics are separated by TLC using various solvent systems.

In the following example, taken from the literature, a significant improvement in the separation and subsequent quantitative evaluation could only be achieved by impregnation with an acid-containing solvent system, i.e., by changing the pH of a layer.

In March 1994, Fulde and Wichtl published in the Deutsche Apotheker-Zeitung a paper on the analysis of greater celandine [23], in which they showed that the main alkaloid in the leaves of the plant was not chelidonine, as was previously assumed, but coptisine. However, on further examination of the data given in the experimental part, no agreement with the chromatograms also shown in the article could be obtained. Also, there was no information in the text about which of the five named reference substances appeared in which part of the chromatogram. Only a reading of the dissertation of Fulde [24] led to a solution of the puzzle: The HPTLC plates were previously subjected to a single development operation using the solvent system chloroform/methanol/water/acetic acid (85 + 15 + 1 + 2 ml) and were then dried at 40 °C for 15 min. Fulde referred to this process as "prewashing of the layer", which it is not, according to the true meaning of the term. The application of the acid to the layer produces a pH gradient in the direction of the development and thereby directly affects the hRf values of the alkaloids under investigation. The chromatographic conditions for this experiment, with and without impregnation, are given in Table 6 (greater celandine), which also includes other alternative TLC systems for greater celandine as well as the data for the DAB system. The corresponding Fig. 23a,b is taken from a poster describing this medicinal plant [25].

▶ **Figure 23:** see Photograph Section.

In connection with this example, it should also be pointed out that publications on TLC only make sense if all experimental conditions are quoted correctly. Nyiredi, in an editorial in the Journal of Planar Chromatography (JPC), which he edits, lists the information that must be provided if scientific contributions to the JPC are to be given any consideration in the future [26].

Further examples of the use of impregnated TLC plates

Normally, impregnated TLC plates are only used for special separations, i.e. rarely, so that commercially produced plates of this type would be too expensive. Thus, A. Koch describes the separation of plant lipids on 5 × 5 cm HPTLC silica gel 60 plates which were impregnated with 0.5 % phosphoric acid especially for this application [27]. The six relevant polycyclic aromatic hydrocarbons are tested for rather more frequently in the monitoring of drinking water. An HPTLC precoated layer impregnated with caffeine is manufactured commercially especially for this application.

Table 6: Greater celandine (*Chelidonium majus L.*)

	DAB 96	**HD [72]**	**Fulde/Wichtl [23]**	**Fulde [24]**	**HD/Koch [25][a)]**
Used parts of the plant	(A) upper parts of the flowering plants, dried			As (A) plus roots	As (A) plus roots
Sample solutions	ff extr. with $CHCl_3$ out of alkaline solution, concentrate	Acc. to DAB, Fulde, with MeOH in USB + water bath	Soxhlet extr. 4 h with MeOH	Soxhlet extr. 4 h with MeOH	Acc. to DAB, Fulde, with MeOH in USB + waterbath
Reference substances	used for all experiments: berberine, chelidonine, coptisine, sanguinarine				
Sorbent	Prescribed: Silica gel GF_{254} R Used: Silica gel for TLC 60 F_{254} GLP 20 × 20 cm (Merck 1.05566)	Silica gel for HPTLC 60 F_{254} GLP, 20 × 10 cm (Merck 1.05613)	Silica gel for HPTLC 60 F_{254}, 20 × 10 cm (Merck 1.05642)	Silica gel for HPTLC 60 F_{254}, 20 × 10 cm (Merck 1.05642)	HPTLC-RP-18 WF_{254s} 10 × 10 cm (Merck 1.13124)
Predevelopment + drying	No	No	No	Yes, in ss, + 15 min 40 °C	No
Solvent system	1-Propanol + water + formic acid (90+9+1)	Toluene + methanol (9+1)	Chloroform + methanol + water + glacial acetic acid (85+15+1+2)	Chloroform + methanol + water + glacial acetic acid (85+15+1+2)	Methanol + water + glacial acetic acid (18+2+0,5)
Applied sample Volume	20 µl	10 – 15 µl	1 µl = 10 µg herb	1 µl = 10 µg herb	2 µl
	20 × 3 mm	10 mm bandwise	Bandwise	Bandwise	7 mm bandwise
Chamber saturation	Yes	No	Yes	Yes	No
Migration distance	10 cm	7,5 cm	[23]: 5 cm, here: 7,5 cm	[24]: 5 cm, here: 7,5 cm	8 cm
Development time	90 min, RT 22 °C	17 min, RT 22 °C	26 min, RT 22 °C	No pred: 26 min, RT 22 °C	Approx. 40 min, RT 22 °C
Detection	UV 365 nm, then approx. 15 min UV 254 nm, once more UV 365 nm				
hRf-values:					
Berberine	17–20	0–3	17–20	13–17	43–46
Chelidonine	39–43 and 43–49	45–49	77–80	43–47	52–56
Coptisine	8–11	0	9–14	8–13	37–40
Sanguinarine	13–34	67–71	84–88	47–53	33–36

a) This system is also suitable for 5 × 5 cm HPTLC plates using the small H-chamber. Development time 8 min for 4 cm migration distance. Especially recommended for schools, because the solvent system is free of CHCs!

ff extr. = fluid-fluid extraction; ss = solvent system; No pred. = time without predevelopment

3.6 Application of Samples

At this point, there usually follows a chapter about the pretreatment of the samples. However, in contrast to HPLC/GC, sample preparation for TLC is not considered to be quite as critical. As well as the use of precoated layers with a concentration zone (e.g. an application zone consisting of silica 50 000 and a separation zone of silica gel 60 or RP-18 material) upon which the matrix constituents can often be held back by suitable choice of solvent system, a chromatogram that is "unusable" for lack of sample preparation is more rapidly rectified (use a different preparation method and a new plate!) than an irreversibly destroyed column. A detailed treatment of the subject of "sample preparation" would exceed the scope of the present book. In Section 9.4, under the title "Examples of GMP/GLP-Conforming Testing Procedures", we describe the extraction of a pharmaceutically active substance from a tablet and the working up of plant components from dry extracts. The reader is referred to other TLC textbooks [2, 21] and to literature and brochures produced by manufacturers of articles for sample preparation [28, 29].

Special care is necessary in TLC when working with light-sensitive samples. As special precautions must be taken when dealing with these throughout all stages of the work from receipt of the material to the documentation of the chromatogram, a special chapter is devoted to sensitive samples (see Section 10.2 "Substances Sensitive to TLC").

In the choice of the solvent system for sample preparation and application, and later in the formulation of solvent systems, there can always be conflicts of conscience: The pharmacopoeias often recommend the use of chlorinated hydrocarbons (CHCs), and benzene is also often specified in modern literature as a solvent. The latter should by now be completely banned from the laboratory because of its carcinogenic effect! The use of chlorinated hydrocarbons should also be viewed with considerable suspicion. Less environmentally hazardous solvents that are almost as effective are often available. The choice of the solvent system mainly depends on the solubility of the substance or mixture of substances. Other criteria as well as environmental compatibility include the volume of the applied sample and the diameter of the initial spot. When applied with *n*-hexane, substances remain exactly at the point of application, while with increasing polarity of the solvent (toluene, dichloromethane, methanol), they are transported towards the edge of the "wet zone", often with formation of a circular chromatogram (anomalous chromatography). The applied substance mainly ends up at the edge of the spot. Then, on chromatographic development, peaks with an approximate Gaussian distribution are obtained whose width increases as the polarity of the solvent increases. Figure 24a–c, in conjunction with [2], shows the substance distribution as a function of the solvent used.

Up to this point in the description of TLC, we have not differentiated between the differing experimental requirements associated with qualitative, quantitative or preparative results. However, very different attitudes lie behind these three concepts and the consequent demands made on the separation techniques used.

Preparative work is not the main topic of this book, so that classification must be in accordance with the other areas of TLC, i.e. identity testing, purity testing and assaying. The demands made on these and hence the differences between them begin with the choice of the layers and end only with the documentation.

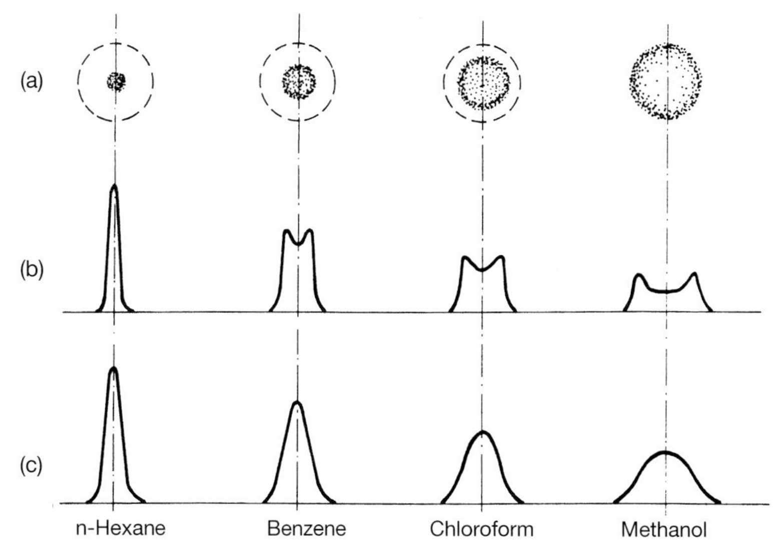

Figure 24. Substance distribution as a function of the solvent system used (from [2, p. 210])
(a) After applying (top view)
(b) After applying (profile)
(c) After developing (profile)
(Note: benzene is carcenogenic! It should be replaced with toluene or a similar solvent)

3.6.1 Manual Application of Samples

You have now carried out the necessary steps of sample preparation, and your samples are now in solution and ready to be applied. Before you take the selected plate material out of the package or the desiccator, please consider how your chromatograms should appear later on, e.g. with respect to the positioning of the samples, migration distance etc.

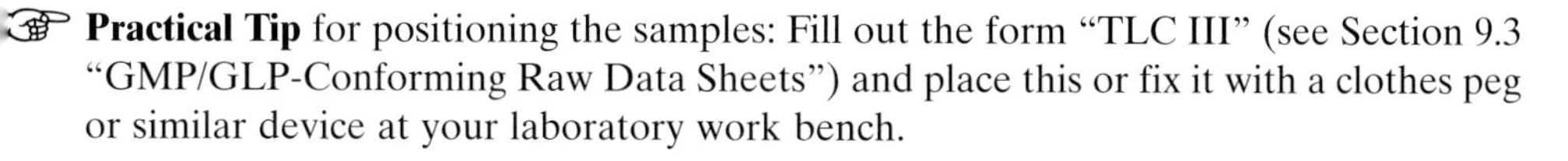

Practical Tip for positioning the samples: Fill out the form "TLC III" (see Section 9.3 "GMP/GLP-Conforming Raw Data Sheets") and place this or fix it with a clothes peg or similar device at your laboratory work bench.

Depending on the number of samples, select the plate size or cut it from a larger plate to the required dimensions. It is a good idea to view every precoated plate horizontally against the light before use. Any faults in the layer such as small air inclusions or cracks are then easily seen, as is the difference between TLC and HPTLC precoated plates, the latter having a more homogeneous shiny surface. Also, the difference can be felt by gently running the fingertip over the surface of the layer close to the edge of the plate.

For beginners or for people less experienced in TLC, it is often difficult to carry out the preliminary work correctly before applying the samples. Beginners have often been seen to mark with a pencil not only the desired position of the front but also that of the application zone by drawing a line across the whole plate! Moreover, other inscriptions such as the number of the solutions and their concentration have been written underneath the application zone. Who teaches young people to behave like this?

Damage to the surface of a layer can lead to errors

Only absolutely essential markings are written on the layer with a soft pencil (e.g. 2B). These can include the starting point and, on the right hand edge (ca. 0.5 cm wide), the desired position of the front. By now you will certainly have made a note of the application scheme on the data sheet, making it is unnecessary to transfer it to the layer.

When applying samples manually, it is advisable not to place the starting point too close to the bottom edge of the plate, as, depending on how much the solutions spread out on application, this can readily cause the starting point to extend into the solvent. Erroneous chromatograms would result. Figure 25a shows an example of a TLC plate in which the starting points have become partially immersed in the solvent. The error is clearly visible compared with Figure 25b, which shows the "correct" chromatogram of the same substance.

▶ **Figure 25:** see Photograph Section.

The commercially available **application templates** are useful aids to the positioning of the starting points, and make it unnecessary to mark the plate directly with a lead pencil. The use of these templates goes back to the time when application of samples for quantitative analysis was done by hand, and where the exact position of the spot was very important. However, modern precoated layers are more abrasion resistant than they were in the early days. Moreover, the distance apart of the spots when using application stencils is inflexible, and some proprietary stencils give a pattern of spots 5 mm apart, while with others the distance is 15 mm.

For manual application, a large number of **instruments** are obtainable. A selection of TLC equipment is shown in Fig. 26. Most of these application pipettes or microcapillaries (also known as microcaps) fill up automatically on dipping them into the sample solution and empty themselves on contact with the sorbent surface. However, their use is limited to relatively low viscosity solvent systems. With aqueous solutions, for example, it is advisable to use microliter syringes with a 90° adaptor for the hollow needle. For in-process control, syringe pipettes are often used because of their speed of operation, although the risk of damage to the layer by the hard tips of the pipettes is extremely great. All the listed instruments are based on the same principle: the application of the sample solutions takes place by direct contact with the layer surface. Care should be taken to keep the instrument at an angle of 90° to the plate, especially when using capillaries.

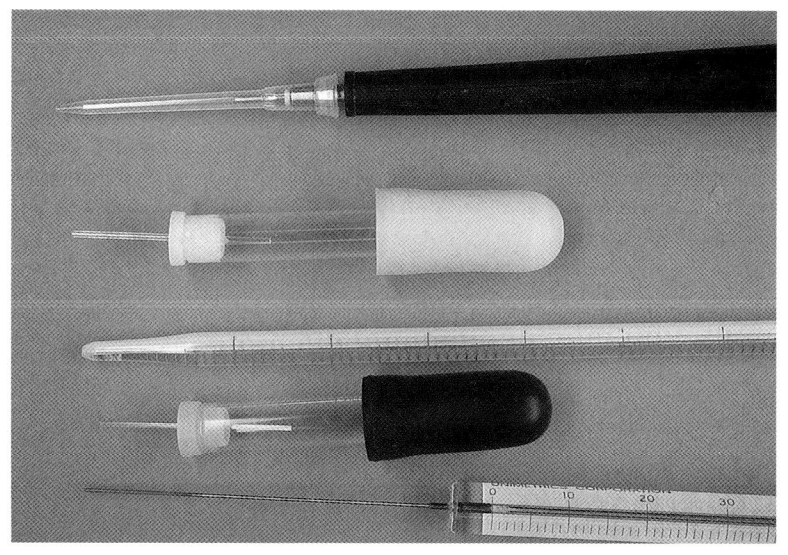

Figure 26. Equipment for sample application
From the top of the picture to the bottom:
1. Eppendorf pipette with 10-µl syringe
2. 5-µl Microcapillary in the holder
3. Application pipette (50 µl)
4. 1-µl Microcapillary in the holder
5. Unimetrics syringe (50 µl)

How is the correct contact with the sorbent obtained?
R. E. Kaiser likes to say in his lectures (see also Fig. 5a):

The stationary phase is like a gravel bed, and should only be found covered with flowers!

An optimal contact of the application system is obtained if the lower surface of the liquid in the capillary comes into contact with the "mountain tops" of the layer, whereupon it begins to flow out. To achieve this contact, gentle pressure is necessary so that the surface of the layer close to the circular outlet of the capillary becomes slightly deformed. This pressure depends on the packing density of the phase and on the binder material used. It must be determined experimentally before the analysis. Too great a pressure (>10 g/mm^2 of the surface) leads to damage to the layer [30].

With manual application at a point, the **sample volume** to be applied depends on the object of the experiment and the concentration of the sample solutions. For identity tests, 0.5–2.0 µl is usual, while to test the purity of a substance a figure of 10 µl (applied in small portions) should not be exceeded. Routine users of TLC apply even 20 µl by a capillary pipette in portions of 4×5 µl with intermediate drying. Thus, for example, when testing samples taken from individual containers of a batch, ca. 120 samples are applied by hand in ca. 4 h on 12 plates [31].

Even today, there are still experts who apply samples for quantitative determinations by hand. Here, the applied volumes are 50–200 nl, and these are applied to the plate either by adjustable microliter syringes (>50 nl) or using a platinum-iridium capillary fused into a glass tube. From our own experience we have become familiar with the problems of quantitative application by hand: one needs very good eyes and a steady hand to position the solutions exactly, one must not allow any disturbance by anyone or anything, and one should take the greatest care of one's own personal Pt-Ir capillary. With syringes and Pt-Ir capillaries, in contrast to single-use capillaries, rinsing after use is extremely important. If you want to find out about the "memory effect", you should apply a solution of a red dye and then check how often the rinsing process must be repeated before no traces of red can be seen on the point of application. Even rinsing ten times may not be enough!

A question often discussed is whether microcaps are disposable items and should be thrown away after use.

Jork thinks it is better to use the same capillary for all application operations in an "experimental run", provided that it is rinsed at least three times between experiments to prevent errors due to carryover [30]. Here, an "experimental run" means the application of several samples to one plate for quantitative analysis, as was usual before the use of automatic application equipment. Today, questions of cost and/or time must be considered when deciding whether microcaps should be disposable items.

☞ **Practical Tip:** The risk of substance carryover in pharmaceutical quality control should be completely excluded by using microcaps once only.

When is it preferable to use syringes rather than capillaries for applying sample solutions?

Mainly if it is difficult to fill the capillary reproducibly. This is the case

- with materials containing a high-density solvent, e.g. chloroform or methylene chloride, so that the liquid tends to run out of the capillary when it is vertical;
- with test solutions that have to be prepared with *n*-hexane, petroleum ether (b.p. 40–60 °C) or diethyl ether. Here, the capillary forces are often insufficient to fill the capillary reproducibly;
- with surfactant solutions in which the surface tension is much reduced, so that here too there can be problems in filling the capillary;
- with viscous liquids that do not easily flow into the capillary (solutions containing albumin or other slimy materials). With narrow capillaries, this problem can even occur with aqueous systems [32].

In quantitative TLC, the analytical results can be affected by the chromatographer's state of mind on the day, and it is therefore usual today to advise against sample application by hand.

Application of samples can lead to unintended **errors**. The commonest of these are described below, and earlier comments in the text are summarized at the same time.

Unsuitable solvents
These include solvents with low boiling points, e.g. acetone, chloroform or ether, which sometimes evaporate from the capillary immediately after it is filled and before it can be used to apply the liquid to the layer. The substance then crystallizes out on the outside of the capillary (creep-back effect). Alternatively, the capillary "loses" the substance in an uncontrolled manner because of the high density of the solvent (e.g. chloroform). The substance then usually appears on the TLC plate in finely divided form. In both cases, less than 100 % of the sample is analyzed.

Solvents with high boiling points (e.g. water) are usually difficult to remove from the layer. The energetic drying needed for this can often lead to decomposition or volatilization of the samples.

Unsuitable concentrations
Written testing procedures often prescribe concentrations close to the saturation point of a substance, simply to keep the applied volume constant when the sample is manually applied at a point. This can cause the substance to crystallize out in the capillary and/or can cause the errors described above for solvents with low boiling points.

☞ **Practical Tip:** Dilute the solutions and apply them in the form of horizontal bands using automatic equipment.

The use of concentrations that are too low combined with application at a point usually leads to the applied volume being too large, which causes spreading of the applied spots and hence poor separation efficiency.

Incorrect concentrations
In purity testing according to pharmacopoeias, substances prescribed as reference substances can have a different counter-ion in the molecule from that in the sample being investigated. In such a case, the difference in the molecular mass must always be borne in mind when determining the amount of reference substance to be weighed out.

Errors associated with capillaries include incomplete filling or emptying of these items. Reproducible operations using them depend, among other things, upon

- the adhesion and cohesion forces (surface tension and viscosity of the solutions being applied, etc.),
- the filling and emptying time,
- the inside diameter of the capillary.

A distinction should be made between nano- and microliter capillaries. Pt-Ir capillaries with a capacity of 200 nl are usually used for application to HPTLC layers. With these, polar and medium-polarity solvents can be applied satisfactorily, but purely aqueous solutions cannot be applied reproducibly. Moreover, as they become blocked very easily, either by crystallized substance or by sorbent picked up from the layer, they are less frequently used today, especially now that they have been largely replaced by automatic equipment.

Microliter capillaries with a capacity of 0.5–10 µl are commercially available, although the use of the 10-µl capillary is not recommended, as substances such as chloroform are quite impossible to handle. Even the 5-µl capillary should be used in exceptional cases only, as the application spots are very large and anomalous chromatography usually occurs. Unfortunately, the pharmacopoeias often prescribe 5-µl application volumes without mentioning whether these should be applied in portions or in one operating step. Therefore, this volume is often used directly in chemical and pharmaceutical laboratories. To obtain good chromatograms, it would be better to apply a correspondingly greater volume of a dilute solution in the form of a band.

Errors in the manual use of syringes can be largely eliminated by the use of application equipment with position control (e.g. the CAMAG Nano-Applikator), equipment controlled by microprocessors and syringes with a 90° attachment. Adequate intermediate rinsing is especially important when using syringes. The number of times this must be performed must be determined experimentally.

Damage to the layer

With all methods of application, damage to the layer and hence changes in chromatographic behavior can occur. This happens very often on contact application using piston pipettes, usually known as "Eppendorf pipettes". The hard tips of the pipettes can easily loosen the sorbent from the support. This often not only leads to damage to the layer but also removal of the substance, so that less than 100 % application is achieved. However, the Pt-Ir capillaries also easily lead to ring-shaped damage to the layer, the sorbent becoming detached within this ring and then blocking the capillary. In this case also, less than 100 % of the sample solution is applied.

☞ **Practical Tip** for beginners: Practice applying the sample a few times to the top part of a precoated layer using the prescribed instrument. You will then discover the filling and emptying time for the sample solution and will have a "feel" for the application pressure to be used with capillaries.

In Sections 3.6.2 and 3.6.3 below, we describe the equipment for semiautomatic and automatic application available from three German suppliers. These are (in alphabetical order) Baron, CAMAG and DESAGA, whose addresses are given in Section 12.5.

3.6.2 Semiautomatic Application

The microprocessor-controlled **Linomat IV** of CAMAG (Fig. 27a) enables the sample solutions to be sprayed onto the plate, preferably in the form of a band, by compressed air or nitrogen, thus requiring no direct contact with the layer (Fig. 27b). Hence, considerably greater volumes of the sample can be applied to the plate, the length of the application zone being in the range 0–180 mm, as desired. A long application zone

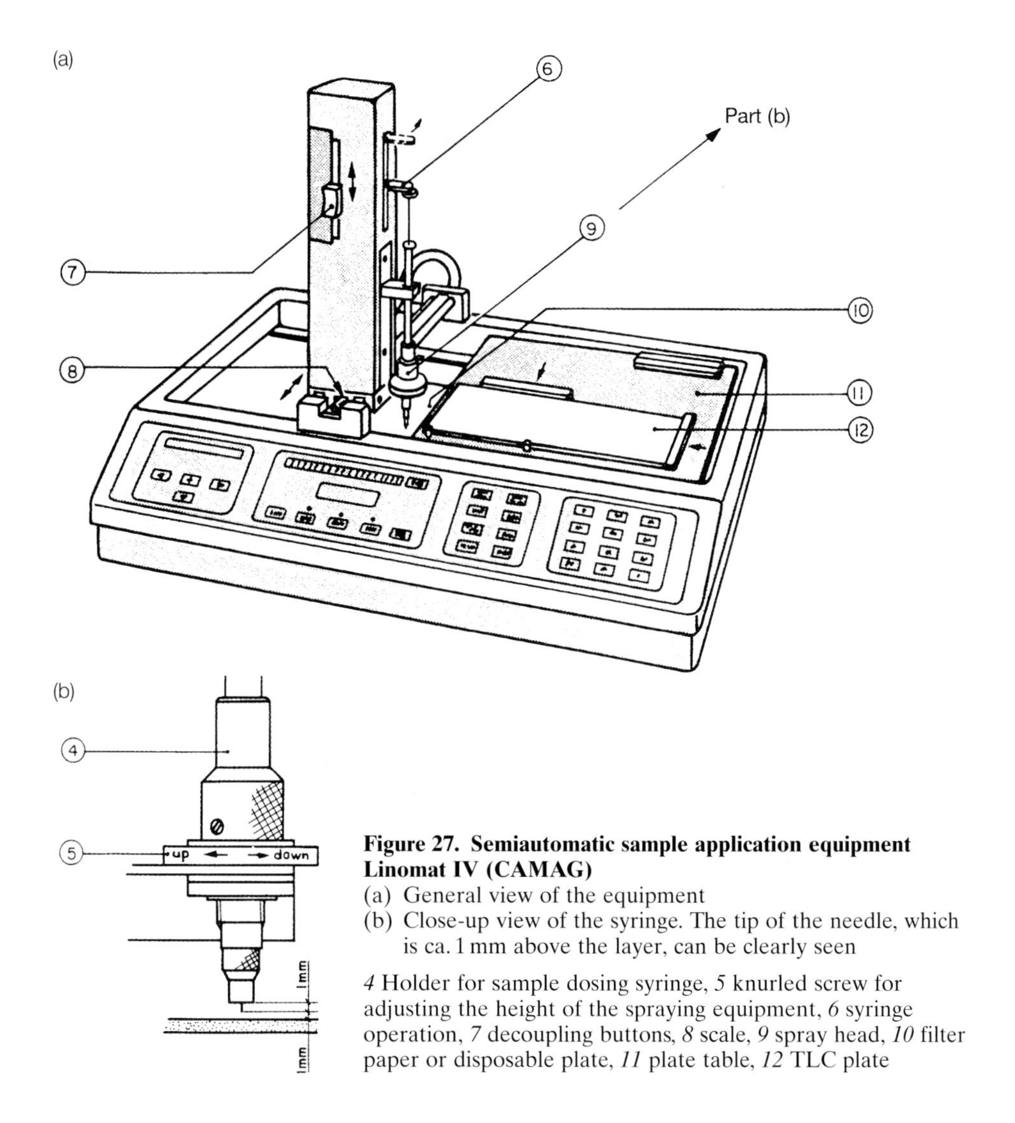

Figure 27. Semiautomatic sample application equipment Linomat IV (CAMAG)
(a) General view of the equipment
(b) Close-up view of the syringe. The tip of the needle, which is ca. 1 mm above the layer, can be clearly seen

4 Holder for sample dosing syringe, *5* knurled screw for adjusting the height of the spraying equipment, *6* syringe operation, *7* decoupling buttons, *8* scale, *9* spray head, *10* filter paper or disposable plate, *11* plate table, *12* TLC plate

is especially useful for preparative work. A 500-μl syringe, with which volumes of 5–490 μl can be applied, is also available. With the more usual 100-μl syringe, 2–99 μl can be applied with good reproducibility.

Band-shaped starting zones give the best possible separation efficiency attainable with the selected TLC system. As an example, Fig. 28 shows the effect of the application method on the separation result. The difference between spotwise and bandwise application can be clearly seen.

▶ **Figure 28:** see Photograph Section.

For quantitative TLC analysis, the Linomat spraying method is claimed by the manufacturer to give further advantages:

- "As the distribution of the substance is uniform over the whole width of the band, densitometric evaluation can be performed by measurement of aliquots.[1] This enables the best possible accuracy to be achieved."
  Our own investigations have shown that if the application is too rapid, e.g. with concentrated methanolic solutions of plant extracts, "dog bone"-shaped application zones can result. The distribution of the substance is then no longer uniform.

- "Facilitation of operation: If the Linomat spraying method is used for calibration purposes, different volumes of the same solution should be applied instead of the same volumes of solutions of different concentrations. This considerably reduces the amount of work required. The standard addition method can be performed with the Linomat by overspraying."
  This advice by the manufacturer should be checked in some cases. We have found from our own experience that, when preparing a calibration curve with volumes of 1–10 μl of the same solution in one series of experiments, the value of the 1-μl volume was too low by a significant amount.

- "In some cases, a prechromatographic in situ derivatization by overspraying the applied samples with reagent solution can be performed." (see also Section 6.4.1 "Prechromatographic Derivatization"). Thus, for example, the application zones of C_6 to C_{14} fatty acids were reacted with dansyl semipiperazide or dansyl cadaverine before chromatography [33]. The result of this can be seen in the title illustration of Vols. 1a and 1b of the book by Jork et al. [34].

The Linomat IV can also be obtained with an inert gas blanket which protects oxygen-sensitive samples from attack by atmospheric oxygen while they are being applied, although only if nitrogen is used as the spray gas. According to the manufacturer's information, the applied substances are protected from oxidation for some time even after the TLC plate is removed, as exchange of the adsorbed nitrogen molecules in the porous surface with the molecules from the air is a gradual process.

In the Linomat IV-Y version, the semiautomatic sample application operation is combined with concentration of the sample solution, preventing damage to the matrix due to the application of large volumes of solution. The samples are applied in a zigzag pattern within a rectangle and are then developed by a solvent system with high elution strength over a short migration distance. After this focusing step, followed by drying of the plate, the actual chromatographic development is performed [35]. Figure 29a illustrates the zigzag application system, and Fig. 29b shows the focussing effect of the initial development step.

In semiautomatic application of samples, the time factor is very important, as the operation should sometimes be performed as quickly as possible to prevent unstable materials from decomposing before the actual chromatography. Therefore, sometimes even an extension of the usual pause times, which are often fixed, can be necessary! To give beginners an idea of the time required for semiautomatic application with the Linomat IV, Table 7 gives some typical examples from our own experience, along with the equipment settings.

---

[1] Authors note: This means that a known fraction of the total zone is measured.

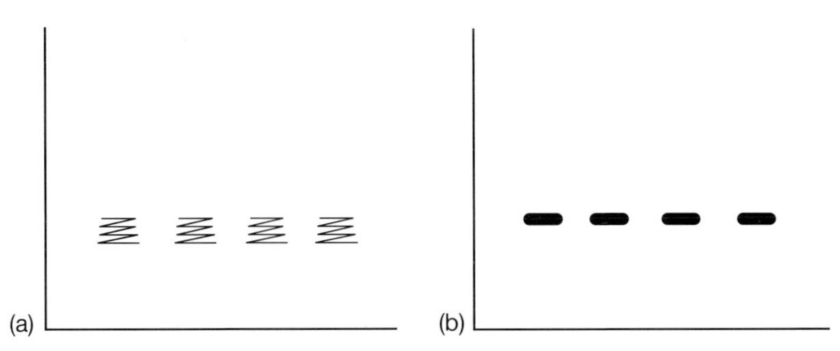

Figure 29. Application of matrix-containing samples to a layer (from [2, p. 228])
(a) After application of the samples
(b) After focussing

Table 7: Semiautomatic application using the Linomat IV

Parameter	Identity	Purity	Assay	Preparative
Class of substance	Herbal extract	Pharm. active ingredient	Pharm. active ingredient	Herbal extract
Dissolved in	Ethanol + water	Acetone	Methanol	Chloroform
Sorbent	HPTLC-RP18 F_{254s} 10 × 10 cm (Merck 113124)	TLC-silica gel 60 F_{254} GLP, 20 × 20 cm (Merck 105566)	HPTLC-silica gel 60 F_{254} GLP, 20 × 10 cm (Merck 105613)	2 mm: PSC silica gel 60 $F_{254+366}$, 20 × 20 cm (Merck 105637)
Quantity of lanes	8	7	20	1
Start [mm]	16	21	22	10
Bandwidth [mm]	5	10	4	180
Space [mm]	4	11	4	–
Applied volume [µl]	8 × 5	5 × 100, 1 × 5, 1 × 10	20 × 10	2000 (4 × 500)
Speed [µl/s]	15	9	7	5
Syringe change	8 ×	6 ×	9 ×	4 × to fill
Total time [min]	10	80	23	ca. 150

☞ **Practical Tips** when working with the Linomat IV:

- Before beginning the application, the distance of the syringe needle from the precoated layer should be checked. This should be ca. 1 mm. If the distance is too small this can lead to damage to the layer, and if it is too large the sample solution will be sprayed over too wide an area.
- When the 500-µl syringe is used, a small warning sign to this effect should be placed in front of the equipment while application is in progress to remind the user that the equipment should afterwards be reset to suit the usual 100-µl syringe. If this is not done, subsequent applications will give five times the expected amount.

- Whenever qualitative or semiquantitative operations are carried out with several plates one after the other using the same reference solutions or sample solutions, it is possible to work with several 100-µl syringes in parallel in order to save time (rinsing operations) or cost (e.g. expensive reference substances). While these are not being used on the equipment, they can be stored in readiness in the bottom part of their styrofoam packing.

- When applying limiting concentrations of various different substances, it is preferable to use different syringes so as to eliminate any possible risk of substance carry-over.

- When applying aqueous solutions to "normal" or water-resistant silica gel layers, the pressure of the spraying gas can be too high, so that there can be a danger of loosening some of the sorbent. In this case, the problem can only be solved by using a different application system if no adjustable pressure-reduction system is available.

- Large amounts of substance are often applied as concentrated solutions. It can then easily happen that particles of substance are spread from the syringe over a large area of the plate (see Fig. 30 "Stability testing of carbamazepine tablets"). In this case, it is preferable to dilute the sample solution to ensure that 100 % of the total amount of sample is applied to the appropriate zone. In addition, the layer should be protected during the application process by placing and securing a plastic film ca. 5 mm above the application zone.

▶ **Figure 30**: see Photograph Section.

- In many years of experience with the 100-µl and 500-µl syringes for the Linomats, an uncontrolled lowering of the plunger was sometimes found and hence a higher sample application rate (per unit of time), causing the application zone to be an irregularly shaped spot. In these cases, it was found to be beneficial to strike the plunger against a flat, hard surface. The PTFE seal at the end of the plunger is thereby slightly widened, so that the plunger fits more tightly in the glass cylinder of the syringe. After several such treatments of the seal, it should be replaced.

- The jets on the application tower should be carefully cleaned after applying samples to one or more plates. However, if a large applied volume of e.g. toluene is used in which a suppository mass is dissolved, it can also be essential to remove any adhering wax in the jets after each individual sample application, using cotton wool "buds" soaked in toluene. Blocked jets lead to defective application zones, as can be seen in Fig. 31, which shows a plate with samples of a drug extract.

▶ **Figure 31**: see Photograph Section.

Figure 32 shows various applied substance zones in three rows. The top and bottom rows were applied with the Linomat IV at various rates. The middle row was applied by hand from various solvents. This is also described in the Figure legend.

The company DESAGA has introduced a semiautomatic system for applying samples to the (somewhat rarely used) TLC round rods ("Chromarods") of the Iatroscan system. In this system, the sample (1–2 μl) is applied to the outside of a fused silica rod (Chromarod), which is coated with thin-layer material by a sintering process. Details of the principle and practice of this method can be found in Section 11.4 "TLC-FID/FTID – Combination of TLC with Flame-Ionization Detector or Flame Thermionic Ionization Detector".

▶ **Figure 32:** see Photograph Section.

The **TLC-Spotter PS 01** (Fig. 33) applies the samples to TLC plates by the contact method. Amounts of 20 nl to 10 μl can be applied spotwise for quantitative TLC. The microliter syringes used have nominal volumes of 0.5, 1, 2 and 10 μl. The positioning of the starting points cannot be freely chosen, but are arranged in patterns at a distance apart of 5, 6, 10, 12 and 15 mm. The distance from the bottom edge of the plate is also fixed. With HPTLC plates, this is 5 mm, while for TLC plates it can be reset to 15 mm. The manufacturer states that the discharge rate is 100 nl/s for the 1-μl syringe, and this can be reduced to 15 nl/s using the "L" function. The filling, emptying and rinsing of the syringe is performed by pressing the appropriate keys [36].

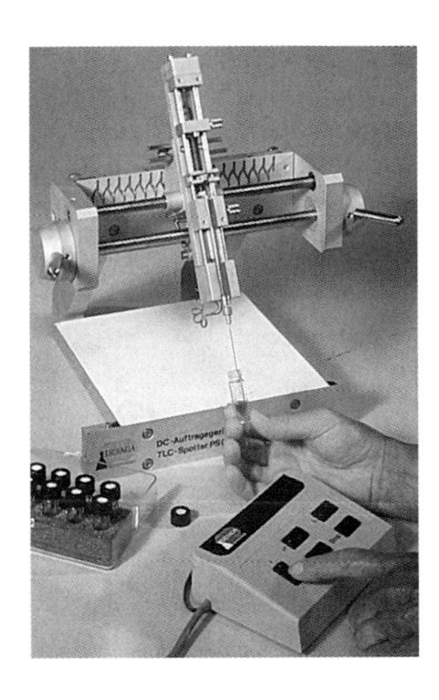

Figure 33. Semiautomatic TLC Spotter PS 01 (DESAGA)

3.6.3 Fully Automatic Application

If you have to perform assays by quantitative TLC, you will find fully automatic sample application indispensable. This will at least require the **Automatic TLC Sampler (ATS 3)** of CAMAG (Fig. 34), in which a PC controls the application and then, in combination with a TLC scanner, the evaluation. The PC also processes the measured data. The samples are applied from a steel capillary, which is connected to a dosing syringe driven by a stepping motor. The samples are applied to precoated layers (on glass or foil), up to the 20 × 20 cm format, either spotwise or bandwise as required. Applica-

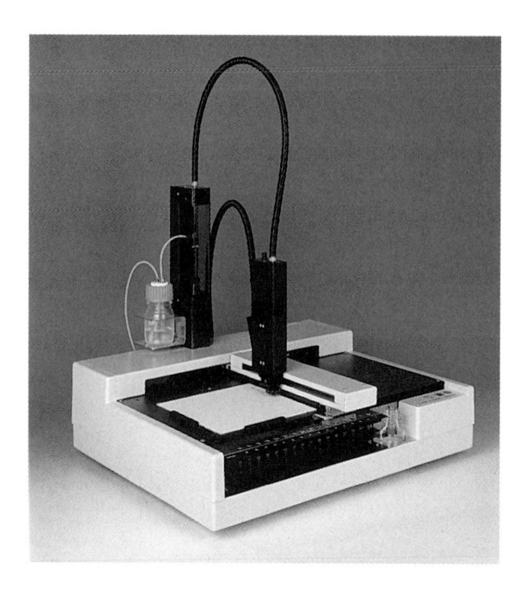

Figure 34. Automatic TLC Sampler III (CAMAG)

tion bandwise is by spraying and spotwise normally by contact transfer, although spraying can also be used for this.

The sample vials, which are closed with normal septa, are placed in the equipment in a rack with 16 positions. Within one application program, up to 32 vials (in two racks) can be addressed.

The arrangement of the samples on the layer can be for normal, double-sided, circular and anticircular chromatographic development.

The Automatic TLC Sampler (ATS 3) is fast. The capillary from which the samples are dosed moves with a lateral velocity of 25 cm/s according to the plotter principle. The dosing rate selected can be in the range 10–1000 nl/s [37].

The DESAGA **TLC-Applicator AS 30** is conceived as a fully automatic application device, but can also be used in semiautomatic mode (Fig. 35). A microcomputer controls the two stepping motors and the gas valve. The equipment operates by the spray-on method in which a stream of gas carries the sample from the tip of the needle onto the plate. This can be as a band or a spot, there being no contact during application. While the sample is being ejected, the application tower moves across the plate. During the filling process, the dosing syringe stays above a collection vessel which removes the rinsing liquid and excess sample. By addition of a sampling device, the AS 30 becomes a fully automatic application system. Here, the Applicator controls the Autosampler via a serial interface and monitors the selection of the sample vessel, the delivery of the required amount of sample and the rinsing processes. A combination of this equipment with the Densitometer CD 60 (naturally PC-controlled) again provides a complete work station for quantitative TLC [38].

Figure 35. TLC Applicator AS 30 (DESAGA)

☞ **Practical Tips** for working with the AS 30 in fully automatic mode:

- Because of the large volume of sample that is pumped into the dosing syringe through tubes, it is advantageous to use pure PTFE septa in the sample vial covers instead of silicone seals with a covering of PTFE. This prevents the formation of a vacuum in the sample vials, which can cause the formation of air bubbles in the dosing syringe and incorrect application volumes.

- Compared with the 100-µl dosing syringe, the 10-µl dosing syringe is very susceptible to warping of the stainless steel plunger. You should therefore avoid applying solutions that are too concentrated, as these are more likely to cause the sample to crystallize. More dilute solutions, which are applied with the 100-µl syringe, are more suitable.

- If "costly" samples are applied from e.g. aqueous solutions (e.g. drugs whose sample preparation has taken several days), the time factor is of minor importance with this fully automatic equipment, as "overnight operation" is always a possibility with the AS 30. Using the slowest delivery rate (120 s/µl), 10 µl takes 20 min. An application time of 16 h was required for e.g. 8 samples of 100 µl each, corresponding to an application rate of 72 s/µl. For all application equipment it is in general advisable to determine the application rates experimentally for each job. Thus, for example, if the application rate with aqueous solutions is too high, the layer can be disturbed or the application zones can be too broad as anomalous chromatography starts to take place.

- After sample application is complete, the system should be flushed out. Here, two rinsing operations over and above the minimum is generally advisable. It is also essential after working with acid solutions to clean the outside of the dosing syringe very carefully. This is because the stainless steel part of the syringe corrodes very quickly, and this can lead to loss of the syringe.

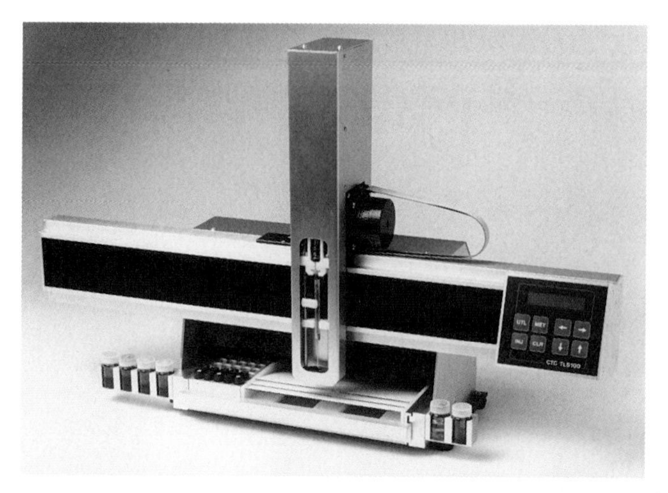

Figure 36. Automatic application equipment TLS 100 (Baron)

Another automatic application system for TLC and HPTLC plates is the **TLS 100** produced by Baron (Fig. 36). This operates with up to 30 samples and 4 standards, which can be combined as required on plates of 10 × 10 cm or 20 × 10 cm format. This is achieved by stacking a maximum of 12 plates of size 10 × 10 cm or 6 plates of size 20 × 10 cm in the equipment. Only one plate of size 20 × 20 cm can be placed in the equipment. A more rapid method of operation is by direct application using easily changed microliter syringes (1, 10 and 100-μl syringes) not connected by flexible tubes. There is therefore no loss of costly test material due to rinsing operations. The application mode can be selected for each plate individually, and 15 methods can be stored [39].

☞ **Practical Tips** when working with the TLS 100:

- As the samples are applied by contact transfer in this equipment, the application pressure of the syringe must be very carefully determined before starting the application process.
- The consumption of sample solution is very small, which is an advantage especially when expensive reference substances are used.

Where our own experience permits it, we describe the operation of the application equipment here in more detail than is found in the information provided by the manufacturers. Our practical advice and tips are also be applicable to comparable equipment produced by other manufacturers.

Before purchasing equipment for semiautomatic or fully automatic application, it is advisable to test its performance oneself. Good information can also be obtained from congresses such as the Internationale Symposium für Planar-Chromatographie which takes place every 2 years in Interlaken (Switzerland) and exhibitions such as InCom, Düsseldorf (annually), Analytica, Munich (every 2 years) and Achema, Frankfurt (every 3 years).

3.7 Positioning of the Samples

In the foregoing Sections on sample application, no recommendations were made regarding the arrangement of the samples on the plate. There are no hard and fast rules for using any particular sequence in either qualitative or semiquantitative analysis, and the user's final assessment of the appearance of the plate as a whole is at best based on experience. However, if a certain sequence for applying samples is adhered to, this considerably facilitates the assessment of thin-layer chromatograms for pharmaceutical quality control.

In **quantitative TLC,** a corresponding number of standards are applied along with the samples and are also determined. The positioning of the samples and standards on the plate is of great importance for the analytical result. Our experience, gained during a course on quantitative TLC with Prof. Jork in Saarbrücken, confirms the superiority of the **Data Pair Method**, and this is therefore the only sample application scheme described in this book. In the data pair method, published by Bethke, Santi and Frei [40], all the solutions are applied twice, but not next to each other, and are separated by half the width of the plate. Figure 37 shows an application scheme of this type.

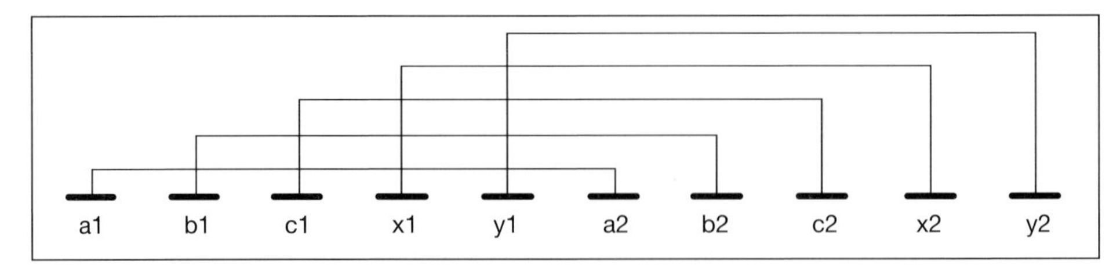

Figure 37. Application scheme using the "Data Pair" method

The following scheme shows an example of an assay performed with 6 samples and 3 standards on 20 lanes of a TLC silica gel 60 WF_{254s} precoated plate (Merck 16485) using fully automatic application by the AS 30:

Lanes 1 + 20:	Edge lanes, *which are not measured*, with the same solutions as lanes 2 and 19
Lanes 2 + 11:	Standard 1, concentration 90 % of target
Lanes 3 + 12:	Sample No. 1
Lanes 4 + 13 :	Sample No. 2
Lanes 5 + 14 :	Standard 2, concentration 100 % of target
Lanes 6 + 15:	Sample No. 3
Lanes 7 + 16:	Sample No. 4
Lanes 8 + 17:	Standard 3, concentration 110 % of target
Lanes 9 + 18:	Sample No. 5
Lanes 10 + 19:	Sample No. 6

Equipment settings used: distance from bottom edge of the plate 15 mm, start 14 mm, lane 5 mm, space between lanes 4 mm.

Apart from the above-described positioning of the samples, as used daily in routine work, there are other possibilities for making a reliable identification of a substance using special application techniques With samples contaminated with matrix, a shift in the hRf values of substances sometimes occurs (Fig. 38a). In such a case, an unequivocal identification can only be made by use of bandwise, overlapping application. The sample (e.g. a drug) is applied in the center of the TLC plate 4–5 cm wide, and reference substances are applied to the right and/or left of this so that they cover ca. 1.5 cm of free layer and extend at least 1 cm into the sample lane. As an example of overlapping application, Figure 38b shows an extract of stinging nettle root and, on the left, scopoletine as a reference substance.

▶ **Figure 38:** see Photograph Section.

A further possibility of the use of overlapping application is shown in Fig. 36. In this example of a sample of complex composition, which contains a large number of zones lying close together, it is shown by overlapping application that two known compounds formed as intermediates in the synthesis route are not present in a sample of the end product that has been affected by exposure to light.

▶ **Figure 39:** see Photograph Section.

Thin-layer chromatograms are often subjected to various methods of detection for identification purposes. For this, the sample is applied in a band several cm wide, up to a maximum of 18 cm, and the plate is cut up after development (see also Chapter 6 "Derivatization"). Samples are also applied at maximum width for preparative analysis. Special application schemes for two-dimensional TLC are shown in Section 4.3.2.

Errors in the positioning of samples are programmed on marking the application points or by the choice of the equipment parameters. The samples are often applied too close to the left or right hand edges of the plate or too near to the surface of the liquid solvent. The application spots or zones are sometimes too close together or vary in distance from the bottom edge of the plate (in the case of manual application). Sometimes expected substance zones are completely missing, and excessive amounts are found on other lanes. The following procedure is helpful in TLC:

☞ **Practical Tip:** Fill in the data sheet (specifying the positioning) and tick off the lanes used on this sheet so that no samples are forgotten and/or other lanes have samples applied to them twice over. (The data sheet "TLC III" is described in detail in Section 9.3.)

Figure 40 shows several of the errors described here. On the left of Fig. 40a on the left of the first lane, a sample has been forgotten at the application point marked (corresponding to lane 3 on Fig. 40b), and the plate was not adequately dried before the development. In Fig. 40b, the second part of a divided 20 × 20 cm precoated plate should be enough for all the chromatograms, but the samples were positioned too close together. Here the drying before the development was adequate.

▶ **Figure 40:** see Photograph Section.

3.8 Drying Before the Development

Every trainee in TLC is told again and again that the solvent used in sample application must be completely removed before the development. Unfortunately no general advice on how to perform this operation can be found in the literature. Kraus recommends a cold current of air, and comments that care should be taken that asymmetrical evaporation of the solvent should not occur, as this can easily lead to shifting of the application zones [21]. Other authors have discreetly avoided this point, as even methods of operation contrary to general common practice can give good results. The properties of the applied samples and the type of solvent system used are very important for the drying.

☞ **Practical Tips** for avoiding errors associated with the drying of plates before development:

- Volatile compounds such as essential oils are usually applied with toluene or *n*-hexane. With these samples it is not advisable to use an electric hair dryer or other type of blower. Instead, the plate is allowed to dry for a few minutes at room temperature in a horizontal position before it is placed in the TLC chamber. However, even using this gentle procedure, very highly volatile substances such as α-pinene can evaporate from the plate before the chromatography. In such cases, the only remedy is to work in a cool room (protective clothing!).
- With thermally stable substances (see also Section 10.1 "Loading with Predetermined Amounts of Substance"), which are applied in amounts up to 1000 µg/lane from chloroform or methanol in purity tests, drying with an electric hair dryer is usually inadequate. Here, uniform heat application (e.g. drying oven with air extraction) for ca. 20 min can be successful. The temperature should be in the region of the boiling point of the solvent. As an example, Fig. 41a shows the chromatographic result of an inadequately dried plate after application of 1000 µg piracetam/lane, and Fig. 41b shows the result of adequate drying of the sample lanes at the same concentrations before the development.

▶ **Figure 41:** see Photograph Section.

- When drying before the development, it is often forgotten that if the plate is exposed to a blower for a long time, large amounts of "laboratory dirt" can be transferred to the plate, ruining any beneficial effects of the prewashing process.
- The following example shows that reproducible results can sometimes only be obtained in TLC by adhering exactly to definite drying conditions: the drying of plates for the quantitative determination of amino acids from dry extract of marshmallow root in an instant tea by applying the samples from solution in aqueous hydrochloric acid fully automatically over a period of 2 h was only successful after drying for 30 min with a warm-air fan heater (grade I). The effect of the drying time on the chromatogram obtained is shown in Fig. 42. Prepared tea samples for the identity of marshmallow root were insufficiently dried on plate A (the last lane was left in front of the warm-air fan heater for only ca. 5 min!), while plate B shows a perfect chromatogram.

▶ **Figure 42:** see Photograph Section.

- With thermally labile or oxidation-prone samples, it is essential to perform a series of drying tests before the development. Our own tests with 400 µg molsidomin/lane showed that a 10-min drying time with the warm-air fan heater (grade I) was too little, while 30 min was too much. For this test for unknown impurities, exactly 20 min drying was therefore prescribed in the testing procedure.

4 Solvent Systems, Developing Chambers and Development

4.1 Solvent Systems

Next to the choice of the stationary phase (precoated layer), the choice of solvent system is the factor with the greatest influence on a thin-layer chromatogram. In only a few cases does the solvent system consist of one component only. Normally, mixtures of up to six components are used, and these must have the appearance of single-phase systems with no sign of cloudiness.

The solvent system performs the following main tasks:
- to dissolve the mixture of substances,
- to transport the substances to be separated across the sorbent layer,
- to give hRf values in the medium range, or as near to this as possible,
- to provide adequate selectivity for the substance mixture to be separated.

They should also fulfill the following requirements:
- adequate purity,
- adequate stability,
- low viscosity,
- linear partition isotherm,
- a vapor pressure that is neither very low nor very high,
- toxicity that is as low as possible [2].

In analogy to "precoated layer = stationary phase", the term "mobile phase" is often used for the solvent system. This is a false analogy in over 90 % of cases, especially when solvent mixtures are used. The liquid (i.e. the solvent system) that is placed in the development chamber can, on migration, release some of its components into the pores of the sorbent where it forms a "liquid stationary phase". In equilibrium with this, the mobile phase then forms, this being depleted with respect to certain components in comparison to the solvent system. As a result, so-called α, β and γ-fronts form along the migration distance. As many fronts can be formed as there are components of differing polarity in the solvent mixture so that, in coordination with a polar stationary phase, the most polar component of the solvent system becomes concentrated near to the starting line and the least polar becomes concentrated near to the front.

In the following three Sections, matters of practical relevance are at the forefront of our description of solvent systems. Theoretical treatments and discussions can be found in the literature [2, 4, 5].

Applied Thin-Layer Chromatography: Best Practice and Avoidance of Mistakes, 2nd Edition
Edited by Elke Hahn-Deinstrop

ISBN: 978-3-527-31553-6

4.1.1 Choice of Solvent Systems

One of the more difficult questions in all TLC is how to pick the most suitable solvent system. Before starting one's own search for a solvent system, such searches usually being laborious and sometimes unsatisfactory, it is advisable to carry out a literature search. However, although specialist publications and monographs of pharmacopoeias offer useful advice, they unfortunately also provide recipes for solvent systems whose components are highly toxic or carcinogenic and therefore should not be used. An example of this is benzene, which regrettably continues to be used, according to recent publications, especially in the Asiatic regions (India). The "eluotropic series" gives an indication of suitable replacements. Trappe introduced this term as early as 1940 to describe the various solvent systems arranged in order of their chromatographic "strength" [41]. Different authors give different sequences for the solvent systems, depending on which sample was used in their investigations and with which sorbents, types of chamber and level of purity of the solvents. Over a period of one year, the eluotropic series of Halpaap became well accepted, in which each solvent system was assessed according to its elution power with silica gel as the stationary phase [42]. Table 8 shows the solvent components most often used in TLC, with their velocity coefficients, arranged in increasing order of elution power. In the example given above, toluene can be used as a substitute for benzene.

In the past, there has been no lack of models for formulating or optimizing solvent systems on paper or, more recently, by computer. Snyder introduced the solvent strength ε_0 as an aid, and calculated this for the most commonly used solvents in chromatography according to their tendency to behave mainly as proton acceptors, proton donors or dipoles [43]. Nyiredy developed the "prism model" proposed by Kirkland and Glaich [44], modifying this for TLC so that the user achieved his or her aim by a trial-and-error method of producing an optimized TLC separation system [45].

Trial and error is also the most practical method of ultimately obtaining a good chromatogram.

Do not believe any recipe for a solvent system if you have not proved it by your own experiments.

☞ **Practical Tips** for the choice of solvent systems:

- If there are no clues to be found in the literature (or from good friends!) for solving a separation problem, the search for a suitable solvent system should always begin with pure solvents of medium elution power. Here, "normal" silica gel 60 is tried first, before testing other (e.g. modified) silica gels.

Table 8: Eluotropic series

Solvent	κ (mm^2/s)
1 *n*-Heptane	11,4
2 *n*-Hexane	14,6
3 *n*-Pentane	13,9
4 Cyclohexane	6,7
5 Toluene	11,0
6 Chloroform	11,6
7 Dichloromethane	13,2
8 Diisopropyl ether	13,2
9 *tert*-Butanol	1,1
10 Diethyl ether	15,3
11 Isobutanol	1,6
12 Acetonitrile	15,4
13 Isobutyl methyl ketone	9,1
14 2-Propanol	2,5
15 Ethyl acetate	12,1
16 1-Propanol	2,9
17 Ethylmethyl ketone	13,9
18 Acetone	16,2
19 Ethanol	4,2
20 1,4-Dioxan	6,5
21 Tetrahydrofuran	12,6
22 Methanol	7,1
23 Pyridine	8,0
Sorbent:	TLC-plate silica gel 60 F_{254} Merck
Type of chamber:	N-chamber with chamber saturation
Room temperature:	22 °C
Migration distance of the solvent:	100 mm

The solvents are listed in the order of Halpaap's eluotropic series with silica gel as the stationary phase.
The complete table with a total of 20 chromatographic and physical parameters can be found in [42].

- Even when single-component systems are used, this can lead to variable results if attention is not paid to the purity and origin of the "pure" solvent. Some of these, e.g. chloroform, methanol, tetrahydrofuran etc., contain added stabilizers. It can easily happen that similar grades from different manufacturers or different grades from the same manufacturer contain different stabilizers. Figure 40a,b shows an example of this in which chromatograms of diosgenine, sitosterol and lanosterin were developed in different grades of chloroform. In the first case a grade of chloroform LiChrosolv stabilized with pentene was used and, in the second, chloroform p.f.a. (pure for analysis) stabilized with ethanol. (See also Section 9.4 "Examples of GMP/GLP-Conforming Testing Procedures.")

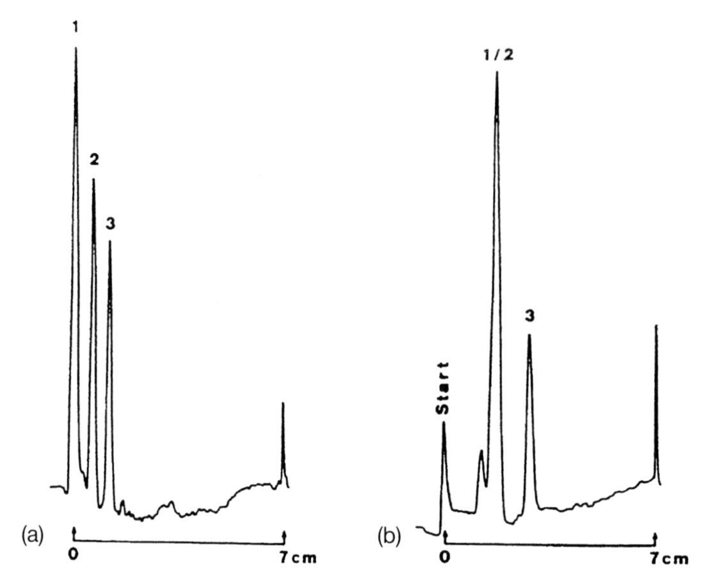

Figure 43. Influence of stabilizers in chloroform on the chromatographic result
The stabilizers present in the chloroform were (a) amylene and (b) ethanol

Sorbent:	HPTLC precoated plate RP-18 WF_{254s}
Solvent:	Chloroform
Migration distance:	7 cm
Chamber:	Normal chamber without chamber saturation
Substances:	1. Diosgenine 2. β-Sitosterol 3. Lanosterol (each 0.2 %)
Applied volumes:	200 nl
Derivatization:	$MnCl_2/H_2SO_4$ (5 min at 110 °C)
Measurement:	UV 265 nm, TLC Scanner II (CAMAG)

- In the DAB method, the identity test for camomile flowers is also performed using chloroform p.f.a. (DAB nomenclature: chloroform *R*) as the solvent. Even the very low percentage of water in the stabilizer present (ethanol) can lead to the formation of a β-front on development (Fig. 44a). If, on the other hand, the chloroform is dried over sodium sulfate shortly before placing it in the developing chamber, a thin-film chromatogram without an interfering β-front is obtained, as shown in Fig. 44b [46]. This example could not be repeated reproducibly, and probably only happens with the 5 × 5 cm horizontal chamber (see Section 4.2.1.2).

▶ **Figure 44:** see Photograph Section.

- If the first results of development tests show that with one solvent system the substance remains at the start and with another it migrates with the front, various mixing ratios should be investigated.
- Tailing often occurs, e.g. with acid and basic compounds, and this can be minimized by suitable choice of pH or by the use of inorganic or organic salts (so-called tailing reducers). An example of this is shown in Fig. 45.

▶ **Figure 45:** see Photograph Section.

- Sometimes only the presence of small amounts of alkaline components enables some substances to be separated at all. Thus, after adding 1 part of concentrated ammonia solution to the solvent system dichloromethane and methanol (90 + 9 ml), theophylline can be separated from the two component substances of dimenhydrinate (diphenhydramine and 8-chlorotheophylline). Figure 46a shows the chromatogram in the above-mentioned solvent system without ammonia, and Fig. 46b the chromatogram in the solvent system with ammonia.

▶ **Figure 46:** see Photograph Section.

- After the solubility of a substance has been determined, nonpolar substances can be developed in nonpolar solvents (adsorption chromatography) and polar substances in polar solvents (partition chromatography).
- It should again be mentioned at this point that it is essential that the layer should be wettable by a solvent, as no development could otherwise occur. This is because the capillary forces within the layer would not be sufficient to overcome the repulsive forces between the stationary phase and the mobile phase. Thus, for example, a very hydrophobic RP-18 phase cannot be used with solvent systems that have a high water content.
- In contrast to HPLC, reproducible results in TLC can be obtained using silica gel as the stationary phase even with additions of basic substances to the solvent systems. It is true that the sorbents are attacked here also, but the damage is not so important as the TLC plate is used once only.
- When choosing between different solvent systems of similar elution strength, those containing lower viscosity components should be preferred, as these give shorter development times.
- If carcinogenic or highly toxic solvents (e.g. benzene, pyridine, etc.) are prescribed in testing procedures, which can happen even today, the attempt should be made to find alternatives, after consultation with your superiors or teachers, and these should then be used consistently.

The following examples illustrate some "tricks" with solvents and their effect on the chromatogram:

Improvement of selectivity
Figure 47a,b shows chromatograms of carbamazepine which were developed in the solvent system prescribed by the DAB. The deformed spot of carbamazepine, indicating the formation of a β front, can clearly be seen. The unknown double zone is here just below the main spot. The choice of a different solvent system gave a spot with a better shape for the carbamazepine and a better separation from the unknown double zone. All the parameters for the chromatography of carbamazepine are listed in Table 9.

▶ **Figure 47:** see Photograph Section.

Better separation efficiency combined with shorter development time
The current DAB monograph on "Birch Leaves" prescribes the use of the solvent system "1-butanol + acetic acid + water". Figure 48a shows the chromatogram obtained with dry extracts of birch and hawthorn leaves viewed by 254-nm UV light. The photograph shows indistinct zones, indicating poor separation efficiency. Many alternative solvent systems have been described in the literature. For example, Wagner proposes in the "Drug Atlas" [47] the by now almost "classical" solvent system for flavone glycosides consisting of ethyl acetate and water with a high content of acids. However, these destroy most of the fluorescence indicator in the lower part of the plate. Pachaly increases the organic content and uses both formic acid and water to adjust the pH [48]. Both use TLC silica gel 60 with a fluorescence indicator as the sorbent. Our own investigations show that a very good separation efficiency can be obtained combined with a relatively short development time if HPTLC silica gel 60 is used with the solvent system "ethyl acetate + formic acid + methanol + water" without use of a saturated chamber. Figure 45b–d shows the chromatograms obtained using the three chromatographic systems described above as alternatives to the DAB system. This clearly shows that the most compact zones are obtained using the HPTLC silica gel layer. This is due to both the solvent system and the high-performance materials used. All the experimental parameters for the TLC of flavonoid-containing dry extracts are listed in Table 10. The sample preparation methods for these extracts are described in Section 9.4.2 "Identity and Purity of Various Flavonoid-Containing Plant Extracts".

▶ **Figure 48:** see Photograph Section.

Table 9: Carbamazepine

	DAB 10	Alternative I	Alternative II	Free of CHC [a]
Sample solution	50 mg/ml chloroform + ethanol 96 % (1 + 1); limit test: 0,005 mg/ml			1 mg/ml methanol
Sorbent	Prescribed: Silica gel G Used: Silica gel for HPTLC 60 F_{254} GLP (Merck 1.05613)	Silica gel for HPTLC 60 F_{254} 10 × 20 cm (Merck 1.05642 or 1.05613)	Silica gel for TLC 60 F_{254} 20 × 20 cm (Merck 1.05715)	TLC-alu sheet RP-18 F_{254s}, 20 × 20 cm (Merck 1.05559)
Solvent system	Toluene + methanol (90 + 14)	Ethyl acetate + methanol + ammonia solution (85 + 10 + 5)	Ethyl acetate + methanol + ammonia solution (85 + 10 + 5)	Methanol + water (72 + 28)
Reference substance	Diluted test solution Iminodibenzyl	Carbamazepine USP Iminodibenzyl		
Applied sample volume	Each 2 µl = 100 µg carbamaz. = 0,01 µg carbamaz. = 0,1 µg iminodibenzyl	Each 2 µl Acc. to DAB Acc. to DAB Acc. to DAB	16 µl = 800 µg carbamazepine as well as 0,08 µg carb. 0,8 µg iminodibenzyl	1 µl
Chamber saturation	Yes	Yes	Yes	No
Migration distance	8 cm	7,5 cm	10 cm	10 cm
Development time	21 min	16 min	21 min	Approx. 54 min
Derivatization reagent	Potassium dichromate-sulfuric acid	Potassium dichromate-sulfuric acid	None	None Optical direct evaluation by TLC-scanner at 254 nm
hRf-values:				
Carbamazepine	31–40	54–63	800 µg: 36–54 0,08 µg: 48–50	Approx. 42
Iminodibenzyl		86–89	77–82	
Iminostilbene [b]		82–84		
Unknown twin zone	25 + 27	35 + 40	23 + 27	

[a] The parameters of this chromatographic system were found in an advertisement of Merck in the Journal of Planar Chromatography, Vol. 7, November/December 1994, p. 489

[b] For validation only

Table 10: Birch leaves

	DAB 10	Wagner: Plant Drug Analysis	Pachaly: TLC-Atlas	Alternative	
Sorbent	Prescribed: Silica gel G Used: TLC-silica gel F_{254} 20 × 20 cm (Merck 1.05715)	TLC- silica gel 60 F_{254} 20 × 20 cm (Merck 1.05715)	TLC- silica gel 60 F_{254} 20 × 20 cm (Merck 1.05715)	HPTLC- silica gel 60 F_{254} GLP 10 × 20 cm (Merck 1.05613)	
Solvent system	1-Butanol + acetic acid + water (66 + 17 + 17)	Ethyl acetate + formic acid + acetic acid + water (100 + 11 + 11+ 27)	Ethyl acetate + formic acid + water (80 + 8 + 12)	Ethyl acetate + formic acid + methanol + water (50 + 2,5 + 2 + 4)	
Applied sample volume	20 µl 20 × 3 mm	25–30 µl 10 mm	20 µl 20 × 3 mm	10 µl 13 mm bandwise	
Chamber saturation	Yes	Yes	Yes	No	
Migration distance	10 cm	Chosen: 10 cm	12 cm	7 cm	
Development time	105 min	40 min	46 min	30 min	
Derivatization reagent	Diphenylboric acid-2-aminoethyl ester reagent and polyethylenglycol 6000 (Flavone reagent acc. to Neu)				
hRf-values:				**Colours of fluorescence**	
Rutoside	45–50	30–34	13–16	15–17	Yellowish orange
Chlorogenic acid	38–42	41–45	28–31	34–38	Pale blue
Hyperoside	55–61	46–50	30–33	38–41	Yellowish orange
Quercitrin	64–70	64–68	50–54	63–65	Yellowish orange
Caffeic acid	77–85	87–91	84–88	90–93	Pale blue
Quercetin	79–85	91–94	87–90	93–95	Yellowish orange

Solvent systems from the upper and lower phase

In the early stages of the development of TLC, especially when used with drugs and their preparations, solvent systems were often used that consisted of the upper or lower phase of a mixture of solvents. Some of these can still be found in current monographs of the DAB. As an example of the use of a solvent system with an addition of acid, we describe here the identity test for primula root. An alkaline solvent system is prescribed for the test for liquorice root and liquid extracts prepared from it. As these upper and lower phases are difficult to handle and their compositions are very temperature sensitive, reproducible hRf values in the TLC can only be achieved at a constant

Table 11: Primula root

	DAB 10	Alternative I	Alternative II	Alternative III
Sample solution	Reflux with ethanol 70 %, 15 min, filter	With ethanol 70 % 10 min ultrasonic bath, centrifuge	Preparation via RP-18 cartridge	Extract with methanol/water (4+1 v/v) 10 min USB, 2 min centrifuge 12000 U/min
Sorbent	Prescribed: Silica gel G F_{264} Used: TLC silica gel 60 F_{254} (Merck 1.05715)	TLC silica gel 60 F_{254} 20 × 20 cm (Merck 1.05715)	HPTLC silica gel 60 F_{254} GLP 10 × 20 cm (Merck 1.05613)	Nano-SIL C 18–100, 10 × 10 cm (Macherey-Nagel 811052) activated 10 min at 110 °C
Solvent system	Upper phase from: 1-butanol + water + acetic acid (50 + 40 + 10)	Ethyl acetate + formic acid + acetic acid + water (50 + 5,5 + 5,5 + 13,5)		2-Propanol + water + acetic acid (40+60+5)
Reference substances	Aescine Saponine	Primula acids or authentic drug, extract		
Applied sample volume	20 µl 20 × 3 mm	10 µl 10 mm bandwise	10 µl 10 mm bandwise	10 µl 10 mm bandwise
Chamber saturation	Yes	Yes	Yes	Yes
Migration distance	12 cm	10 cm	6 cm	8 cm
Development time	170 min	45 min	28 min	78 min
Derivatization reagent	Anisaldehyde-sulfuric acid	Vanillin-sulfuric acid	Vanillin-sulfuric acid (Autosprayer)	Vanillin-sulfuric acid
hRf-values:				
Primula acids	16–18; 20–23	13; 16 and 19	13; 16 and 19	33–36; 39–42
Aescine	22–26	23–29		
Saponine	9–11			
Application for (recommendation)		Identity and purity	Assay semiquantitative: estimation by eye Assay quantitative: TLC-Scanner or Video-Densitometer	

laboratory climate. For the identity test of the primula root extract, the alternatives listed in Table 11 have been found to be suitable, and these also give a considerably shorter development time (30 min compared with 170 min). Figure 49a,b shows the chromatograms of dry extract of primula root in these two solvent systems.

▶ **Figure 49:** see Photograph Section.

Table 12: Liquorice root

	DAB 10	Pachaly: TLC-Atlas	Alternative	Testing for Aglykon
Sample solutions a	Extract with chloroform, org. phase = 0, dissolve R in $CHCl_3$/MeOH	Extract with dichloromethane, org. phase = 0, dissolve R in CH_2Cl_2/MeOH	Extract with acidified ethyl acetate, org. phase = 0, dissolve R in MeOH	Extract with acidified ethyl acetate, org. phase = 0, dissolve R in MeOH
Acc.to DAB and Pachaly Sample solutions b	R of a: 1 h hydrolyze with H_2SO_4, Extract twice with CHC			
Sorbent	Prescribed: TLC silica gel G 60 F_{254} Used: TLC silica gel 60 F_{254} 20 × 20 cm (Merck 1.05715)	TLC silica gel 60 F_{254} 20 × 20 cm (Merck 1.05715)	TLC silica gel 60 F_{254} 20 × 20 cm (Merck 1.05715)	TLC silica gel 60 F_{254} 20 × 20 cm (Merck 1.05715)
Solvent system	Upper phase from: ethyl acetate + ethanol + ammonia solution 1,7 % (60+13+27)	Ethyl acetate + ethanol + water + ammonia solut. conc (65+25+9+1)	Ethyl acetate + formic acid + acetic acid + water (50+5,5+5,5+13,5)	Toluene + ethyl formate + formic acid (50+40+10)
Reference substance	18β-Glycyrrhetinic acid		Glycyrrhicinic acid	18β-Glycyrrhetinic acid
Applied sample volume	10 μl 20 × 3 mm	20 μl 20 × 3 mm	10 μl 10 mm bandwise	10 μl 10 mm bandwise
Chamber saturation	Yes	Yes	Yes	Yes
Migration distance	15 cm	15 cm	10 cm	10 cm
Development time	54 min	70 min	45 min	19 min
Derivatization reagent	Anisaldehyde-sulfuric acid	Anisaldehyde-sulfuric acid	Vanillin-sulfuric acid	Vanillin-sulfuric acid
hRf-value:				
Glycyrrhicinic acid	Start	Start	21–24	Start
18β-Glycyrrhetinic acid	13–17	37 (acc. to literature)	92–95	40–43

org. phase = 0 : organic phase is evaporated to dryness, R : residue, CHC : chlorinated hydrocarbon

The alternative system can also be used in the identity testing of liquorice root and its preparations if, in contrast to the DAB, the test is not for aglycon 18β-glycyrrhetinic acid obtained by previous hydrolysis, but for the component glycyrrhizinic acid without any previous reaction. Here, care must be taken that the water content of the sample solution should be ca. 40 % (→ care: slow application!), as considerably smaller quantities of glycyrrhizinic acid are transferred into the purely organic solvent during the sample preparation [49]. The complicated preparation method prescribed by the DAB requires an additional ca. 100 min per sample compared with the alternative method. The purity test for the aglycon can be performed separately with a different lipophilic solvent. All the parameters for this example are listed in Table 12, and Fig. 50a–c shows the relevant chromatograms.

▶ **Figure 50:** see Photograph Section.

4.1.2 Preparation and Storage of Solvent Systems

Every TLC plate has the right to its own fresh solvent system!

This instruction should be clearly displayed in every TLC laboratory, so that it may ultimately put an end to the bad habits of using solvent systems many times over, adding used solvent to fresh solvent, or storing solvents in bottles for many months and then helping oneself at irregular intervals. Stahl has pointedly remarked on p. 70 of the 2nd edition of his 1967 laboratory handbook: "Mixtures of different solvents should only be used a small number of times for development, and the chamber should never at any time be opened for any longer than necessary." Thirty years have now gone by, but **not** without leaving their mark on TLC! Objections are sometimes made to the systematic use of fresh solvent for each plate, with the claim that reproducible hRf values should be obtainable even with multiple use of the solvent. Facts in support of these assertions have never been produced, so that the best principle must always be that, at least in experiments performed in compliance with GMP/GLP, solvent systems should be used only **once**. The exception to this, of course, is the single-component system, but these are seldom used in TLC.

An example of the harmful effects of the multiple use of the solvent system is shown in Fig. 51. Four plates, each with two 10-mm lanes, each containing 20 μg metoclopramide hydrochloride, were developed successively with the solvent system "chloroform + methanol + concentrated ammonia solution" (56 + 14 + 1 ml). The continuous downwards shift of the hRf value can be clearly seen, combined with an increasing broadening of the spots of the substance during the chromatography. These effects were caused by the reduction in the amount of base in the remaining solvent.

▶ **Figure 51:** see Photograph Section.

☞ **Practical Tips** for the preparation and storage of solvents:

- Single components should be measured out separately and then well mixed. However, this should not be done in the TLC chamber itself! Suitable mixing vessels are conical flasks with ground-glass stoppers, or mixing cylinders, which have the advantage that the main components can be measured out into these.
- The term "mixing cylinders" means glass vessels with ground glass stoppers and not "measuring cylinders", where the palm of the hand can replace the stopper, a common occurrence!
- Further solvent components are added by pipette, and this can cause heating of the mixture. In this case, the mixture should be cooled before adding acidic or volatile components.
- As a rule it is not important to follow any particular order of addition of solvent components. However, every rule has its exception! The DAB gives an example in which the order of addition of solvent components starts from that with the smallest volume and highest polarity and ends with the component with the largest volume and lowest polarity. In this case, the prescribed order of addition should be strictly adhered to, as otherwise a cloudy solvent system will be obtained (as we have demonstrated many times!). In the DAB monographs for fructose, glucose, lactose and sucrose, the following solvent system composition is given:

  10 volumes of water
  15 volumes of methanol *R*
  25 volumes of anhydrous acetic acid *R*
  50 volumes of dichloroethane *R*

It is recommended in the monographs that the solvent systems should be measured accurately, as a small excess of water can cause the mixture to be cloudy. Chromatography with this solvent system is described in Section 4.3.1.1 "Vertical Development".

- Many methods prescribed for TLC include the addition of concentrated ammonia solution. Here, the following is recommended: Only use ammonia solution of the highest purity (e.g. Suprapur, Merck No. 1.05428), order the smallest package size (250 ml) and always store bottles that have been opened in the refrigerator. Leftover amounts of <ca. 20 ml should be used for purposes other than TLC; it is much better to open a fresh bottle of solvent. If there is too little ammonia gas dissolved in the aqueous phase, experimental results will not be reproducible.
- If several plates have to be developed with the same solvent system in one day, the total amount required can be prepared in one batch provided that the solvent compositions are uncritical. This is recommended even for quantitative work in order to maintain reproducible hRf values. Critical solvent systems are discussed in Section 4.1.3 "Problematical Solvent System Compositions".
- Solvent system compositions are often given in the literature such that the total equals 100, e.g. 60 + 38 + 2. This can facilitate both the comparison of one development method with another and the calculation of amounts required in a given ex-

Table 13: Solvent requirements for different TLC separation chambers

Manufacturer of chamber	Type of chamber	Dimensions [cm]	Bottom	Lid	Volume of solvent [ml]	
					Without CS	With CS
Baron	Automatic		Solvent box of PTFE		Max. 20	Additionally 10–20[a]
CAMAG	Vertical	20 × 20	Flat bottom	Metal	110	110
	Vertical	20 × 20	Twin trough	Metal	35	70
	Vertical	20 × 20 Light chamber	Twin trough	Glass	40	80
	Vertical	20 × 10	Twin trough	Metal	35	70
	Vertical	10 × 10	Twin trough	Metal	10–12	20–25
	Horizontal	10 × 10	PTFE	Glass		each side approx. 1,5[a]
	Horizontal	20 × 10	PTFE	Glass		each side approx. 3[a]
	Automatic ADC	20 × 10				max. 12[a]
	Automatic ADC	20 × 20				max. 25[a]
	Automatic AMD				8 /development, gradient with 20 steps approx. 200 incl. rinsing steps[a]	
DESAGA	Vertical	20 × 20	Flat bottom	Glass with knob		155
	Vertical	20 × 10	Conical	Metal		70
	Vertical	10 × 20	Flat bottom	Glass		90
	Round chamber	∅ approx. 12				
	Horizontal	5 × 5	PTFE	Glass	3	3 + 2 for CS
	Horizontal	10 × 10	PTFE	Glass	6	6 + 3 for CS
Macherey-Nagel	Vertical	4 × 8 Glass with screw cap	Slightly curved	Plastic material		Approx. 10
Merck	Vertical	20 × 10	Twin trough	Glass	35	70
Schott Glaswerke	Vertical	7 × 7	Flat bottom	Glass	Approx. 10[b]	
	Staining trough acc. to Hellendahl					

[a] Acc. to the manufacturer
[b] Acc. to literature
In all "vertical" chambers the quoted quantity of solvent is per 1 cm height of solution in the tank.

periment. To minimize the consumption and hence the cost of solvents as well as the amounts of used solvents to be disposed of, only the minimum amount of freshly prepared solvent system should be used. The amount of solvent required depends on the type of developing chamber in which the development is to be performed. TLC developing chambers have been discussed in Section 4.2, and the amounts of solvent needed for the various types of chamber are listed in Table 13.

- Storage of solvents is unnecessary if they are used in a TLC chamber immediately after they have been prepared. However, it is sometimes stated in the literature that certain solvent systems can be stored for several months [50]. In this case, the best advice is to store them in a dark bottle in a cool place. The "daily quota" of a solvent system should also be kept cool in the summer, e.g. if laboratory temperatures exceed 25 °C. Care must be taken to adjust the temperature to room temperature before the development.

- The storage behavior of solvent systems, as described in the last paragraph, was checked by Wagner using the example of the "classical" mixture used for flavonoids. The mixture of ethyl acetate, formic acid, glacial acetic acid and water (100 + 11 + 11 + 27 ml) was used, and a reference plate was prepared using the fresh solvent. After 17 days, the same standard solution and sample solution were applied to two plates and developed in parallel, on one hand in the stored solvent and on the other hand in freshly prepared solvent. The development time for a migration distance of 7.5 cm was 43 min with the stored solvent and only 37 min at 23 °C room temperature with the freshly prepared solvent. The chromatograms obtained in this experiment were different, as shown in Fig. 52a,b. The recommendation only to use freshly prepared solvents is therefore well justified.

▶ **Figure 52:** see Photograph Section.

4.1.3 Problematical Solvent System Compositions

Mixtures With Ether

The use of diethyl ether, known simply as "ether", requires special care by the user because of its very low b.p. and flash point. Before use, the ether should be tested for absence of peroxides; also, any solvent residues left over after the chromatographic process must be disposed of separately from other solvents. The reason why the DAB (1993, Appendix 2) prescribed the solvent system "ether + methanol" (90 + 10 ml) for the identity test for stinging nettle root is difficult to understand, as, in summer months, the composition of this solvent system after charging it into the TLC chamber would be the same as that of the starting mixture only if the work was performed in a climatic laboratory. Use of this solvent system will in fact kill two birds with one stone, as the detection of scopoletine (Fig. 53a$_3$) under 365-nm UV is performed **before** the derivatization with vanillin-phosphoric acid and the determination of β-sitosterol (Fig. 53a$_1$). However, depending on the quality of the starting drug, the scopoletine content can be very low, so that a clear identification becomes more like a guessing game. Separate preparations and TLC systems for these two substances have been performed as follows. For β-sitosterol an identity test is performed on a sample taken from every container of a batch (Fig. 53b), and for scopoletine the purity test is performed on a mixed sample, i.e. a homogeneous mixture of the samples taken from the containers of one batch, after concentrating the substance, using a solvent mixture suitable for coumarins (Fig. 53c) [51]. For β-sitosterol, the solvent system "toluene + ethyl acetate" (50 + 20 ml) is quite satisfactory, not only for qualitative detection but also for assay after appropriate validation (Fig. 53d). All the experimental conditions for the chromatography of the dry extract of stinging nettle root are given in Table 14. The chromatogram obtained using the DAB solvent system is also given for comparison. Here, the development was performed at a temperature of 5 °C. The result is shown in Fig. $53a_2$.

▶ **Figure 53:** see Photograph Section.

Table 14: Dry extract of nettle root

	DAB 10	Test for β-sitosterol	Test for scopoletine
Sample solutions	Reflux with methanol + ethyl acetate + toluene, filter, concentrate	With methanol ultrasonic bath, centrifuge: 100 mg DE/ml	Dissolve in water, extract with HCl-acidified $CHCl_3$, org. phase = 0, dissolve R in $CHCl_3$: 500 mg DE/ml
Sorbent	Prescribed: Silica gel G Used: TLC silica gel 60 F_{254} 20 × 20 cm (Merck 1.05715)	TLC silica gel 60 F_{254} 20 × 20 cm (Merck 1.05715)	TLC silica gel 60 F_{254} 10 × 20 cm (Merck 1.05729) Used transversely
Solvent system	Ether+ methanol (90 + 10)	Toluene + ethyl acetate (50 + 20)	Acetone + water + ammonia solution conc. (45 + 3,5 + 1,5)
Reference substances	Scopoletine + cholesterol (1 + 30 mg/20 ml)	β-Sitosterol 1 mg/ml	Scopoletine 0,1 mg/ml
Applied sample volume	20 µl 20 × 3 mm	20 µl 10 mm	10 µl 10 mm
Chamber saturation	Yes	Yes	Yes, not less than 30 min
Migration distance	10 cm	10 cm	7 cm
Development time	16 min/5 °C 11 min/22 °C	16 min	12 min
Derivatization reagent	Vanillin-phosphoric acid	Vanillin-phoshoric acid	Detection at UV 365 nm
hRf-values:			
β-Sitosterol	67–73	39–43	91–96
Scopoletine	51–58	11–16	38–42

org. phase = 0: organic phase is evaporated to dryness, R: residue

⚠ Chlorinated Hydrocarbons (CHCs)

CHCs must always be treated as hazardous solvents!

Following a decision of the Conference of the German Ministers of Culture, Education and Church Affairs, CHCs must no longer be used in school chemistry lessons, and must be excluded from the laboratory as soon as possible. Unfortunately, this will not happen very quickly, as these substances are often prescribed in the DAB monographs and in pharmacopoeias of other countries, e.g. as solvents for ibuprofen and haloperidol or in the preparation of samples of drugs and drug preparations, e.g. chloroform for greater celandine and dichloromethane for liquid extract of thyme. They are also often used for the subsequent chromaphotography, e.g. chloroform and its mixtures with toluene for steroid hormones and dichloromethane for liquid extract of thyme.

Some modern precoated layers have been developed with the object of avoiding CHCs in solvent systems [52]. For example, separations have been performed on RP-18 layers using solvent systems without any CHCs, which in contrast are used for chromatography on silica gel layers. A comparison of CHC-containing and CHC-free TLC systems is given in two examples below. Figure 54a,b shows chromatograms of theophylline, theobromine and caffeine, and Fig. 55a,b the scanned chromatograms of furosemide and spironolactone. Tables 15 and 16 give the experimental conditions.

▶ **Figure 54:** see Photograph Section.

The replacement of CHCs prescribed in official publications will take some years, as past experience shows. The question therefore arises why every user should not be free to use alternative environmentally friendly solvents after appropriate validation, without the necessity for audits or discussions with FDA inspectors or other authorities, bearing in mind the fact that the pharmacopoeia permits the use of alternative testing procedures provided that these lead to the same result as that obtained by the prescribed procedure (DAB, Chapter IV, General Instructions).

☞ **Practical Tips** for the replacement of CHCs:

- In many cases, it is possible to replace CHCs by other solvents. Here, it can happen that a larger volume of the new solvent is required to dissolve a given amount of substance. The volume of solution applied must be correspondingly larger.
- If there is not enough time available to devise a CHC-free chromatographic system, an attempt should at least be made to replace the chloroform in the solvent system by dichloromethane. However, the method must then be revalidated.

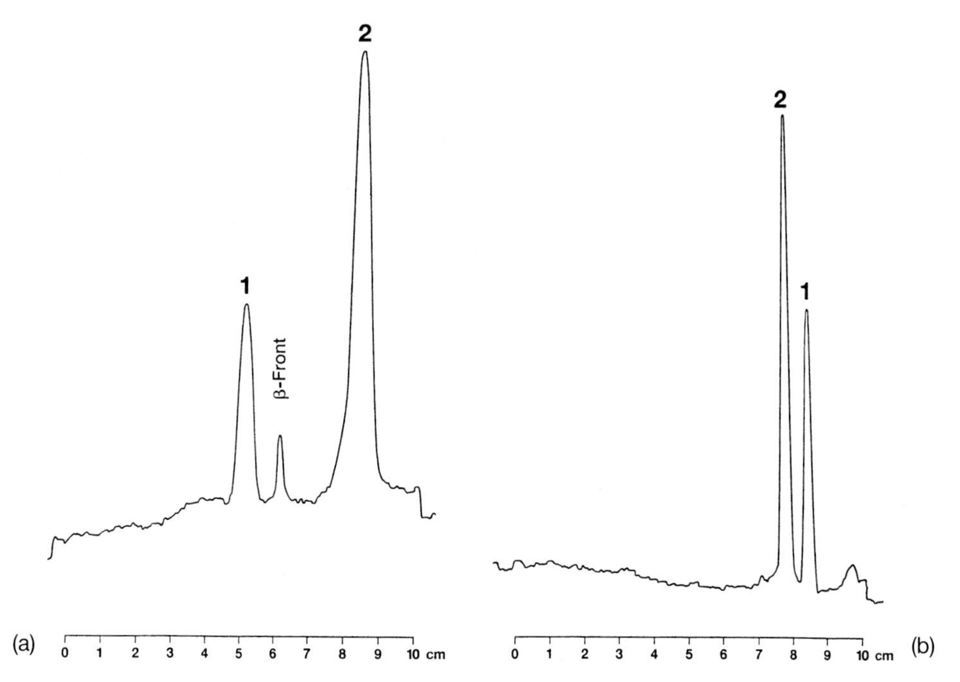

Figure 55. Direct optical evaluation of spironolactone and furosemide from the contents of a capsule measured by the TLC Scanner II (CAMAG) at 254 nm
(a) Chromatography on TLC silica gel 60 F_{254} with CHC-containing SS according to the DAB
(b) Reversed-phase chromatography on RP-18 TLC foil with CHC-free SS
1 1.6 μg furosemide/spot,
2 4.0 μg spironolactone/spot

Table 15: Spironolactone and furosemide

	DAB 10		Alternative (free of CHC)	
Sample solution	2 mg each of spironolactone CRS and furosemide/ml methanol			
Sorbent	TLC silica gel 60 F_{254} 20 × 20 cm (Merck 1.05715)		TLC alusheet RP-18 F_{254s} 20 × 20 cm (Merck 1.05559)	
Solvent system	Chloroform + methanol + acetic acid (70+30+1)		Acetone + water (90+10)	
Applied sample volume	10 μl 10 mm bandwise		10 μl 10 mm bandwise	
Chamber saturation	No	Yes	No	Yes
Migration distance	10 cm	10 cm	10 cm	10 cm
Development time	50 min	24 min	33 min	16 min
hRf-values:				
Spironolactone	Front	83–87	56–58	68–72
Furosemide	93–95	53–56	70–73	82–85

Table 16: Theophylline, theobromine, caffeine

	DAB 10	Renger et. al. [144]	Alternative (free of CHC)[a]
Sample solutions	200 mg/4 ml MeOH + 6 ml $CHCl_3$	100, 120, 140 µg/ml Methanol + water (1+1v/v)	50 mg/10 ml methanol
Sorbent	Prescribed: Silica gel G Used: TLC silica gel 60 F_{254} 20 × 20 cm (Merck 1.05715)	HPTLC silica gel 60 F_{254} GLP 20 × 10 cm (Merck 1.05613)	TLC alusheet RP-18 F_{254s} 20 × 20 cm (Merck 1.05559)
Solvent system	1-Butanol + chloroform + acetone + ammonia solution conc. (40 + 30 + 30 + 10)	Toluene+ 2-propanol + acetic acid (16 + 2 + 1)	Methanol + water (60 +40)
Applied sample volume	Purity test: 10 µl = 200 µg 20 × 3 mm Limit test: 1 µg = 0,5 %	Assay: 1 µl = 120 ng (100 %-value) 4 mm bandwise	Purity test: 40 µl = 200 µg 10 mm bandwise Identity test: 10 µg/10 mm
Chamber saturation	Yes	No	No
Migration distance	10 cm	3 cm	10 cm
Development time	40 min	15 min	63 min
hRf-values:			
Theophylline 10 µg	15–19	Approx. 27[b]	49–53
Theobromine 10 µg	43–47		56–59
Caffeine 10 µg	73–78		36–39

[a] The chromatographic system is very environmentally friendly, therefore recommended for schools.
[b] Acc. to a private communication of the authors.

⚠ "Critical" Solvent Compositions

The following two examples are not quite so problematical with respect to handling of the solvents, but are critical with respect to the reproducibility:

- It is not recommended to store solvent systems that contain alcohols and acids, as these components tend to esterify, and this eventually leads to an altered chromatogram. An example of this is the frequently used mixture "1-butanol + glacial acetic acid + water".
- Saponification in ester-containing solvent systems should also be regarded as a problem, as the free acids and alcohols cause a shift in the polarity, which again strongly influences the chromatographic results.

- Solvent systems in whose phase diagram there is in any danger of a miscibility gap are also critical. Even with small changes in temperature, these can lead to emulsion formation, i.e. separation of the mixture (example: the DAB solvent system for the sugars fructose, glucose, lactose and sucrose).

4.2 TLC Developing Chambers

In the early days of TLC, before the advent of HPLC, researchers experimented in their laboratories with developing chambers of glass, sometimes V4A steel or, for less aggressive solvent systems, plastics. Various chambers were used for the following purposes:

- ascending development,
- descending development,
- horizontal development.

After some years, two of these development methods are still used, but new techniques of performing them have come along (see Chapter 11 "Special Methods in TLC"). Ascending development is referred to in the present book as the "classical" method, as in the author's opinion it offers a wider spectrum of possibilities than horizontal development. Lively discussions continue in "TLC expert circles" on the subject of which of these two methods of development is the "better". But what does "better" mean? Every user must find this out for himself or herself at his or her workplace, often doing this afresh for each task.

4.2.1 What Types of TLC Developing Chambers Are There?

Between a simple screw-top jar and an automated multidevelopment system (AMD) there is a very wide range of TLC developing chambers, differing from each other in materials of construction, but especially in price, and finally also in the chromatographic result obtained in the various fields of application. The chambers for "classical" ascending chromatography are of glass and have glass or stainless steel lids, while the bodies of the horizontal chambers are of PTFE ("Teflon") and have glass lids, as can be seen in Fig. 56, which shows a selection of developing chambers. All the automatic chambers commonly used today in analytical equipment are built into a metal housing packed with electronic circuitry.

Figure 56. Selection of developing chambers for TLC

Back row from left to right:
Screw top jar (brown glass) for maximum plate size 5 × 8 cm
Double-trough 10 × 10 cm chamber with stainless steel lid (CAMAG)
Double-trough light-weight 20 × 20 cm chamber with glass lid (CAMAG)
Flat-bottomed 20 × 20 cm chamber placed on its side for drying
Chamber, 20 × 10 cm, with V-shaped bottom and stainless steel lid (DESAGA)
Cylindrical 10 × 20 cm developing chamber with glass lid (DESAGA)

Front row from left to right:
Horizontal 5 × 5 cm chamber (DESAGA), common name: "baby chamber"
Horizontal 10 × 10 cm chamber (DESAGA)
Screw top jar (clear glass) for maximum foil size 4 × 8 cm (Macherey-Nagel)

4.2.1.1 TLC Chambers for Vertical Development

- For teaching in schools, the principle of TLC can be demonstrated using equipment never intended to be used for this analytical method and in the lowest price bracket (up to 50 DM). Hellendahl's staining trough recommended by Surborg in 1981, like the screw-top glass jars in the TLC "micro-set" of Macherey-Nagel, also falls into this category. Both these types of chambers are suitable for precoated layers on foils up to 4 × 8 cm in size. Necessity is the mother of invention, and even preserving jars from mother's pantry have been used with great success in teaching [54].

- The circular development chambers now commonly used for the TLC plate formats 5 × 20 and 10 × 20 cm (DESAGA) stem from the era of paper chromatography. These containers may often be misappropriated for keeping flowers fresh, but in general their preferred use is for in-process control. They are not particularly heavy, and are still in the lower price bracket (up to 180 DM). Also in this price range are the flat-bottomed "light-weight chambers" for plate sizes of 20 × 20 and 20 × 10 cm obtainable since 1996 and the flat-bottomed chambers for 10 × 10 cm plates

(CAMAG). According to the manufacturer's information, the latter are not available in light-weight material, as chambers of this size do not weigh very much even in the normal material.

- Most classical TLC developing chambers (also known as "tanks") are in the medium price bracket (up to 500 DM). For the plate sizes 10 × 10 cm, 20 × 10 cm and 20 × 20 cm, these are available in rectangular form and as double-trough chambers with V-shaped bottoms or with side grooves for simultaneous blank chromatography for prewashing. One gets the impression that the heavier the chamber the dearer it is. The only exceptions to this are the "light-weight chambers" in the double-trough form for plate sizes 20 × 10 and 20 × 20 cm. These weigh only ca. 25 % of the usual tanks, and even the larger chambers can be carried comfortably in one hand. Breakages caused by dropping heavy tanks when cleaning them in the sink should now be a thing of the past!
- New development chambers for small UTLC plates (60 × 36 mm) are available: CAMAG markets a small double-trough chamber. Baron has modified a staining trough for microscope slides with a tailor-made cap and a plate holder as a small development chamber.

If they are used correctly, all the TLC developing chambers described after this point in the book fulfill the requirements for identity testing, purity determinations and assays.

- Automatic developing chambers were developed especially for work in accordance with the GMP/GLP guidelines, and with their aid the user is relieved of all supervisory functions (and can go to lunch with a clear conscience!). At ca. 5500 DM for the simple version of Baron (produced by DESAGA) and ca. 10 000 DM per set of ADC (automatic development chamber) equipment manufactured by CAMAG, these are in the upper price bracket and are therefore usually found only in the laboratories of the pharmaceutical industry, where they are used for high-quality **quantitative TLC**. However, it should not go unremarked at this point that ADC offers not only comfort at the workplace but also certainty with respect to reproducibility, which is today regarded as standard in pharmaceutical quality control. Without automatic operation, the daily volume of work could not be coped with.
- The AMD method of Burger is also based on vertical TLC. This system for automated multiple development of thin-layer chromatograms is discussed in Chapter 11 "Special Methods in TLC". This equipment falls into the top price bracket and is included here only for completeness.

4.2.1.2 TLC Developing Chambers for Horizontal Development

With developing chambers for horizontal development, the chamber type again determines the price.

- The 5 × 5 cm horizontal chamber of Kraus is in the bottom price bracket [55]. To obtain good thin-layer chromatograms with this "mini" apparatus, sometimes also known as a "baby chamber", the user first needs to gain some experience with it. With this proviso, the Kraus chamber can be used in teaching and for identity testing in pharmacies with good results. Kraus recommends the 1/16 TLC plates with normal silica gel 60, but the pre-scored 10 × 10 cm HPTLC silica gel 60 plates (Merck Article No. 1.05644), which can readily be broken into four plates of the 5 × 5 cm format, are more suitable. With these, all work, including assays, can be performed, provided the appropriate fittings are available and the work is performed correctly [56, 57].
- The "baby chamber" has now been adopted by the DAC, and the World Health Organization (WHO) has accepted it for monitoring medicinal plants and natural products [58].
- The manufacturer of the Kraus chamber (DESAGA) also markets a horizontal chamber of similar type for the plate size 10 × 10 cm. This also falls into the medium price range, and, after a suitable period of familiarization (as with the "baby chamber"), can be used for identity testing. For purity testing using the large amounts of substance often recommended, the suitability of this developing chamber should be tested. With these horizontal developing chambers, the chromatographic development has to be performed in one direction only.
- A different manufacturer (CAMAG) produces two horizontal developing chambers for the plate sizes 10 × 10 cm and 20 × 10 cm, mainly for use with HPTLC plates. These types of chamber enable the samples to be developed from the two opposite sides towards the middle, giving an available migration distance of only 4.5 cm. With optimal operation, ca. 70 samples can be analyzed in parallel on the 20 × 10 cm plate. These developing chambers are mainly used where a large number of samples must be analyzed quantitatively in a short time. However, this type of chamber is not suitable for use at low temperatures, as in this case liquid can easily condense out and cause problems. The price per set of equipment is over 100 DM.
- The Vario-KS chamber supplied by Geiss and Schlitt [59] and marketed by CAMAG is a special variation of the horizontal development technique. As well as five separate solvent troughs, there are also four different interchangeable conditioning trough inserts. This enables up to five solvent systems to be investigated simultaneously under the same conditions on the same TLC plate. Moreover, this type of chamber enables the relative atmospheric humidity to be kept constant and different types of conditioning of the layer to be carried out. This type of chamber is recommended for development work, as a great deal of time can be saved in the optimization of separating problems by skillful selection of the experimental conditions. The cost of the Vario-KS chamber, including accessories, exceeds 3000 DM (1996).

- Developing chambers for horizontal development were originally conceived as sandwich chambers (S-chambers) in which the plate, with the layer underneath, is placed above the counter-plate at a distance of 0.5 mm. Without this counter-plate, a filter paper soaked in solvent can be placed at the bottom of the chamber, enabling development to be performed in a saturated chamber. Chromatography can be performed with only a few ml solvent in horizontal chambers, which is, of course, economic and reduces disposal problems. However, the question arises how much of a four- or five-component solvent system is actually used if only 2 ml is consumed in the development process in the 5 × 5 cm horizontal chamber?

4.2.2 Influence of the Chamber Atmosphere

After the layer of sorbent and the solvent system, the next most important influence on the result of a thin-layer chromatogram is the gas space of a developing chamber. However, the development processes that can take place are of so many kinds that they are discussed together in this book and mainly described for the "classical" normal developing chamber (N-chamber). For completeness, the influence of the chamber atmosphere is also illustrated using examples of other chamber types.

Detailed theoretical treatments on the theme "Gas/Vapor Space and Sorbent Layer" can be found in the literature [2, 4, 21].

Some important chromatographic terms are explained below and illustrated by a schematic diagram based on a diagram by Geiss [4] (Fig. 57):

- Chamber saturation is the controlling factor when all components of the solvent system *before* and *during* the development are in equilibrium with all the zones of the gas space. If there is a solvent inside the chamber which is not present in the solvent system (e.g. ammonia solution in a glass beaker or extra trough), this additional component must also be in equilibrium with the gas space.
- Pre-loading means, in general terms, i.e. irrespective of the degree of saturation of the chamber and layer, any sorptive uptake of gas molecules from the chamber atmosphere by the unwetted TLC layer (see also Section 3.4 "Conditioning").
- Sorptive saturation means that condition in which the chromatographic layer is in equilibrium with all components of the *saturated* gas space. This is a special case of "pre-loading", namely its upper limit.
- Capillary saturation means the process of capillary filling of any free volume still remaining in the sorbent layer following pre-loading *and* the condition of the layer after completion of the development process.

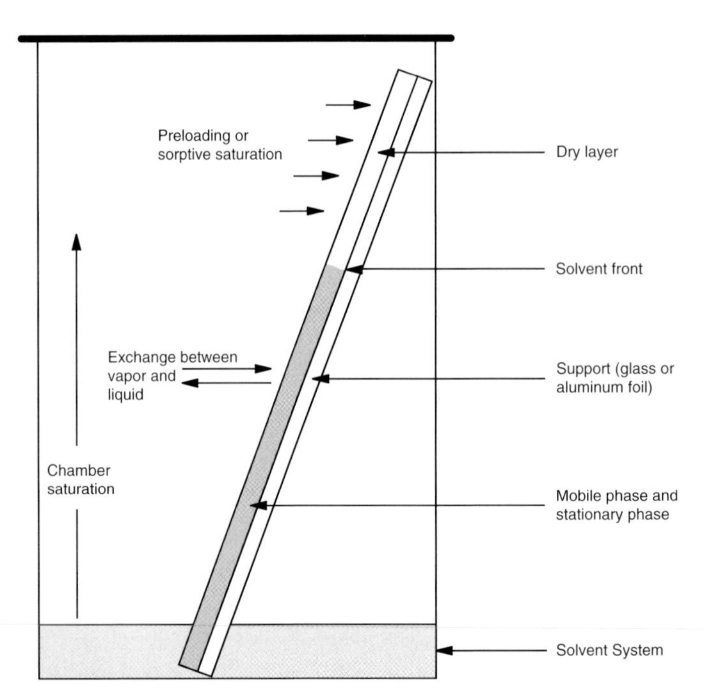

Figure 57. Terms used in connection with chamber saturation
Schematic representation of the exchange processes between the gas space, the solvent system reservoir and the TLC plate before and during development.

In contrast to the N-chamber, completely different conditions exist in a **"sandwich chamber"** or **S-chamber**. This type of chamber, introduced by Stahl, is hardly used in TLC nowadays, but is often described in the literature as an example of the influence of the chamber atmosphere.

For the purpose of comparative investigations, an S-chamber can be quickly constructed in the laboratory. One wall of the chamber consists of the back of the TLC plate to be developed. The counter plate is a clean glass plate of the same width, whose height should be ca. 2 cm less than that of the TLC plate. This is so that the samples can be positioned at a maximum distance of 15 mm from the bottom edge of the plate. After applying the samples, a strip of sealant made of cardboard or any other suitable material 5 mm wide and 1 mm thick is applied to the three remaining edges. This is covered with the counter plate so that the two plates are sealed at the top, and the whole assembly is then stabilized with transparent adhesive tape. This can now be placed in an N-chamber containing a solvent system.

The special feature of this type of chamber is that the dry part of the layer above the ascending front is unsaturated during the entire development process.

To illustrate the influence of the chamber atmosphere, four different solutions containing samples of greater celandine were each applied to three 10 × 10 cm HPTLC silica gel 60 plates, and these was developed in an unsaturated N-chamber, a saturated N-chamber and an S-chamber (Fig. 58 a–c). Table 17 shows the hRf values for sanguinarine together with the chromatographic conditions and also the chromatographic results obtained by development in 10 × 10 cm and 5 × 5 cm H-chambers, both unsaturated and saturated. The chromatograms obtained in the S-chamber showed the narrowest zones but also the lowest hRf values. Those of the saturated N-chamber have broader zones and higher hRf values than the chromatograms after vertical development in the unsaturated N-chamber. For comparison, Fig. 58d also shows the 5 × 5 cm plate developed in the unsaturated H-chamber.

▶ **Figure 58:** see Photograph Section.

Table 17: Influence of the chamber atmosphere as shown with different samples of greater celandine

For all tests:	
Sorbent:	HPTLC silica gel 60 F_{254} (Merck 1.05629 or 1.05635)
Solvent system:	Toluene + methanol (9 + 1 ml)
N-chamber:	Normal, vertical
S-chamber:	Small, vertical
H-chamber:	Horizontal

Type of chamber	Dimensions [cm]	Migration distance [cm]	Time [min]	hRf-value sanguinarine
N (unsaturated)	10 × 10	7,0	23	64,3–70,3
N (saturated)	10 × 10	7,5	18	62,9–68,5
S	10 × 10	7,5	31	29,3–30,7
H (unsaturated)	10 × 10	7,0	23	50 R 62 M[a)]
H (saturated)	10 × 10	7,2	15	54,6–59,7
H (unsaturated)	5 × 5	3,7	6	42,3–50,3
H (saturated)	5 × 5	3,8	4	32,9–36,3

a) R = position of the spot in the edge zone, M = in the middle of the plate (result cannot be evaluated)

4.2.2.1 The Unsaturated N-Chamber

In the early days of TLC, unsaturated chambers without any absorbent lining were used. This was because on the one hand there was little knowledge at that time of the influence of the gas space, and on the other hand the manual operations involved are simpler. The latter factor is the reason why preliminary investigations are often performed even today using this type of chamber. If a TLC plate is placed in the development chamber immediately after placing the solvent system in it, the capillary saturation commences and the front begins to migrate. The mobile phase is released from the upper part of the wet layer, beginning to saturate the surrounding gas space, and this leads to increased transport of the solvent system. This, expressed simply, means that more available development steps are used, which usually leads to an improvement in the separation of spots that lie close together. An example of an improved development result without chamber saturation is the chromatography of four sugars on HPTLC-amino-modified silica gel (Fig. 59a,b). The hRf values from the developments with and without chamber saturation are given in Table 18 together with the other experimental conditions. It can be stated in general that in an unsaturated chamber the hRf values are *higher* and the development times *longer* than those in a saturated chamber.

▶ **Figure 59:** see Photograph Section.

4.2.2.2 The Saturated N-Chamber

In order to avoid the supposed edge effect, Stahl in 1959 introduced chamber saturation by lining the chamber with filter paper. In the technique described in the first handbook (1962), he describes a paper size of 15 × 40 cm, which is placed, bent into a U shape, in the glass trough, and which becomes saturated with the solvent system already present. However, this technique cannot be used with the double-trough chambers that use only small amounts of solvent; moreover, the chamber saturation would not be sufficiently reproducible for a quantitative determination. The DAB states that, in principle, chamber saturation should be performed for 1 h with a fitted filter paper (**CS paper**) unless an unsaturated chamber type is expressly prescribed in a monograph. This time specification would seem to be a precautionary measure, as, according to Geiss, chamber saturation in an N-chamber lined with filter paper and in which the solvent system has a b.p. below 100 °C is achieved in 5 min at the most [4, p. 167]. In our own experience, a period of at least 30 min to give adequate chamber saturation can be necessary for certain solvent systems, and this should be expressly stated in a prescribed testing procedure.

A particularly impressive example of an improved result after development in a saturated N-chamber is shown by the chromatogram of the separation of spironolactone and furosemide on normal silica gel 60, where without chamber saturation the spironolactone is in the front and the furosemide is slightly below this (Fig. 60a,b). However, in this example the development is performed with a CHC-containing solvent system, and an alternative system for separating these two substances is therefore given in Table 15. Other information on the conditions is also given.

▶ **Figure 60:** see Photograph Section.

☞ **Practical Tips** for the chamber atmosphere:

- The lid of the TLC chamber must on no account be opened during the chromatographic development process! It is therefore a good idea to leave a viewing slit in the CS filter paper lining to check the height of the solvent system.
- To produce saturation of the chamber, circular filter papers must *never* be used, as these would not take up solvent over the whole width of the chamber.
- To line the developing chambers for this purpose, filter papers with good absorption properties for the solvent system should always be used, although if the filter paper is too thick it can become separated from the chamber wall, so that the TLC plate is no longer firmly supported. Suitable filter paper is marketed in sheet form by Schleicher & Schüll (Article No. 604, 58 × 58 cm).

- For **qualitative** work, the sheets of filter paper should be cut into pieces of size ca. 19.6 × 19.6 cm. To do this, several sheets are stacked together and folded twice, each fold being 19.6 cm from the edge. The paper is then cut twice with a guillotine to produce 19.6 × 58 cm sheets, each folded twice. These are cut along the folds to give 19.6 × 19.6 cm sheets, which can be conveniently stored at the TLC workplace and can be cut up as required to obtain smaller formats.
  To line a double-trough chamber, two sheets of CS paper are cut to size and fitted in the troughs, and the developing chamber is tilted forward so that one sheet rests on the front wall and the other can be held onto the back wall with thumb and forefinger. Approximately half of the solvent system to be used is now carefully poured onto the front glass wall over the paper and into the front trough. Care must be taken to avoid the formation of any air bubbles between the paper and the wall of the chamber. The chamber is then turned through 180° and carefully tilted forward, and the operation is repeated.
  It is here worth repeating our advice about the maximum amount of solvent to use in a particular chamber. For example, if 100 ml solvent is placed in a 20 × 20 cm double-trough chamber and the samples are positioned on the plate in the normal way, this can lead to errors in the development!

- For **quantitative** work, 20-cm wide plates are normally used. As the development chamber should be lined on all four sides in this case, the sheets of filter paper folded as described above can be cut at the folds with a sharp knife unless a suitably large guillotine is available. In the latter case, the 58 × 58 cm sheets can be marked out with a lead pencil and cut into six strips just under 10 cm wide or three ca. 19.6 cm wide.
  Charging of a developing chamber with solvent for an assay determination is as a rule performed as follows: A 58 × 19.6 cm strip of CS paper is fitted into a flat-bottomed chamber so that one end of the strip is placed in one corner of the chamber. A ruler is then used to smooth down the paper strip and "fold it into" the next two corners. The other end of the paper strip then does not quite reach the fourth corner in the case of the 20 × 20 cm chamber of DESAGA, so that a small viewing slit remains. With other types of chamber, the length of the CS paper strip must be adjusted accordingly. Alternatively, the start of the paper strip can be placed in the middle of a narrow side, so that the viewing slit is then obtained on the outside. Using a pipette, a few milliliters of solvent are then applied onto the CS paper at the top of the two narrow sides, at least enough to wet the paper completely. The solvent is then added in a similar way to that used for qualitative work, i.e. by tilting the chamber towards the two long sides.
  It is even more important than with the "qualitative chamber" to avoid air inclusions between the paper and the chamber wall, as otherwise the development may not be reproducible. Formation of these air bubbles cannot always be avoided, but they must be completely removed before the chamber is allowed to stand in order to saturate the gas space. A long (but sturdy) glass rod can be useful for this.

- Instructions are sometimes found in monographs, especially in the USP, in which chamber saturation with ammonia solution is prescribed although this component is not present in the solvent mixture. To achieve this, the USP states that two beakers of ammonia solution should be placed on the bottom of a chamber in which the solvent system is already present. Our own investigations have shown that a trough of width ca. 2 cm produced from a block of PTFE of the same width as the chamber gives better saturation of the sorbent with ammonia vapor, as the NH_3 molecules are then transferred to the layer over the whole width of the chamber and not just from the two beakers, whose diameter is only ca. 3 cm.

- The sorptive saturation of the layer with solvent molecules can also be achieved using the time-controlled conditioning equipment of Baron (Fig. 61). At the end of the selected conditioning time (maximum 60 min), the plate is automatically lowered into the solvent.

- The question whether “better” thin-layer chromatograms are obtained with or without chamber saturation cannot in general be answered. According to Geiss [4], there are about 25 different factors that can influence a separation by TLC, so that the parameters must be determined empirically for every task.

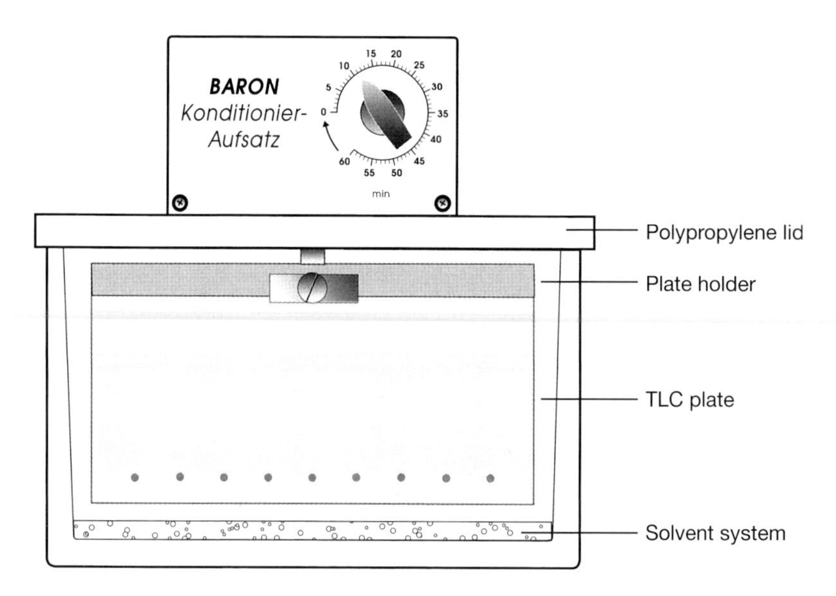

Figure 61. Conditioning device as an aid to the sorptive saturation of the layer by solvent molecules (Baron)

4.2.3 Influence of Temperature in Chromatography

In TLC, development is usually performed at room temperature. If samples are volatile or solvent systems have a low b.p., better results can be obtained by cooling the chamber and the solvent. In the analysis of dry extract of stinging nettle root, for example, the use of the DAB solvent system (diethyl ether + methanol, 90 + 10 ml) can enable an additional violet-colored zone to be separated from the β-sitosterol if the chromatography is performed at 5 °C instead of 22 °C (Fig. $53a_2$). To separate polycyclic aromatic hydrocarbons, development at a temperature as low as –20 °C may be necessary [60]. However, such extreme deviations from room temperature are not always necessary. It is usually sufficient to set the temperature for the chromatography at the temperature used in the validation. To achieve this, Baron in 1996 developed the TLC Thermo-Box, which gives the desired temperature with an accuracy of ±1 °C using a Peltier element. This technique can give temperatures of 15 °C below or 25 °C above ambient temperature; cooling from 25 °C to 10 °C takes ca. 1.5 h.

Refrigerators for chromatography are also commercially available (NUNC). These are obtainable in various sizes, but can only be operated in the region of 4 °C, which greatly limits their usefulness.

TLC is also performed at higher temperatures. Thus, Lichius analyzed the sugars from the nectar of digitalis (foxglove) by one-dimensional twofold development on HPTLC precoated plates Si 50 000 (Merck Article No. 1.15134) without chamber saturation in a drying oven at 80 °C (!) with the solvent system acetonitrile + water (17 + 3 ml) [61] (N.B.: these chromatographic conditions were positively confirmed on further inquiry.)

4.2.4 Location and Labeling of TLC Developing Chambers

The charging of a TLC chamber with solvent is usually performed at the place where the development will later take place. This is in order to avoid moving the chambers back, as they are sometimes quite heavy. However, "sensitive noses" like to have a fume extractor for the charging operation. There can be no objection to this, especially when concentrated ammonia solution is used, if the TLC chamber is then replaced on the level (!) laboratory bench. However, if the chromatography is to be performed in an unsaturated chamber, it is necessary to wait long enough for the solvent to become static after insertion of the plate. On no account should the TLC chamber be in a fume cupboard during development. The continuous current of air leads not only to the formation of a temperature gradient on and in the chamber, but also to consequent phenomena that are uncontrolled and hence not reproducible during the development process. Thus, for example, 9 plates, each with 8 samples of a pharmaceutical substance, were prepared by a laboratory assistant in one day for purity testing. The developing chambers were kept in a fume cupboard during the whole of the development period. Figure 62a,b shows a section of two of these 9 plates in which undesired fluctuations of the hRf values can clearly be seen. Each of the remaining 7 plates had a different appearance.

▶ **Figure 62:** see Photograph Section.

In rare cases, however, problems can arise from the continuous circulation of atmospheric air even when the work is performed on the laboratory bench. The quality and reproducibility of the separation in TLC analyses can be improved by placing a styrofoam hood of suitable size over the TLC chamber during the development [62].

Direct sunlight should not be allowed to fall on the TLC chamber, as this can lead to temperature effects and catalytic interactions between the samples and the sorbent. It may therefore be necessary to work in subdued light for some hours. The treatment of sensitive samples is described in Section 10.2.

For teaching in a pharmacy or in laboratory areas where TLC is demonstrated only occasionally, only one experiment usually has to be performed, i.e. only one developing chamber is used in a given period of time. In this case the chambers do not need to be marked, as there is no possibility of a mix-up. This is not so if several TLC plates are to be developed in parallel or within a limited time period, when several chambers are standing on the laboratory bench at the same time. In this case it is important to mark them properly. If the chambers (and therefore solvent systems) become mixed up, all the previous preparative work is futile, and an entire working day can be wasted.

☞ **Practical Tips** for marking TLC developing chambers:

- On no account should the lids of the TLC chambers be marked, as these can easily be mixed up.
- The front side of a TLC chamber should not be marked with a felt-tipped pen, as it is often forgotten to remove this marking when the chamber is cleaned after completion of the development.
- Write on a strip of paper the daily serial number of the chamber or, better, the solvent mixture composition, and wedge the label half under the TLC chamber.
- Labels, which can be kept in small plastic holders, are very suitable for marking TLC chambers by sticking them to the sides.
- For solvent mixtures used over and over again, fairly permanent labels can be produced. For these, there are no limits to one's own fanciful abbreviations for solvent systems. Moreover, such labels can include other information, e.g. about the migration distance and development time. For the TLC of flavonoids on HPTLC plates, a permanent label can, for example, contain the following information:

SS Flavonoids HD
EA + FA + MeOH + H_2O
40+2.5+2+4 [ml]
7 cm; 35 min

This means:

SS Flavonoids HD	Solvent system for flavonoids (in-house development)
EA	Ethyl acetate
FA	Formic acid
MeOH	Methanol
H_2O	Water, HPLC quality
7 cm	Migration distance
35 min	Developing time (at 23 °C)

4.3 Development of Thin-Layer Chromatograms

All the operations described above are in general only preliminaries to the actual process of TLC, i.e. the development of the TLC plates by a liquid with the object of separating mixtures of substances. "All roads lead to Rome" is an old saying. The same applies to TLC, in which the results can be obtained by various kinds of development:

1. One-dimensional development
   - Single development
     - vertical
     - horizontal, in one direction
     - horizontal, in opposite directions
     - circular
     - anticircular
   - Multiple development
     - separate runs over the same migration distance
     - stepwise, increasing
     - stepwise, decreasing
     - automated multiple development, stepwise with solvent gradient

2. Two-dimensional development
   - Two dimensions, one solvent system
   - Two dimensions, two solvent systems
   - SRS (separation in 1st dimension → chemical reaction → separation in 2nd dimension)

3. Development with forced flow
   - OPLC: overpressured layer chromatography
   - HPPLC: high pressure planar liquid chromatography
   - RPC: rotation planar chromatography

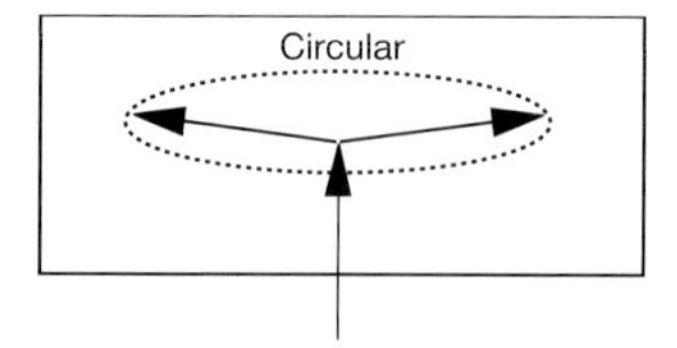

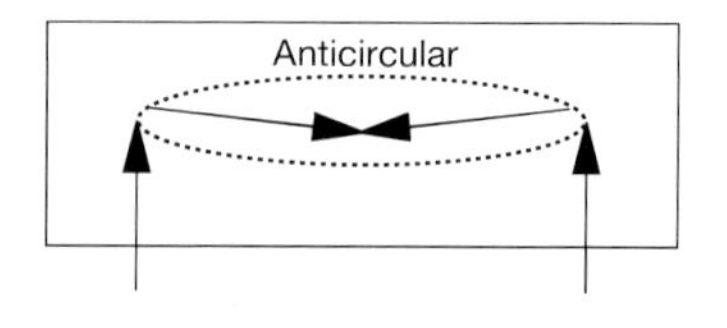

Both circular techniques were used in the 1980s and are of the horizontal development type. Special equipment is required not only for application and development of the sample but also for its evaluation. This equipment is not discussed in this book, especially as it is now hardly ever used. The simple horizontal and counter-current horizontal development techniques are described in Section 4.3.1.2. The special development techniques AMD, OPLC, HPPLC and RPC are discussed in Chapter 11 "Special Methods in TLC". In all the other development techniques mentioned above, the TLC plate is chromatographed one or more times in a vertical position. The separations described in the next Section were achieved in various trough chambers. Chromatography performed in automatic development chambers, all of which are also constructed as trough chambers, is not described here, as these have so far not attained great importance. However, we would point out that, in our experience, when TLC plates are wetted with a fluorescence indicator and the automatic Baron chamber is used overnight, there should be no acid present in the solvent system, as this destroys the fluorescence indicator by the next morning and no substances can then be detected by short-wave UV light.

4.3.1 One-Dimensional Thin-Layer Chromatography

Most thin-layer chromatograms are produced in one dimension, and in fact even today it is very difficult to obtain quantitative results from plates developed in more than one dimension. All present-day commercially available TLC scanners therefore operate on the principle of a one-dimensional chromatographic lane.

Whether working with N-chambers (vertical development) or H-chambers (horizontal development), experience with the appropriate equipment is necessary. The personal preference of the user is a very important factor when deciding which of these development techniques to use in quantitative TLC. This question is superfluous for purity tests using the procedures prescribed in the pharmacopoeias, as neither the capacities of HPTLC plates nor the available migration distances in the 5×5 cm H-chamber are sufficient for such analyses. Experiments of this sort have been reported [63, 64], but the quality of the chromatograms obtained was not impressive. Our own investigations showed that in some cases a reduction of the amount of sample applied to ¼ of that prescribed in the DAB enabled the chromatography to be performed even in the "baby" chamber, but reduction of the limiting concentration in the same ratio was not practicable (as this meant that the limit of detection was not reached), so that the conditions prescribed by the DAB were not complied with.

4.3.1.1 Vertical Development

In most cases, a single development is all that is necessary to obtain a good thin-layer chromatogram. For this, the precoated TLC plate is carefully placed in the prepared development chamber with its back surface leaning against the back wall, and the chamber is immediately covered with the lid, upon which a lead "doughnut" is then placed for extra weight. The development by the solvent mixture starts as soon as the plate is immersed in the solvent. When the latter has reached the previously marked front line, the lid is removed and the plate is quickly taken out. The solvent that runs off is absorbed in a Kleenex cloth and the plate is placed in the fume cupboard to dry.

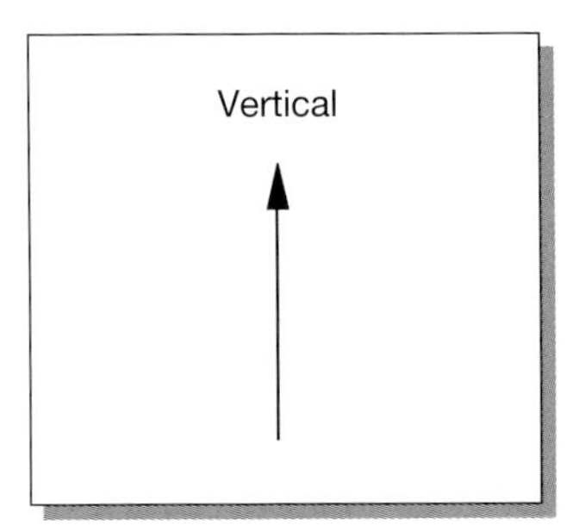

And that's it! "Is all this work really necessary?" I hear you ask. The answer to this can only be: only **if** you have put this much effort into it will you obtain good chromatograms and not the "abstract art" shown in our cartoon.

☞ **Practical Tips** for vertical development:

- During development, the development chamber should be standing on a flat, horizontal surface and should be left completely undisturbed.
- During development, do not under any circumstances turn the chamber to observe the solvent front.
- It must again be emphatically repeated here that the chamber must under no circumstances be opened during the development, even to have a quick peep!

If certain separation problems cannot be solved by a single development, multiple developments can be the answer. Here, the chromatogram is developed two or more times with intermediate drying of the layer. There are five types of repeated development:

Table 18: Sugar

	DAB 10	Pachaly: TLC-Atlas
Sample solution	10 mg/20 ml water + methanol (2+3)	10 mg/4 ml water + methanol (1+1)
Sorbent	Prescribed: Silica gel G Used: TLC silica gel 60 F_{254} (Merck 1.05715)	TLC silica gel 60 F_{254} 20×20 cm (Merck 1.05715)
Solvent system	1,2-Dichlorethane + methanol + acetic acid + water (50+15+25+10)	Acetone + ethylmethyl ketone + boric acid sol. 3 % (40+40+25)
Reference substance	Fructose Glucose Lactose Sucrose	Lactose Glucose
Applied sample volume	2 µl Pointwise	4 µl Pointwise
Chamber saturation	Yes	Yes
Migration distance	2×15 cm	12 cm
Development time	2×95 min intermediate drying and cooling: 20 min	52 min
Derivatization reagent	Thymol-sulfuric acid	Diphenylamine-aniline-phosphoric acid
Colour after derivatization		Acc. to literature
Lactose	Light red	Grayish blue
Fructose	Light red	Brownish red
Glucose	Light red	Light blue
Sucrose	Light red	Brownish green
hRf-values:		
Lactose	21–24	34–39
Sucrose	27–30	43–52
Glucose	33–36	44–48
Fructose	36–38	16–22

Alternative I	Alternative II	Alternative II with chamber saturation
5 mg/10 ml water + methanol (2+3)	5 mg/10 ml water + methanol (2+3) for fructose, glucose, lactose, 50 mg/10 ml for sucrose	
HPTLC-silica gel 60 F_{254} 10×20 cm (Merck 1.05642)	HPTLC NH_2-modified silica gel 60 F_{254s} 10×10 cm (Merck 1.05647)	
1,2-dichlorethane + methanol + acetic acid + water (50+15+25+10)	Acetonitrile + water + phosphoric acid-buffer (pH 5,9) (85+15+10)	
Fructose Glucose Lactose Sucrose	Fructose Glucose Lactose Sucrose	
1 µl Pointwise	10–13 µl 10 mm bandwise	
Yes	No	Yes
2×7 cm	7,5 cm	7,5 cm
2×36 min intermediate drying and cooling: 20 min	18 min	14 min
thymol-sulfuric acid	Heating not less than 15 min at 160 °C	
	Evaluation in UV light for all sugars	
Light red	Fluorescence quenching at UV 254 nm	
Light red	Light blue fluorescence at UV 365 nm	
Light red		
Light red		
16–21	7– 9	4– 6
23–27	11–13	8–10
30–33	18–20	13–15
33–36	23–25	17–19

Repeated development

- with the same solvent system over equal migration distances,
- with the same solvent system over increasing migration distances,
- with different solvent systems over equal migration distances,
- with different solvent systems over increasing migration distances,
- with different solvent systems over decreasing migration distances.

The aim of the first technique is improved separation of neighboring substances.

As an example, the DAB gives a method for the identification of the four sugars lactose, sucrose, glucose and fructose by twofold development on TLC silica gel 60 over a migration distance of 15 cm. The evaluation must only be performed if, after derivatization with the thymol-sulfuric acid reagent, four clearly separate spots can be seen. The question arises here whether a single chromatographic development over a migration distance of 10 cm is sufficient for an identity test for the lactose and sucrose, even though the glucose and fructose are not completely separated from each other. Twofold development on HPTLC silica gel 60 precoated plates gives good results even over a migration distance of 7 cm. The relevant chromatograms are shown in Fig. 63 a–c, and comparative data on the time requirements and other parameters for these developments are given in Table 18. It may be noted, however, that it was not in fact possible to perform a successful separation of the above-mentioned sugars in the "baby" chamber.

▶ **Figure 63:** see Photograph Section.

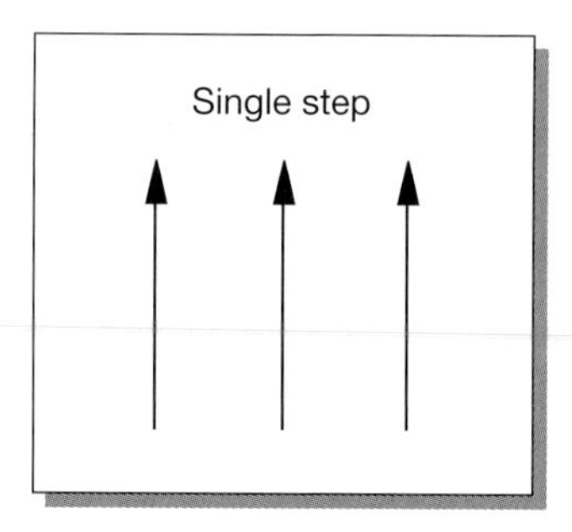

Even threefold developments with the same solvent mixture are described in the literature. For example, Kennedy describes a solvent mixture for the separation of hexoses, deoxyhexoses and some disaccharides on cellulose [65]. The threefold development over the same migration distance is performed with a mixture of ethyl acetate, pyridine and water (100 + 35 + 25 ml). Madaus uses this type of threefold development for the sugars obtained by hydrolysis of mucilages. The separation is performed on TLC silica gel 60 using acetonitrile + glacial acetic acid + water (85 + 14 + 1 v/v) as the solvent system [65a].

The developments performed using "progressively increasing migration distances" are nowadays favored for the stability testing of some samples held in a matrix in order

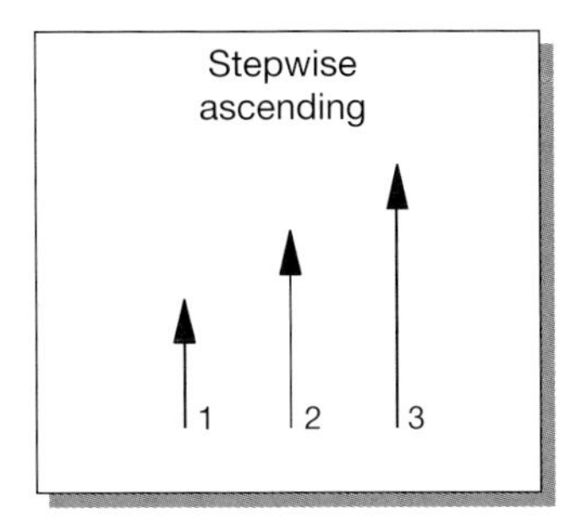

to release the substance to be analyzed from the application zone. Auxiliary substances in tablets, which cannot be chromatographed, can represent a matrix of this type. In these cases, a chromatographic development is performed over a few centimeters by a strongly polar solvent mixture. After complete removal of this solvent mixture, observation of the plate under short-wave UV light often shows that the substance, after its release from the matrix, is in the form of an ill-defined "smear". The desired result is only obtained after development by the actual developing solvent. If the initial development in the polar solvent mixture is omitted, cloud-like shapes are often obtained, as shown in Fig. 64.

▶ **Figure 64:** see Photograph Section.

We now describe an example of an analysis of natural products published by Stahl and Kaltenbach in 1961 [66]. This was also described by Ebel and Völkl in 1990 [67] as an example of repeated development with different solvent mixtures in the same direction with increasing migration distances. The following is a quotation: "The Separation of Podophyllum-Lignanes and their Glycosides. A strong eluent is used for the first development, and the solvent front is allowed to migrate for about half the planned migration distance. This separates the polar glycosides, while the less polar aglycons are transported with the solvent front without being separated. The aglycons are separated in the second stage, using the total migration distance, by a solvent mixture of lower elution strength. The glycosides do not migrate on the plate in this mobile phase, and remain in their original position."

The procedure for development with progressively decreasing migration distances is the exact opposite.

An example from pharmaceutical quality control can serve as a model here. In the analysis of suppositories, the matrix components (e.g. wax) are displaced by a nonpolar solvent for a few centimeters **beyond** the solvent front of the subsequent chromatography of the substance, and this can then proceed through the matrix without further problems. Figure 65 shows the chromatogram of Gastrosil® suppositories (active substance: metoclopramide base) after derivatization in an iodine chamber. The matrix components are clearly visible in the upper part of the plate.

▶ **Figure 65:** see Photograph Section.

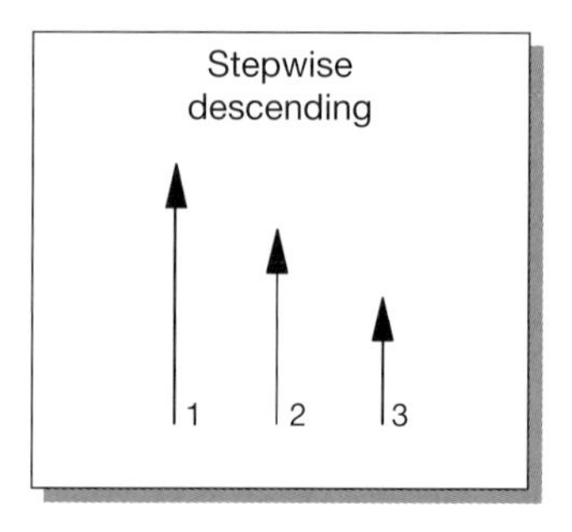

4.3.1.2 Horizontal Development

In contrast to vertical development, in which the solvent system is always placed in the development chamber first, in horizontal development the TLC plate must be positioned on the bottom of the chamber first, i.e. before adding the solvent system. In both types of H-chamber of DESAGA, as soon as the solvent is added to the chamber the process of development begins by wick transportation through a sintered glass insert, while in the linear development chambers of CAMAG the chromatographic development starts as soon as a glass strip is tilted into a position such that the solvent is transferred to the layer through a capillary gap. In the latter chambers, development from the two sides towards the center is possible, and this is also known as "countercurrent horizontal development".

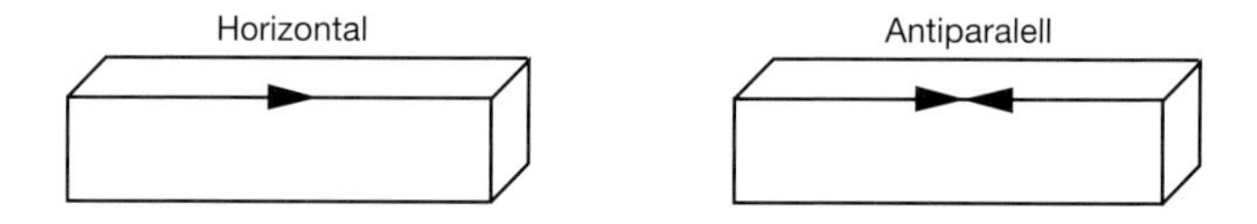

Development stops automatically when the two fronts meet. If a longer migration distance is desired, only one side of these CAMAG chambers can be used, and in this case some solvent usually has to be added during the chromatography to prevent "starvation" of the plate.

Professor Veit of the Department of Pharmaceutical Biology in the University of Würzburg describes his glass "Micro-TLC" apparatus [68], shown here in Fig. 66, which combines the dimensions of Kraus's "baby" chamber (for 5 × 5 cm plates) with the development technique used for CAMAG chambers.

During the students' practical training course in Würzburg, 59 drugs were investigated of which 39 are suitable for "micro-TLC". Of these, 14 gave even better results than those obtained by the TLC test of the DAB. For 6 drugs, no micro-TLC methods have so far been developed. Veit confirmed the comment made in Section 4.3.1, that in general a direct change from the DAB TLC method to the micro-TLC method (and hence to the 5 × 5 cm H-chamber) is not automatically possible, even if the volume applied is reduced.

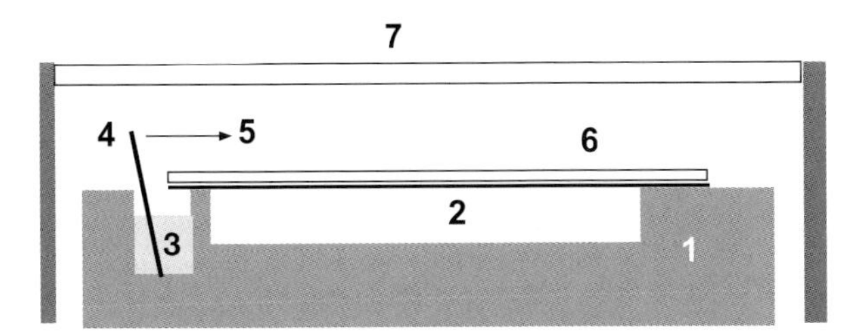

Figure 66. "Micro-TLC" chamber in three parts (from [68])
The base (*1*) (6 × 7.4 × 1.5 cm) includes two reservoirs filled with the solvent system, one of which (*3*) is for development, the other (*2*), in the central part, being for chamber saturation if required. For development, the prepared TLC plate (*6*) with the coating facing downwards is placed on the central reservoir. The glass plate (*4*) (1 × 5 cm) in the left-hand solvent system reservoir is tilted (*5*) to cause development of the TLC plate to commence. The anti-evaporation hood (*7*) with a transparent top is placed over the equipment after the start of the process.

Figure 67 shows, as an example, the purity test of cimetidine on 10 × 10 cm or 5 × 5 cm HPTLC plates. The amount applied to the 10 × 10 cm plate was reduced to half the amount prescribed in the DAB, and the amount applied to the 5 × 5 cm plate was 1/4 of this. Whereas the limiting concentrations on the larger plate are just detectable after derivatization with iodine, a zone can be seen only on lane 2 of the smaller plate.

▶ **Figure 67:** see Photograph Section.

4.3.2 Two-Dimensional Thin-Layer Chromatography

In an attempt to maximize the utilization of the space available on an existing layer, Consden et al. [69] experimented with a two-dimensional technique as early as the era of paper chromatography, both the direction of development and the solvent being changed after the first development.

This gives a theoretical increase in the capacity of the spots, so that this method should be ideal for the identification of mixtures with a large number of components. Even in the very early years of TLC on precoated plates, two-dimensional TLC **(2D-TLC)** was of great importance in the analysis of metabolic diseases. With the aid of this TLC technique it was possible to carry out screening tests for a large number of patients in a relatively short time, to test urine for anomalies in the amino acids present, and to detect hyperaminoacidity in urine [70].

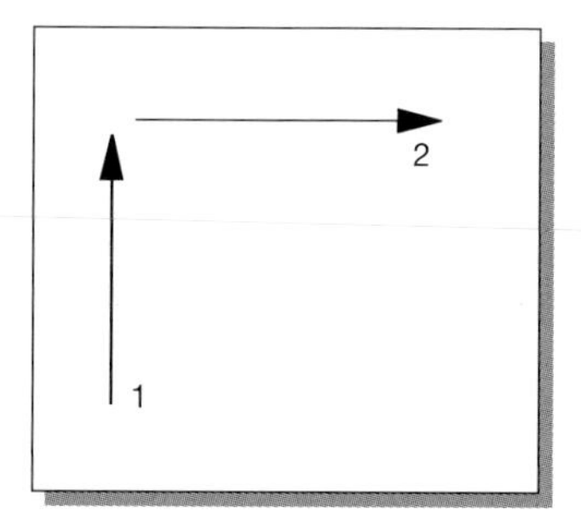

The principle of the procedure is described below:
An amino acid standard and a small amount of urine are applied in turn to the same point on a cellulose-precoated plate. This is then developed two-dimensionally using two different solvent systems, causing all the amino acids to be separated from each other. These are then visualized by derivatization with ninhydrin. The amino acids are identified from the known pattern of spots produced by the standard which is applied at the same time. Anomalies are detected by the significant intensification of certain spots compared with the chromatogram of a normal urine [71]. Figure 68 gives a schematic diagram of the positioning of the urea amino acids after a 2D-TLC.

Substance identification by 2D-TLC is also often performed in the investigation of phytopharmaceuticals, which usually have complex compositions. In the analysis of ginseng roots, for example, development is performed in the 1st dimension with *n*-butanol + ethyl acetate + water and in then in the 2nd dimension with chloroform + methanol + water. After derivatization with the vanillin-sulfuric acid reagent, it was found that the critical substance pairs had been separated [71a].

From a logical point of view, 2D-TLC using the same solvent in two directions should be the best system. However, this does not usually lead to any additional information, as all the substances would lie on the diagonal, as shown in the diagram in Fig. 69. This method of 2D-TLC only becomes interesting if a reaction has occurred between the two developments, and deviations from the diagonal line can be observed after the second development.

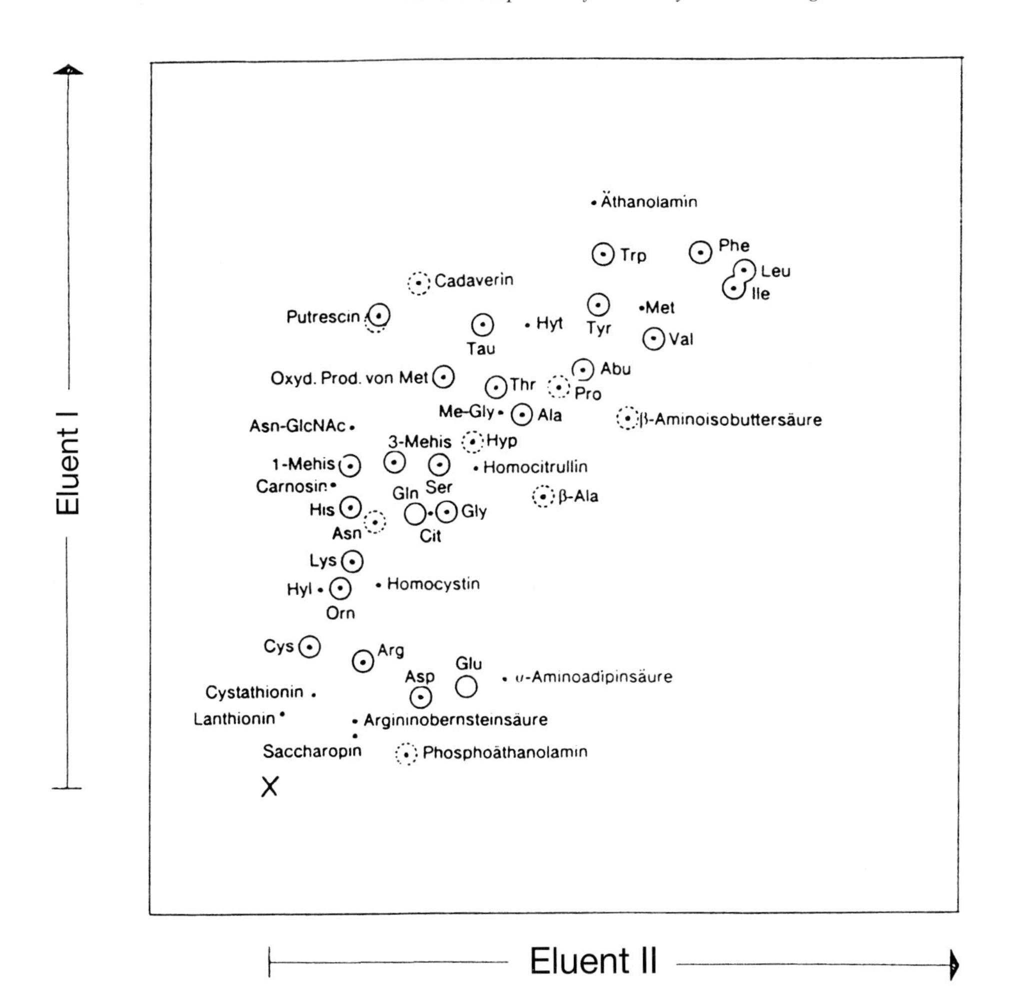

Figure 68. Position of the urine amino acids on a 20 × 20 cm TLC plate after a two-dimensional development and subsequent derivatization with the ninhydrin reagent (from [71])

× Start point

⊙ Position of substances present in the standard solution and not visible after the first heating process

(⋅) Position of substances present in the standard solution and in general only visible after the second heating process

○ Position of substances that become visible in most urine chromatograms but are not present in the standard solution for technical reasons

• Position of substances that do not usually occur in urine and are also not present in the standard

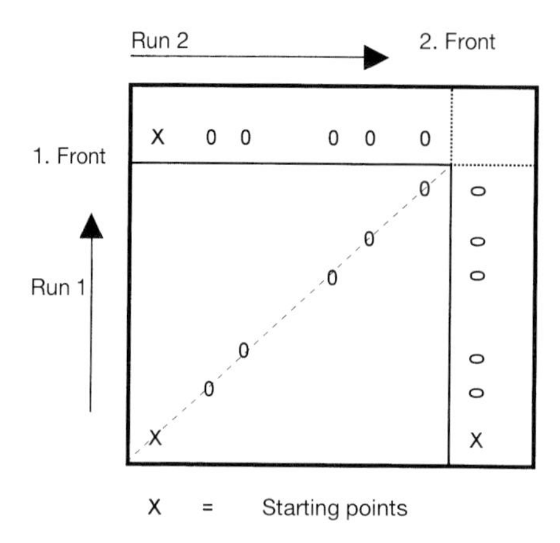

Figure 69. Schematic diagram of a 2D-TLC in which with the same solvent system was used for developing in both dimensions
The direction of flow in the two runs is indicated by arrows.

An example of this is described in a paper on greater celandine [72]. To demonstrate that the action of high-energy light on pure chelidonine leads to a chemical change, the development in the 1st dimension was followed by exposure of part of the plate to 254-nm UV light for 15 min, and the plate was then developed in the 2nd dimension using the same solvent system. After removal of the solvent, several decomposition products with a yellow fluorescence could be seen in 365-nm UV light. The plate was then exposed to short-wave UV light for a further 5 min. Only after this were zones with a yellow fluorescence detected at the height of chelidonine on the reference lanes and the diagonals produced by development in two dimensions. The detection steps described

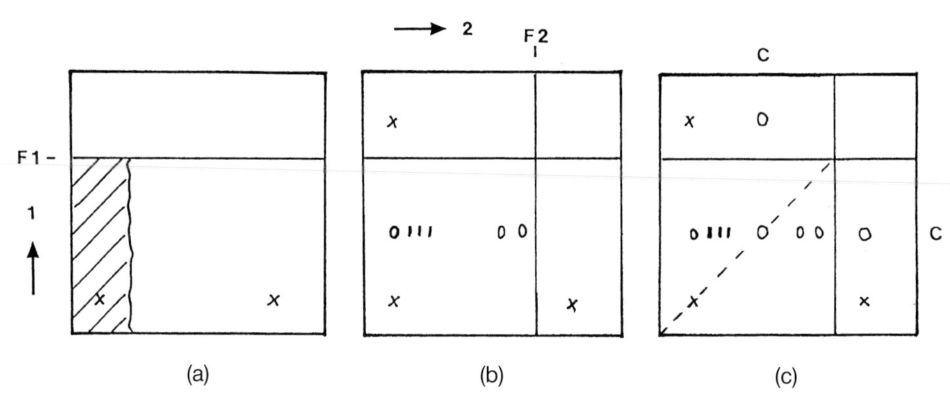

Figure 70. Scheme of a 2D-TLC in which part of the plate (chelidonine) was irradiated with UV light (254 nm) after the first run (showing all points fluorescing in 365-nm UV light)

× Application points
C Chelidonine
F1 and F2 Solvent system fronts
The flow direction in the two runs is indicated by arrows.
(a) The dark area shows the part of the plate irradiated with short-wave UV light after the first run
(b) After the second run, before further irradiation with UV light
(c) After further irradiation with UV light

here are represented graphically in Fig. 70a–c. Further details on the analysis of greater celandine are given in Section 6.2 "Irradiation with High-Energy Light". See also [72a].

In analogy to greater celandine, the TLC described below is also a so-called **SRS technique** (separation-reaction-separation). Whereas in the first example there is a photochemical reaction in which the sorbent layer is unchanged, there is now a chemical reaction with decomposition of any fluorescence indicator present. Heisig and Wichtl describe the use of this technique in a TLC reaction chamber [73]. In the determination of plant glycosides, using the example of marigold flowers (*Calendula officinalis* flos), this very selective identification by 2D-TLC with an intermediate chemical change of the substance is described. As the procedure can be used with all flavonoid-containing drugs, it is briefly described below:

A sample is applied in spot form to the bottom edge of a TLC plate and in band form to the right-hand edge. The plate is first developed in a solvent system suitable for flavone glycosides (shown from bottom to top of Fig. 71). The left-hand chromatogram is hydrolytically decomposed by treatment with 5 N hydrochloric acid followed by heating (8 min at 120 °C) in the TLC reaction chamber. Rotation through 90° gives the starting band. A reference solution containing the appropriate aglycons is now applied to both sides of this and is then chromatographed in a solvent system suitable for these substances. Residues from the incomplete hydrolysis can be seen on the left-hand side of the plate. The solvent front of the second development run is vertical and is on the left of the right-hand chromatogram. If one imagines horizontal lines between the two chromatograms of the first run, the aglycons lie on these lines on the right of the corresponding glycosides. Based on the reference substances applied before the second development, aglycon can be assigned to glycoside with certainty. In Fig. 71a sample of birch leaves is used to illustrate the SRS technique described here

A further example of the technique is the two-dimensional separation of the ginkgo flavonol glycosides with intermediate glycoside decomposition [73a].

▶ **Figure 71:** see Photograph Section.

4.4 Drying After Development

After the chromatography, the solvent must be removed from the layer

- to prevent any subsequent deterioration of the chromatogram obtained due to diffusion effects and
- so that the solvent will not have any harmful effects on the subsequent detection process or even make it impossible.

For this, the plate is placed in a level position (!) in the fume cupboard until the odor of the solvent has disappeared. The plate can sometimes be carefully blown with compressed air, but care must be taken to avoid blowing the solvent across it, as this could lead to an undesired chromatographic effect. With powerful solvent systems containing water or acids, drying with warm air is usually necessary, and the last residues of solvent should therefore be removed using a fan heater. If a densitometric measurement is to be performed, it is advisable to specify this drying time in the testing procedures to ensure that the measured values will be reproducible.

☞ **Practical Tips** for drying after development:

- With volatile compounds such as essential oils, it is advisable to avoid the use of high temperatures and simply to allow the TLC plate to dry for a time in a fume cupboard before treating it with a reagent.
- If the solvent system already contains the derivatization reagent (e.g. in case of the quantitative determination of amino acids), the drying of the plate and the simultaneous reaction with the reagent are performed either in a drying oven (with air extraction!) or by placing the plate on a cold plate heater and then switching it on to start the heating (see also Section 6.5: "Further Treatment of Derivatized Chromatograms").
  The end of this Chapter on the subject of development is an appropriate point to offer some hints on the removal of residual solvent and the cleaning of TLC chambers:

☞ **Practical Tips** on tidiness:

- When the chromatography is finished, it is good practice to remove the residual solvent as soon as possible and to clean the developing chamber, especially if there are several people working in the TLC laboratory. Nothing is more unpleasant at the start of the day than always finding a used developing chamber on the laboratory bench, when it could have been dried overnight ready for use.
- Solvent residues are disposed of in special waste containers, keeping halogen-containing solvents separate from the rest.
- Crucible tongs (if available) are used to take the CS paper out of the chamber. This is then allowed to dry by evaporation in the fume cupboard before disposing of it as waste paper.
- The developing chamber is now kept in the fume cupboard for a time, before rinsing it with tap water and then with distilled (or deionized) water. It is not usually necessary to clean it with detergent, except occasionally the outside.
- The sintered glass inserts from DESAGA H-chambers should be carefully cleaned, especially if they have been used with acid- or alkali-containing solvents. They can then be rinsed with alcohol and dried in a drying oven.
- Glass trough chambers are dried by placing them on their sides on a clean (!) cloth. Under no circumstances should a chamber be turned upside down, as the flat ground glass top of the chamber gives such a good seal that air circulation is almost impossible, and the inside of the chamber then takes much too long to dry. The chamber lid can be placed half underneath the chamber to prop it up at an angle.

5 Evaluation Without Derivatization

When the solvent residues have been removed from the TLC plate, this is the moment to take the first look at it to find out whether the work done so far has given the desired result.

5.1 Direct Visual Evaluation

TLC plates are still evaluated with the naked eye, a process which requires minimum expenditure on apparatus and which is still prescribed by the pharmacopoeias. The evaluation can be performed in daylight, using either reflected or transmitted light, or can be assisted by the use of UV equipment providing short- and/or long-wave illumination.

5.1.1 Detection in Daylight

The simplest method of detecting substances on the TLC plate after chromatography is by the visual detection of spots caused by substances with a color of their own. Here, precoated layers without fluorescence indicators can be used. Dyes, which are analyzed for the identification of gelatin capsules, are a good example of this, being visible in daylight. Dyes are very often added to foodstuffs, e.g. wine gums, effervescent powders (sherbet), fruit juices, yoghurt and margarine. As well as legally permitted dyes, banned additives are often found, at least in Germany, including sandalwood and its extracts, in meat, sausages and fish products [74]. In doubtful cases, an analytical determination of added sandalwood must be performed semiquantitatively using amino-modified HPTLC-silica gel layers [75].

5.1.2 Detection With 254-nm UV Light

Nonselective detections can be performed in short- and long-wave UV light. For this, UV lamps of the appropriate excitation wavelengths are necessary, and these are supplied in various forms by, e.g., CAMAG, DESAGA and Merck. Darkrooms are very suitable locations for UV lamps combined with documentation equipment, as the eval-

Applied Thin-Layer Chromatography: Best Practice and Avoidance of Mistakes, 2nd Edition
Edited by Elke Hahn-Deinstrop

ISBN: 978-3-527-31553-6

uation of TLC plates by daylight would not be done here. Certain types of evaluation equipment intended for use in daylight rooms are provided with weak UV lamps. These are used for visualization only and not for photographic documentation. The closed systems for video documentation described in Section 7.3 can be used in daylight rooms, but precautions must be taken.

☞ **Practical Tip** for working with UV lamps: dark glasses must always be worn to prevent eye damage.

The fluorescence indicators show their usefulness in short-wave UV light, as all UV-active substances show fluorescence quenching when irradiated by UV light with an excitation wavelength of 254 nm, appearing as dark spots on a bright background. The latter fluoresces bright blue with acid-stable fluorescence indicators and bright green with the indicators normally used (see also Section 2.2.3 "Additives").

An example of a semiquantitative evaluation in short-wave UV light without derivatization is the purity testing of a pharmaceutical substance (Fig. 72). Here, 100 % of the sample under investigation was chromatographed next to 100 %, 0.25 % and 0.5 % of the reference substance (Z 4), and 0.25 %, 0.5 % and 1.0 % of the known impurity was developed on the same plate (Z 1). The evaluation is not completely straightforward, although it can be seen that the content of known impurities is in the region of 0.5 %. Other unknown additional zones are determined from the limit values of the substance. The top additional zone (Z 5) is difficult to evaluate, as it is spread by diffusion, but is probably just below the 0.5 % level. The situation is exactly the opposite in the case of the bottom additional zone (Z 2). This is very narrow and concentrated, and lies just below the 0.25 % level. Zone 3 has a similar spread to that of the limit value zones of the substance under investigation, and its concentration can be estimated at just below 0.5 %. If all the additional zones of lane 7 are now added together, this gives a total of a little below 1.5 %, so that the sum of the unknown impurities is only just 1.25 %. No limit value for the sum of all unknown additional zones is given in the pharmacopoeia. Only the concentration of an individual spot due to an additive is limited to 0.5 % and the concentration of the known impurity to 1 %. This substance would therefore pass the purity test according to the DAB. However, any manufacturer of medicaments is free to draw up his own specification for a raw material, including the sum of unknown impurities. If in the above example a specification of the limit value for the sum of the unknown impurities had been set at 1 %, the substance could **not** have been released. The evaluation of this plate is also described in Section 7.3.2 "Quantitative Video Evaluation" (see also Table 21).

▶ **Figure 72:** see Photograph Section.

The assessment of the additional zones is very risky if the substance under investigation shows different absorption behavior from that of the impurities. Figure 70a shows how this difference can be recognized on the TLC plate. Here, two pharmaceutical substances were applied pointwise and bandwise in amounts of 400 µg and 8 µg. After the development, strong fluorescence quenching of the second substance can be seen in short-wave UV light, while this is much weaker for the first substance, so that the 8-µg amount only appears as a trace. After derivatization with iodine, no difference between the intensities of the spots of the two substances can be detected (Fig. 73b).

▶ **Figure 73:** see Photograph Section.

Figure 74 shows another example of evaluation in 254-nm UV light from the area of pharmaceutical process development. Here, starting materials, in-process control, mother liquor, wash water and the end product are monitored by TLC. With in-process control, TLC gives rapid and reliable results with an accuracy that is quite adequate for the chemical production process.

▶ **Figure 74:** see Photograph Section.

☞ **Practical Tip** for documentation with 254-nm UV light: 35-mm cameras or video equipment is used to produce the required images, which are in black-and-white because these give better contrast of the spot with the background. The color of the fluorescence indicator cannot therefore be seen on these images, so that no conclusions can be drawn about the precoated layers used.

5.1.3 Detection With 365-nm UV Light

If a UV lamp with an excitation wavelength of 365 nm is used for the detection, the background of the layer appears dark, and substances that absorb in the UV region and are thereby excited to emit fluorescence or phosphorescence (luminescence) can be seen as bright spots against this dark background. Examples of self-fluorescence with various colors are given in Figs. 58 and 78, which show images obtained by the action of 365-nm UV light on greater celandine without additional derivatization. Fluorescent constituents are often present in samples of plant origin, e.g. sinensetine in javatea (*Orthosiphonis folium*) and scopoletine in nettle root (*Urticae radix*), both of which give a bright blue fluorescence [76]. Many self-fluorescent substances are often found in the analysis of perfumes, which often contain essential oils in addition to the carrier (ethanol) [77]. Of the essential oils used as medicaments, only the coumarins obtained from the oils expressed from citrus rind (oils of lemon, bergamot and orange) fluoresce under 365-nm UV light. The detection of these substances is specified in the European Pharmacopoeia as a test for adulteration with oil of lemon [78].

Theoretical and practical aspects of visual evaluation using various light sources can be found in Vol. 1a of the books on TLC analytical methods by Jork, Funk, Fischer and Wimmer [34].

5.2 Direct Optical Evaluation Using Instruments

Unlike visual evaluation of a chromatograms before derivatization, which can only give qualitative or semiquantitative results, direct optical evaluation using instruments enables quantitative results to be obtained. For this, a traditional TLC scanner, diode-array scanner or video equipment, either alone or in combination with a flat-bed scanner, is used. Quantitative evaluation with these instruments is described in more detail in Sections 7.2–7.4. However, the limits of this book would be exceeded if we gave a detailed description of **all** the commercially available equipment that can be used to quantify substances on TLC plates. Training in the use of TLC scanners can be obtained in company seminars (e.g. CAMAG) and detailed instructions are provided by the manufacturer when the equipment is purchased.

5.2.1 Principle of Operation of a Traditional TLC Scanner

Quantitative analyses are obtained by densitometry. The TLC scanner used for this is linked to a personal computer (PC) and is controlled by an evaluation program. The PC performs the calculation of the result, supports the protocol, and provides an archive of all the parameters of the equipment and evaluation program, the raw data and the numerical and graphical results.

The measurement consists of a direct photometric evaluation. The TLC plate is scanned lane by lane by monochromatic light of the appropriate wavelength. The measuring slit is adjusted to give a light beam whose dimensions suit the size of the spot. Spots obtained by pointwise application are scanned over their whole diameter. With bandwise application, which is preferable for quantitative evaluation, an aliquot of 50–75 % is taken from the homogeneous central part of the band.

The light that is diffusely reflected (re-emitted) by a blank part of the TLC plate is measured by a photomultiplier (PM) and is set equal to 100 % (0 % absorption). When the absorbing substances are scanned, the absorption increases with increasing amount of substance. A plot of the absorption signal against the migration distance gives the so-called absorption-position curve (representation of the chromatogram).

5.2.2 Direct Optical Evaluation Above 400 nm

When TLC scanners are used for absorption measurements in the 400–800 nm range, illumination is by tungsten filament lamp. To determine the optimum wavelength for measuring a substance (the absorption maximum), the absorption spectrum is measured at the centre of the spot obtained by chromatography. This spectrum can then be kept in a library of spectra and later used for identifications of other samples.

Dyes that absorb in the visible range, e.g. those found in the analysis of writing inks, can be quantified.

5.2.3 Direct Optical Evaluation Below 400 nm

A deuterium lamp is used for absorption measurements in the 190–400 nm range. Here also the optimum wavelength for measurements is determined by obtaining an absorption spectrum. At an excitation wavelength of 254 nm, UV-active substances can be directly measured on TLC plates. Alternatively, it is possible to use a high-pressure mercury vapor lamp, which gives an extremely intense line at 254 nm (see also Section 5.2.4). Examples of measurements at an excitation wavelength of 254 nm using a high-pressure mercury vapor lamp can be found in the literature [79]. The chromatograms of the sulfonamides in Section 7.2.2.3 were measured with deuterium lamps using an irradiation wavelength of 285 nm.

5.2.4 Direct Optical Evaluation With 365-nm UV Light (Fluorescence Measurement)

Substances that themselves fluoresce (self-fluorescence) are determined by measuring this fluorescence. A high-pressure mercury vapor lamp, which is usually used as the light source, emits a line spectrum in the range 254–578 nm. The fluorescence light of longer wavelength emitted by the substance on the TLC plate after it has absorbed the light at the selected excitation wavelength is recorded by the PM. To prevent the reflected light which is also emitted from the TLC plate at the excitation wavelength from reaching the PM, this is eliminated by a cut-off filter or narrow-band filter. For fluorescence measurements in general it is standard practice to use an excitation wavelength of 366 nm combined with a cut-off filter which is not transparent to light of a wavelength less than 400 nm. As with absorption measurements, the fluorescence intensity is plotted against the migration distance, giving a fluorescence-position curve [80].

The chromatograms of greater celandine shown in Section 10.2.3 "Effect of Light" are examples of fluorescence measurements without previous derivatization.

☞ **Practical Tips** for working with traditional TLC scanners:

- **Determination of the optimal measurement wavelength**
With older equipment which cannot be used to obtain spectra, the optimal measurement wavelengths are determined as follows: If it is known, for example, that the maximum is at a wavelength less than 300 nm, the chromatogram lane containing the substance to be determined is scanned several times using a scheme such as the following:

  (a) From 220 to 300 nm in 10-nm steps (9 ×)
  Result: Maximum is in the medium range

  (b) From 250 to 270 nm in 5-nm steps (5 ×)
  Result: Maximum is between 260 and 270 nm

  (c) From 260 to 270 nm in 1-nm steps (11 ×)
  Result: Maximum is at 265 nm

- **Measurement with incomplete separation of the substance**
If two substances on a chromatogram lane are not separated at the base line, but only one of the two substances is needed for the analytical result, the spectra of both substances are obtained and the measurement wavelength is selected such that there is a minimum for the substance not used. In the ideal case, however, the substance to be determined has a maximum or at least gives a satisfactory signal.

- **Measurement with multiple wavelength scan (MWLS)**
The use of MWLS gives additional information enabling compounds to be identified or at least classified as members of a group of substances. An example of the use of MWLS is the analysis of drinking water for pesticides by AMD using Burger's method [81]. Here, the chromatograms are measured successively at the wavelengths 190, 200, 220, 240, 260, 280 and 300 nm. The measurement curves of the chromatogram at the various wavelengths are all printed together, i.e. superimposed, using a multicolor plotter to enable them to be visually evaluated (Fig. 75). In UV multidetection, each substance is thus characterized by its chromatographic migration distance and its behavior.
Multiple wavelength measurements are also performed in the analysis of natural substances, which have complex compositions, e.g. in the determination of flavonoids after AMD separation. Here, absorption measurements are performed at wavelengths of 235, 286, 300 and 314 nm [82].

► **Figure 75:** see Photograph Section.

- **Advantages of fluorescence measurement over absorption measurement:**
  - The limits of detection are lower by about 1–3 orders of magnitude.
  - The calibration curves are linear over a greater range of concentrations.
  - The selectivity of the determination is usually greater [80].

- **Intensification and stabilization of a fluorescence signal**
  The intensity and the stability of a fluorescent spot can be increased by Triton X-100 in chloroform or paraffin in *n*-hexane. For example, in the determination of testosterone, the TLC plates are dipped twice for 1 s in a solution of Triton X-100 in chloroform (1 + 4). Between the two dipping operations, the plate is air dried for 30 min in the absence of light until the chloroform has completely evaporated. This causes an increase in the spot intensity by a factor of 2.5 [83].
  In the determination of polycyclic aromatic hydrocarbons (PAHs), the separation is performed either on caffeine-impregnated HPTLC silica gel 60 plates, on HPTLC-RP-18 plates with a concentration zone or on precoated HPTLC silica gel 60 plates with a concentration zone (Merck Item Nos. 1.15086, 1.15037, 1.13728). Subsequent dipping in paraffin/*n*-hexane (1:4) gives not only stabilization of the PAHs on the plate but also an increase in the fluorescence intensity by a factor of 4–5 [83a].

- **Sensitive samples** are often destroyed when subjected to measurement by fluorescence. This sensitivity should be borne in mind when performing a validation and, if appropriate, it should be stated in the testing procedures that the chromatogram should be measured only once. Our own investigations into the determination of dried extract of nettle root showed that a second determination of scopoletine on the same chromatogram lane gave a value that had decreased by ca. 5 %.

- **Replacement of lamps**
  If the sensitivity of a measurement decreases, e.g. because of loss of intensity, the lamps should be replaced. Here, the operating instructions should be followed strictly. It is advisable to practice the changing of the lamps, which are usually preset, on delivery of a TLC scanner.

- **Measurement at 190 nm and 200 nm in a nitrogen atmosphere**
  To obtain reliable results at wavelengths of 190 nm and 200 nm, it is strongly recommended to purge the monochromator housing with nitrogen (ca. 0.5 L/min). This is because short-wave UV light causes oxygen to be converted into ozone, the shorter the wavelength the greater being the extent of conversion. Consequently, local "clouds" of ozone escape intermittently through the ventilation holes. Ozone absorbs in the short-wave UV range with a maximum at 254 nm. Purging with nitrogen prevents ozone formation and hence prevents fluctuations in the intensity of the short-wave UV light [84].

5.3 Diode-Array Detection

The evaluation of thin-layer plates by optical methods is based on a differential measurement of light reflected from the sample-free and the sample-containing zones of the plate. Two different scanner systems are on the market. In the main, a mono-wavelength scanner is used to evaluate a TLC-plate (see Section 5.2.1). This system was introduced by *H. Jork* in 1966 [84a]. Mono-wavelength scanners illuminate the plate with nearly monochromatic light of variable wavelengths to scan an absorption spectrum directly from a sample spot or to measure densitograms at different wavelengths.

In contrast, diode-array scanners use white light for illumination purposes. Diode-array scanners register absorption and fluorescence spectra and densitograms at different wavelengths simultaneously from the plate. This kind of scanner was first introduced in 2000 by Spangenberg [84b].

In diode-array scanning the light source and detector are both placed 450 μm above the surface of the HPTLC plate. Commonly used scanners need an angle of 45° between the light emitting device and the bulky detector. According to the Lambert cosine law ($J = I_0 \cos \theta$), this angle (θ) is responsible for a reflected light intensity reduction of nearly 30 %. A light fiber array overcomes this limitation. For dense light intensities the light emitting and the light detecting fibers are arranged parallel to each other, because only in this arrangement does the Lambert cosine law predict an optimum response.

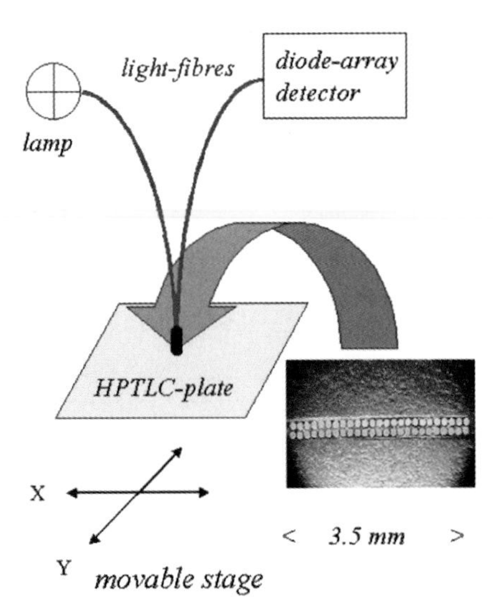

Figure 76. The principle of diode-array TLC scanning. A row of 25 glass fibres transport light from a lamp to the surface of a HPTLC-plate and another row of 25 fibres transport the reflected light to the diode-array detector. The plate is fixed on a mechanical stage, which moves with constant velocity.

A bundle of 25 glass-fibers with a diameter of 100 μm each, which are highly transparent in a wavelength-range from 190 to 1100 nm are fixed side by side in one line to transport light from a combined light source of a deuterium and a tungsten lamp to the surface of the plate. In this way a small scanning slit is formed on the plate surface. Another bundle of 25 fibers in slit arrangement transports the reflected light from the plate surface to the diode-array detector. The intensity distribution of this reflected light is measured.

Another 25 fibers in slit arrangement illuminate the plate for fluorescent measurements by use of an LED, emitting dense light at 365 nm. A reduced attachment for absorption measurements in the UV-range consists of two 25 fibber arrays each, to illuminate the plate by use of a deuterium lamp and to transport the scattered light back to the diode-array detector.

For evaluation purposes, the reflected light from a clean TLC-plate (J_0) is combined with the reflected light of a TLC-track (J). The combination of the wavelength dependent light distributions are combined according to expression (1).

$$R = \frac{J}{J_o} \tag{1}$$

To scan an absorption spectrum directly from a sample spot, a mono-wavelength scanner has to illuminate the plate with light of different wavelengths. Using a diode-array scanner the absorption spectrum of a TLC-plate spot can be directly extracted from the measured scan.

For absorption evaluations expression (2) can be used,

$$A = \frac{(1-R)}{R} \tag{2}$$

where a fluorescence spectrum can be extracted from the same raw data using formula (3).

$$F = (R-1) \tag{3}$$

The well-known *Kubelka-Munk* equation (4) is the combination of equation (2) and equation (3) and describes both, absorption and fluorescence spectra [84c].

$$KM = \frac{(1-R)^2}{2R} \tag{4}$$

▶ **Figure 77:** see Photograph Section.

5.4 Coupled Methods for Substance Identification

TLC is an ideal sample preparation method prior to further analysis for substance identification. Using suitable TLC systems, especially those involving optimal precoated layers, it is possible not only to separate mixtures of unknown substances but also to use various spectroscopic evaluation methods directly on the layers. By this combination of two different analytical methods, it is often possible to perform an unequivocal identification of a substance even when the samples are extremely small. These spectroscopic evaluation methods include the following:

- **FTIR (Fourier Transformation Infra-Red)**
  Although the use of silica gel precoated layers in the so-called fingerprint region of the IR spectrum leads to strong interfering bands caused by the silanol groups and the water molecules on the surface of the stationary phase, reliable information about the molecular structure of the sample being investigated can be obtained. This method has been used with great success for the identification of drugs of abuse [85–87].

- **Raman**
  Raman spectroscopy is a type of molecular spectroscopy similar to FTIR, but in this case there are no interfering bands even when silica gel precoated layers are used. HPTLC-aluminium sheets silica gel F_{254} (Merck no. 1.05548) are suitable ready-to-use plates for this purpose. TLC/Raman coupling is described in [88, 89], and a new publication describes the routine use of TLC-Raman, taking the analysis of hydroxybenzenes as the example [89a].

- **SERS (Surface-Enhanced Raman Spectroscopy)**
  Compared with normal Raman spectroscopy, the SERS technique gives a considerably lower limit of determination of the substance to be analyzed. This is achieved by bringing colloidal noble metal atoms into close proximity to the sample molecules after they have been separated by TLC. Further information can be found in [90, 91].

- **MS (Mass Spectrometry)**
  MS is a widely used analytical method which enables a substance to be identified from the size of its molecules and characteristic molecular fragments. It has so far been mainly used in combination with GC, but attempts have been made for some time to couple TLC with this spectroscopic method [92, 93].
  Many publications exist in the field of food analyses. *Gerda Morlock* described in 2004 the determination of heterocyclic aromatic amines after a 6-step AMD-separation. She used an HPTLC-MS online extractor, developed by Luftmann [93a]. In 2004 a patent application was made for the mode of operation of this so-called "ChromeXtrakt" device [93b]. In her latest paper (2006), the quantification of isopropylthioxanthon (ITX) in food using HPTLC/FLD coupled with ESI-MS and DART-MS was reported. The prepared samples were separated on a HPTLC-plate and determined by a fluorescence detector (FLD). Positive results have been verified by ESI-MS (Electrospray-Ionisation-Mass-Spectrometry) and DART-MS (Direct Analysis in Real Time-Mass-Spectrometry) [93c].

5.5 Documentation Without or Before Derivatization

This subject is brought up in this chapter deliberately in order to emphasize the importance of step-by-step documentation with no gaps. The evaluation of chromatograms is often described or demanded only after a derivatization. However, this method of working can often lead to the loss of important information. With few exceptions, in routine analysis (the results of which would not be expected to provide any surprises), it is important that a plate should be "pictured" after the development stage and **before** any further treatments are carried out. This can be in the form of Polaroid photographs, drawings on paper, diapositives, video shots or video prints. Of course, the hard copy outputs produced by plate scanners are also included here. These various techniques and the results that they provide are described in detail in Chapter 8 "Documentation".

6 Derivatization

For all TLC users who perform derivatizations, the book of reagents (in two volumes) by Jork, Funk, Fischer and Wimmer [34], which has already been cited many times in the present book, is an absolute must! Vol. 1a contains 80 reagent monographs and also gives a detailed presentation of the fundamental principles of the physical and chemical of the methods of detection. Vol. 1b deals with special methods of detection, including photochemical, thermochemical and electrochemical activation reactions; reaction mechanisms are described and a further 65 reagent monographs are also included. In Sections 6.1–6.3 of the present book, the practical execution of the most important methods of derivatization is described, advice on the avoidance of mistakes is given, and each case is illustrated by a practical example.

The subject of "Prechromatographic Derivatization" would have been described at the end of Chapter 3, this being an operation **before** the development, but because the technique is not widely used it is included in Section 6.4 "Special Cases of Derivatization". The subject of "Simultaneous Derivatization and Development" is discussed in detail in Section 6.4.2.

Chapter 5 contains a description of the evaluation of substance classes that can be detected from their own color or fluorescence or from components that absorb UV light. Compounds that do not show such properties must be converted into other substances (derivatives) that are detectable in order to evaluate a TLC separation, i.e. they must be derivatized. These can be universal reactions or, if suitable functional groups are present, they can be selective. Substance-specific derivatizations are practically impossible.

The aim of every postchromatographic derivatization, which, unlike prechromatographic reactions, should be performed as quickly as possible, is therefore principally:

- Detection of the chromatographically separated substances, giving improved visual assessment of the chromatogram.

However, the following are equally important:

- Improvement of the selectivity, which is often linked to this.
- Improvement of the detection sensitivity, with consequent optimization of the subsequent in situ evaluation [94].

Applied Thin-Layer Chromatography: Best Practice and Avoidance of Mistakes, 2nd Edition
Edited by Elke Hahn-Deinstrop

ISBN: 978-3-527-31553-6

6.1 Thermochemical Reaction

A simple heating process without the use of reagents will convert a large number of substances, after the chromatography on amino-modified silica gel layers, into stable compounds that fluoresce with varying intensities and at various wavelengths [95]. Here, extremely complex reactions take place, which can sometimes be described as Maillard reactions, in which the reducing sugar reacts with amino acids. The authors of this paper give an overview of the substance classes so far investigated. This shows that mono-, di-, tri- and oligosaccharides, creatine, creatinine, uric acid, catecholamines and related substances, steroid hormones and other compounds such as lipids, amino sugars, fruit acids, sugar acids, amino acids, bilirubin, choline, digitalis glycosides and hemoglobin are derivatized by heating and can be evaluated in UV light at 365 nm. This derivatization can be performed using infrared heaters, plate heaters or drying ovens. As a general indication, 3–4 min at 150 °C is sufficient to convert most substances. However, figures as high as 45 min at 200 °C can be found in the literature.

In Vol. 1b of the reagent books, Jork et al. describe and hypothesize about the reactions that take place in thermochemical activation, the term used for this physical process, and give the conditions for the thermal treatment of a large number of substances on aluminum oxide gel, silica gel and amino-modified silica gel precoated layers.

☞ **Practical Tips** for the thermochemical reaction:

- For each substance class, the optimum conditions such as reaction time and reaction temperature should be determined experimentally and then incorporated in a testing procedure.
- As well as the evaluation in UV light at 365 nm, it is advisable also to view the treated chromatogram in UV light at 254 nm, as it is always possible that the reaction products formed will be more easily detectable in short-wave UV light.

Example of a thermochemical reaction: sugar

The identification of the four sugars lactose, sucrose, glucose and fructose by thermochemical reaction is not only an elegant method (as by its use the spray reagent thymol-sulfuric acid prescribed in the DAB can be dispensed with), but by using another chromatographic system the time required according to the DAB method is considerably reduced. Table 18 in Section 4.3.1 shows all the parameters for the TLC investigations of these sugars. Figure 59a shows the fluorescence-quenching zones of the derivatives in UV light of 254 nm and Fig. 59b the fluorescent compounds formed in UV light of 365 nm. It should be noted that sucrose can only be determined at concentrations ten times higher than those of the three other sugars; it can just be seen in Fig. 59a as a trace in lane 2.

6.2 Irradiation With High-Energy Light

The reaction of substances with light in order to derivatize them is seldom used, as these reactions often cannot be performed reproducibly. Also, the reaction mechanisms are not fully understood; oxygen and the active surface of the sorbent play a part as well as the light. Reaction with high-energy light, also known as **photolysis,** is mainly used in qualitative detection. Here, the 254-nm UV lamps in the illumination equipment of the documentation systems of CAMAG and DESAGA and the "Suntest" apparatus of Atlas are used. The light source is a xenon lamp, which gives UV radiation with a wavelength of 270 nm. The manufacturer states that irradiation for 31 days in this apparatus gives a dose approximately corresponding to 1 year's natural light in the open air in central Europe [96].

In Vol. 1b of the above-mentioned series "Reagents and Methods of Detection", Jork et al. dedicate a chapter to photochemical activation, giving numerous examples, and try to throw some light on the theoretical background. For example, Funk et al. used a low-pressure mercury vapor lamp without a filter to liberate inorganic tin ions from organotin compounds separated by TLC, and then treated these with 3-hydroxyflavone to give chromatogram zones with a blue fluorescence on a background that emitted yellow light [97]. A quantitative evaluation was also possible in this case (γ_{exc} = 405 nm, γ_{fl} = 436 nm, monochromatic filter).

Using the same irradiation unit (Heraeus, Hanau, Osram STE 501; UV lamp TNN 15–3200/721) it was possible to activate 15 β-blockers photochemically and then to evaluate them quantitatively [98].

☞ **Practical Tips** for irradiation with high-energy light:

- It may be assumed that when detection is performed using light-induced reactions, light-sensitive substances are present in the samples, and therefore the entire operation as far as the reaction with light should be performed **in the absence of light**. See also Section 10.2.3 "Effect of Light".
- Simple UV viewing apparatus does not normally have enough energy to react with substances. For teaching purposes, a quartz-mercury vapor lamp can be used (wearing sunglasses!) or, if it is not actually raining, direct sunlight.
- The intensity and time of illumination must be determined individually for each substance. The time can vary between a few seconds and several hours.

Example of a chemical change caused by high-energy light: chelidonine

A monograph in the DAB describes a medicinal drug consisting of the above-ground parts of *Chelidonium majus* L. (greater celandine) collected while the plant is in flower and then dried. On the other hand, the German homeopathic pharmacopoeia (HAB) uses the fresh root stock of the same plant. Both pharmacopoeias prescribe the same solvent system for the identity test, for which the DAB specifies evaluation after chromatography, first with UV 254 nm and then with UV 365 nm, whereas the HAB specifies only the longer-wave UV light for the evaluation. It is expressly stated in the DAB that chelidonine shows a fluorescence-quenching zone at 254 nm but does **not** fluoresce at 365 nm. Our own investigations have shown that chelidonine does not fluoresce **only if** operations are carried out in the complete absence of light and the chromatogram is observed **first** at UV 365 nm and **afterwards** at UV 254 nm. In a repeated evaluation at UV 365 nm, chelidonine then gives an intense yellow-green fluorescent chromatogram zone, and, in the lane with the drug, further intensely fluorescent new chromatogram zones (previously visible with other colors or not visible at all) can be seen (Fig. 78a–c). For fluorimetric evaluation, excitation by γ_{exc} = 313 nm is used and the fluorescence emission is measured at γ_{fl} >400 nm (cut-off filter). Figure 79 shows the fluorescence-position curve of a greater celandine extract before (A) and after (B) irradiation for 1 h with short-wave UV light; two new and hitherto undetected zones can be clearly seen. The parameters of the TLC of greater celandine are given in Table 17 in Section 4.2.2.

▶ **Figure 78:** see Photograph Section.

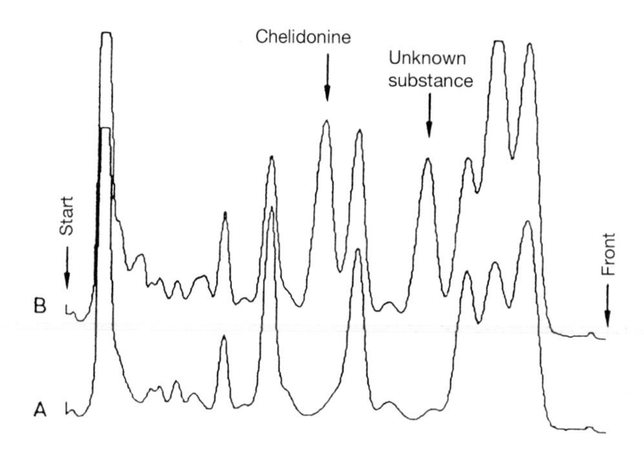

Figure 79. Fluorescence-position curves of a greater celandine DE lane with ca. 5 µg chelidonine
Before (A) and after (B) irradiation with short-wave UV light for 1 h

6.3 Reaction With Reagents

If the substances separated on a TLC plate are not directly visible and do not react to UV light, they can be caused to absorb light or to fluoresce by reaction with suitable reagents. In the case of biological reactions, the result can be obtained, for example, as the hemolytic index (HI) on coating the plate with blood-gelatin or as the measured values found from an enzymatic inhibition test.

The various commonly used techniques for applying the reagents to the TLC plate are described below.

6.3.1 Spraying of TLC Plates

6.3.1.1 Manual Spraying of TLC Plates

In qualitative TLC, the commonest technique is to spray the chromatogram. For this, all-glass sprayers can be used in which the upper part, the spray head, is connected by a ground glass joint either to a conical flask or to a ca. 15-ml capacity test tube containing the reagent (Fig. 80). The nozzle is designed such that on application of a pressure of 0.6–0.8 bar a mist of reagent solution can be applied homogeneously to the chromatogram. However, in many laboratories neither the necessary pressure tubing nor a nitrogen cylinder is available. Here, the TLC spraying equipment of Merck, also supplied by CAMAG, can be used. This consists of an electropneumatic spraying system whose motor is powered by an inductively rechargeable accumulator. This TLC spray-

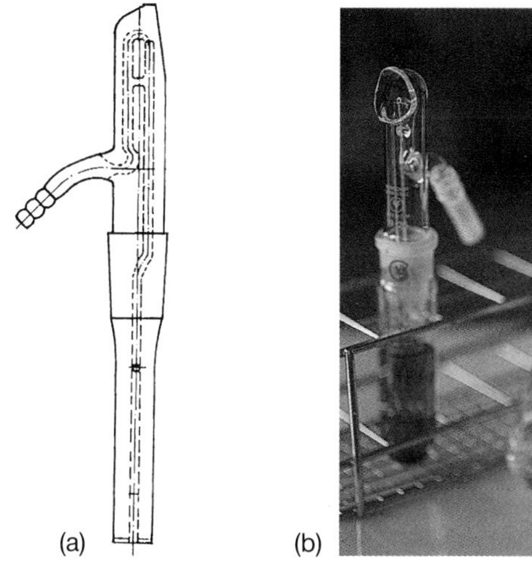

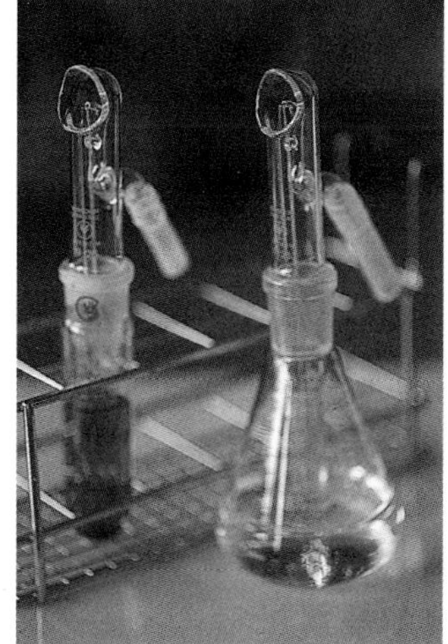

Figure 80. All-glass spray head
(a) Cross-section
(b) Left: spray head with 15-ml test tube
Right: spray head with 100-ml conical flask

ing equipment consists of the electrical charging station, the spray holder, and two spray heads with capillaries of different diameters. To spray, e.g., alcoholic reagent solutions, the smaller-diameter (0.8-mm) spray head is used. For higher-viscosity spray solutions, e.g. sulfuric acid reagents, the spray head with the larger-diameter (1.25-mm) reagent capillaries is used. The spray heads are screwed onto reagent bottles (50-ml or 100-ml capacity) containing in-house prepared or proprietary spray solutions. The latter are more commonly used, changing the spray head being a more user-friendly procedure than attaching a new reagent bottle, as in the former case it is not necessary to rinse the bottle when the reagent is changed, thereby saving time and solvent. Figure 81 illustrates the operations to be performed before spraying, i.e. screwing on the spray head, fixing the bottle and pressing the button to start the motor [99]. The spraying equipment "Sprayer SG1" of DESAGA is of similar construction, and can also produce an extremely fine mist without having to being plugged in to an electrical power supply.

Figure 81. Method of setting up TLC spraying equipment (from [99])

The spraying of TLC plates should always be performed in a fume cupboard with good air extraction to avoid any health hazards due to the inhalation of harmful or aggressive aerosol mist and to prevent contamination of the workplace. In addition, a device to trap and siphon off any excess reagent is absolutely necessary. These functions are performed by the TLC spray cabinet of CAMAG or the similarly designed "spray box" of DESAGA. Both of these are designed so that both the spray mist produced by the sprayer and the particles that rebound from the plate are safely contained without losing any of the spray stream before it reaches the plate, as can easily happen if the plate being sprayed is simply placed in a fume cupboard. Both the spray housings feed excess spray mist through a connecting duct back into the waste air shaft of the fume cupboard. Any reagent lost in the form of liquid drops is collected in a tray at the bot tom of the housing, which can easily be removed for cleaning. Devices containing built-in radial fans are also provided in which the motor is external to the suction pipe. Here also, the spraying of TLC plates is possible outside permanently installed laboratory fume cupboards.

The TLC plate, which is positioned vertically in the housing, is sprayed from a distance of 20 to ca. 30 cm (Fig. 83). On starting the motor of the spray equipment or pressing the trigger of the spray pistol, compressed air is "fired" through the fine nozzle and sucks up the reagent solution, which is then carried with the air (a similar principle to that of the water jet pump). Spraying should be performed in a meandering to-and-fro pattern, as described by Waldi [100], with the changes of direction occurring outside the chromatogram (Fig. 82).

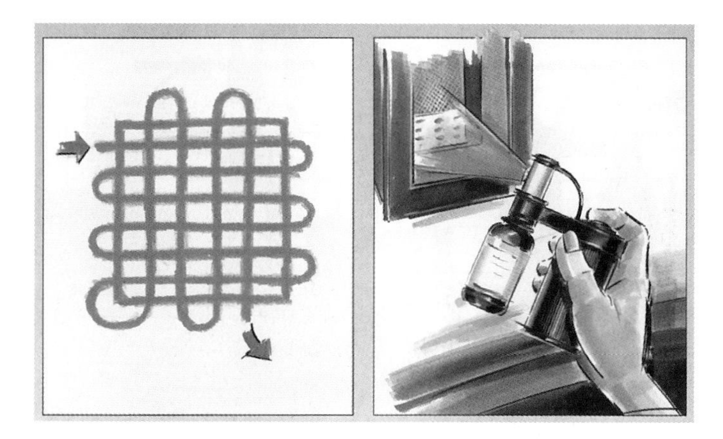

Figure 82. Spraying scheme recommended by Waldi

Figure 83. Spraying a TLC plate with the TLC spraying equipment (from [99])

Practical Tips for spraying TLC plates:

- It must always be assumed that the aerosols formed during spraying are toxic. This means extra precautions! There should be a good extraction system, and protective goggles, disposable gloves and a dust mask should be worn.

- The spraying of TLC plates is a simple and rapid method of derivatizing substances. However, the coating produced is usually more inhomogeneous than that obtained by dipping. Moreover, it is more difficult to apply controlled amounts of the reagent.

- Spray reagents should be as fresh as possible and should be available only in the amount necessary for that day. CHCs should be avoided! The storage behavior of spray reagents over a longer period of time should be tested and documented in a testing procedure specification. For example, a vanillin-sulfuric acid reagent kept in an ice bath can be used until the solution develops a yellow color. If the reagent is stored in a refrigerator, this color can sometimes only begin to appear after ca. 6

months. In contrast, the vanillin-phosphoric acid reagent can be kept in a refrigerator for only a few days and should always be freshly prepared for quantitative determinations.

- The nozzles in the all-glass sprays can vary greatly. If they are too narrow, the spraying process can take too long, and if they are too wide the plate becomes spattered with large drops of reagent, which can make the evaluation very difficult. The only solution to the problem is to order several spray heads, to test them and to reject the unsuitable ones. The number of spray heads kept in a pharmaceutical or teaching laboratory should be at least two in the first case and six in the second.
- When assembling the all-glass spray equipment, care must be taken to keep the spray head clean and dry!
- It is advisable to place a Kleenex cloth under the plate before spraying to absorb excess reagent.
- Before starting to spray, always carry out a test by spraying an area outside the developed part of a plate. This is because unwanted spots can be produced if the spray head is not quite dry,
- Plates in the 5×5 cm format can be attached with adhesive tape to the middle of an old larger plate to facilitate uniform spraying.
- With some reagents, it is advisable to spray a **heated** plate to cause the solution to evaporate more quickly, as a further spraying process with a different reagent is often subsequently performed, e.g. with the flavone reagent according to Neu [101]. In these cases, the plate is placed in front of a hot-air dryer after the documentation and before the derivatization. The derivatization can then proceed more rapidly and the layer can take up more solvent without causing the substances to run.
- Spraying until the plate becomes transparent seldom occurs, e.g. with aqueous starch solution after treatment with iodine vapor. Here, the user should have some experience, and this can be gained by first trying out the spraying technique.
- On completion of the spraying, the spray heads and housings should be thoroughly cleaned. Contaminated spray heads can lead to uncontrolled reactions with reagents that are sprayed at a later stage.
- Reagents that contain manganese heptoxide and perchloric acid and also solutions of sodium azide and iodine azide should never be sprayed, as these can cause explosions in the air extraction ducting.
- The formerly used cans of propellant gases, usually containing fluorinated hydrocarbons, are never used today. Alternatives are described above.

Examples of derivatizations by spraying: flavonoid-containing phytopharmaceuticals

A large number of drugs mentioned in pharmacopoeias contain flavonoids or similar classes of substance that can be converted into fluorescent compounds by complexation with the diphenylboric acid 2-aminoethyl ester reagent (the flavone reagent according to Neu). As described in the Practical Tips, the warm plate is first sprayed with a 1 % solution of natural substance reagent A in methanol and then with a 2 % solution of magrocol in ethanol to intensify and stabilize the fluorescence. The evaluation is performed after derivatization by treatment with UV light at 365 nm for ca. 15 min, after which the intensity of the fluorescence colors is very high. Figure 84 shows the chromatogram observed in 365-nm UV light after spraying of three types of coneflower (Echinacea) and a hybrid of *Echinaceae angustifolia* and *Echinaceae purpurea*. Table 19 shows the parameters for the TLC of Echinacea according to the pharmacopoeias, from the literature [102] and from our own results. The chromatographic detection of the hybrid by TLC and HPLC is described in detail in [102a].

► **Figure 84:** see Photograph Section.

6.3.1.2 Fully Automatic Spraying of TLC Plates

The ChromaJet DS 20 (Fig. 85), produced by DESAGA, is a completely new concept for spraying reagents with the greatest precision onto TLC plates or films. This, of course, is achieved with the aid of computer and microprocessor control. Advantages compared with hand spraying include the considerably reduced amount of reagent used, the reduced amount of aerosol formation and the uniformity of the sprayed area. As many spraying methods as necessary can be created, stored and called up as required, depending on individual chromatography schemes. Operating programs to facilitate the positioning of the lanes for the DESAGA TLC Applicator AS 30 can be directly loaded from the AS 30 list of methods.

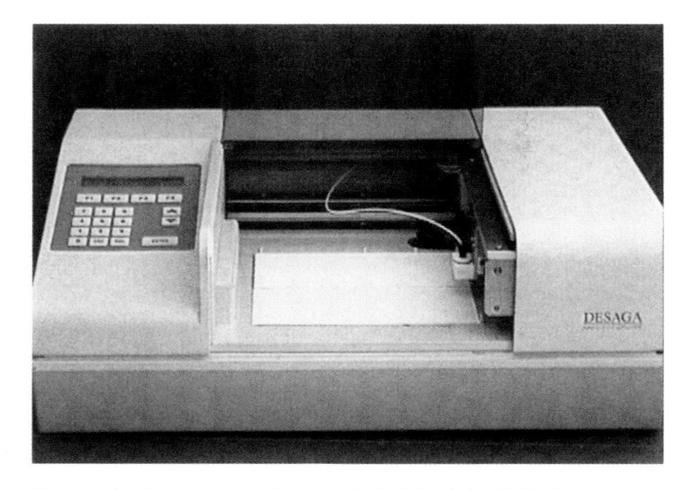

Figure 85. Autosprayer ChromaJet DS 20 (DESAGA)

The individual operating steps of the spraying process, including the date and the continuous numbering, are stored with the spraying method, and for the first time enable GMP/GLP-conforming operations to be performed. This is because of the exact reproducibility and the documentation now possible with the derivatization process.

All the important parameters such as spray rate, sprayed volume and reagent selection are included in the spraying method. The integrated sample changer can also select the desired spraying medium out of four possible reagents according to the spray program. Infinitely adjustable *x/y* coordinates enable individual lanes or any areas up to 20 × 20 cm to be sprayed uniformly. A built-in rinsing system prevents carryover. Any aerosols formed are removed by air extraction [102b].

At the time of preparation of the present manuscript, comparative investigations of hand spraying, dipping and automatic spraying are in progress. The first results have been presented at a symposium [102c], and are available on request from DESAGA.

6.3.2 Dipping of TLC Plates

The following Section on the dipping of TLC plates was written at the same time as Section 6.3.1.1 on manual spraying. There may be some discrepancies between these Sections and the discussion on fully automatic spraying in Section 6.3.1.2, which was written later.

The process of dipping TLC plates in suitable reagent solutions is being used increasingly for the following reasons [103]:

- The application of the reagent solution to the sorbent layer is more uniform than that achieved even by the most careful manual spraying.
- The distribution of the reagent is not influenced by the manual skill of the user, the capacity of the spraying equipment, the viscosity of the reagents or the droplet size of the spray mist.
- The exact quantitative evaluation of chemically reacted TLC chromatograms only became possible using this method. This is because considerably less structured baselines result from the uniformity of the coating. Therefore, the limits of detection are considerably lower than those with sprayed chromatograms. Also, the reproducibility of the results is much improved because of the more homogeneous application of the reagent.
- The consumption of reagent is very low, especially with series investigations.
- Contamination of the workplace with aggressive reagents which are sometimes harmful to the health is very much lower when dipping is used rather than spraying.
- Complex spraying equipment with integrated air extraction is not required.

Equipment for dipping TLC plates can be obtained in various forms. Until recently, dipping chambers suitable for manual dipping of 10 × 10 cm, 20 × 10 cm and 20 × 20 cm plates were often used. However, instructions in personal terms such as “brisk”, “rapid” or “careful” dipping cannot usually be exactly duplicated by another operator, and the transverse marks that often appear when a dipped plate is not raised at a steady rate can sometimes make the work of one or even several days worthless. For this reason, automatic dipping equipment for lowering and raising the plate at a uniform rate is nearly always used today. This prevents the formation of the above-mentioned tranverse marks which resemble a solvent front and which have a harmful effect on the densitometric evaluation. The dipping conditions can be standardized to give a definite residence time.

☞ **Practical Tips** for dipping TLC plates:

- Dipping solutions are in general less concentrated than spraying solutions. The solvents used must be suitable for the special requirements of dipping. Water can on the one hand form droplets which stay on the surface of RP phases and do not penetrate into the layer, but it can on the other hand lead to loosening of the layer if this is incompatible with water. It is therefore usually replaced by alcohol or other lipophilic solvents.
- When choosing solvent systems, one should in general ensure that neither the substances to be separated chromatographically nor the reaction products are soluble in the solvent system of the dipping reagent.
- In each individual case, the dwell time in the dipping chamber should be determined experimentally and documented in a testing procedure.
- With large amounts of substance or if the dipping time is too long, “comet tails” can easily form after dipping. In such cases the possibility of obtaining a better analytical result by using spray application of the reagent should be investigated. Figure 86a shows the chromatogram of components of plant medicaments in 365-nm UV light after dipping in the flavone reagent according to Neu, and Fig. 86b shows the same chromatogram obtained after spray application of this reagent.

▶ **Figure 86b:** see Photograph Section.

- For a dipping bath, the amount used is often up to ten times the amount required for spraying a plate. Although the concentrations of the reagents are usually lower in the former case, if the substances used are costly (e.g. hexachloroplatinic-IV acid hexahydrate: 1 g ≅ 86.25 DM, Merck No. 807340, VWR 2005–2007 catalog), it should be determined whether dipping may not be justified on cost grounds even though it is more environmentally friendly.
- For the first dipping “run”, it is advisable to perform a test using an old plate with water as the solvent to check the liquid level needed for each plate size and to mark the outside of the glass container accordingly. If reagent solutions overflow, this will lead not only to a cleaning problem but also to a health hazard!
- After each dipping operation, the dipping chamber should be closed with a lid to prevent loss of solvent and to protect the environment.
- After the dipping operation, the back of the plate, which is wet with reagent solution, should be carefully cleaned (Kleenex cloth), especially before it is placed on a hotplate, on a laboratory table, on the coordinate table of a TLC scanner or on documentation equipment.
- Dipping equipment can also be used for prewashing (see Section 3.2) and for impregnation of TLC plates (see Section 3.5).

Table 19: Coneflower (*Echinacea*)

Parts of the Echinacea species used for pharmaceutical products:

(a) *E. pallida*:	roots (acc. to Kommission E of the German official administration)
(b) *E. angustifolia*:	roots (acc. to DAB 9), whole plant (acc. to HAB 1)
(c) *E. purpurea*:	upper parts of the flowering plants (acc. to HAB 1), (acc. to Kommission E)

	DAB 9[a)]	Bauer/Wagner [102]
Used parts of the plant	Dried roots	Roots of all species
Sample solutions	With methanol, 5 min water bath 65 °C.	
Reference substances	Rutoside	Echinacoside, cynarine, cichoric + chlorogenic acid
Sorbent	Prescribed: Silica gel G R Used: TLC silica gel 60 F_{254} GLP 20 × 20 cm (Merck 1.05566)	(Probably TLC) Silica gel 60 F_{254} (Merck) Used: (Merck 1.05566)
Solvent system	Ethyl acetate + water + formic acid (67+20+13)	Ethyl acetate + formic acid + acetic acid + water (100+11+11+27)
Applied sample volume	20 µl 20 × 3 mm	10–20 µl sample Bandwise
Chamber saturation	Yes	Yes
Migration distance	9,8 cm	11 cm
Development time	107 min, RT 21 °C	65 min, RT 20 °C
Derivatization reagent	Flavone reagent acc. to Neu and polyethylenglycol 6000	
Detection	UV 365 nm	UV 365 nm
hRf-values:		
Rutoside	52–55	38–41
Echinacoside	36–40	20–24
Chlorogenic acid	61–64	47–52
Cichoric acid	96–100	89–92
Hyperoside	69–72	59–62
Caffeic acid		

[a)] When the 1st supplement of the DAB 9 took effect on the 1.1.1990, the monograph was repealed.

Bauer/Remiger/Wagner [102e]	Alternative	Bauer/Wagner [102]
Herb of all species	Dried parts of all species	Herb + roots of all species
Methanol extracts, after concentration: 1 g/ 10 ml	Acetone/water extracts 50 mg + 3 ml	Chloroform extracts, after conc.: 1 g/ 1 ml ethanol
Perhaps rutoside, isochlorogenic acids	Rutoside, chlorogenic acid, caffeic acid (each 1mg/ml)	β-Sitosterol
(Probably TLC) Silica gel 60 F_{254} (Merck) Used: (Merck 1.05566)	HPTLC silica gel 60 F_{254} GLP, 20 × 10 cm (Merck 1.05613)	(Probably TLC) Silica gel 60 F_{254} (Merck) Used: Merck 1.05566
Toluene+ ethyl formate + formic acid + water (5+100+10+10)	Ethyl acetate + formic acid + methanol + water (50+2,5+2+4)	Hexane + ethyl acetate (2 + 1)
10–20 µl sample Bandwise	40 µl sample, 1 µl ref. soln.. 10 mm bandwise	10–20 µl sample Bandwise
Yes	No	Yes
10,4 cm	7,5 cm	
25 min, RT 20 °C	35 min, RT 21 °C	
for all systems evaluated at 365 nm		Anisaldehyde-sulfuric acid
UV 365 nm	UV 365 nm	White light
8–10	12–14	
22–25	29–32	
47–50	77–84	
76–79	94–96	

Example of a derivatization by dipping: identity testing and assay of polidocanol in Gastrosil® suppositories

Polydocanol is a mixture of monoethers from dodecyl alcohol and Macrogols, which contains an average of 9 oxyethylene units and is used as a local anesthetic. As this mixture of substances is not UV active, visualization by derivatization is performed with **a modified Dragendorff reagent**:

Solution A: Bismuth oxide (600 mg) and hydriodic acid (3 ml) are dissolved in water and made up to 10 ml. Ascorbic acid is slowly added with stirring until the color due to excess iodine disappears; this gives a ruby red solution.

Dipping solution: Barium chloride (10 g) and calcium chloride (6 g) are dissolved in 220 ml 30 % acetic acid, mixed with 4 ml solution A and placed in the dipping chamber.

The TLC test for identity and the assay of polidocanol are performed with various solvent systems. In the very elegant assay procedure, the active substance of the suppository, metoclopramide base, is scanned at a measuring wavelength of 307 nm **before** the derivatization, the plate is then dipped, and the polidocanol is then determined at a measuring wavelength of 510 nm.

For the identity test, the plate is dipped for 3 s in the derivatization reagent, dried for 15 min in a cold current of air, allowed to stand for 2 h at room temperature and then documented photographically or measured with a TLC scanner.

For the quantitative determination, the results of the metoclopramide determination should be checked **before** dipping the plate, and the measurement repeated if necessary. The plate is then dipped for 3 s also and dried for 15 min in a cold current of air. It is then covered with a clean glass plate and kept in a desiccator overnight. Further information on the measurement of this plate can be found in Section 7.2 "Evaluation using a TLC Scanner".

Figure 87a,b shows the dipped plates in solvent mixtures 1 and 2 and Fig. 87c the scanned chromatogram of polidocanol in solvent mixture 2.

► **Figure 87:** see Photograph Section.

Remark: In the German language the word „tauchen" means "dipping" and "diving"

6.3.3 Vapor Treatment of TLC Plates

Homogeneous application of the reagents to the layer can also be achieved by applying vapor to the chromatogram. For this, trough chambers of the appropriate size are used. Small porcelain dishes or glass beakers can be placed at the bottom of a flat-bottomed chamber and filled with the reagent. One or two plates, with the layer facing towards the middle, are placed in the chamber and left there for up to 20 h. Using a double-trough chamber, the reagent can be placed in one trough and the TLC plate in the other. Jork et al., in Vol. 1a of the series "Reagents and Methods of Detection" describe a large number of vaporization reactions that can be performed in trough chambers. Also, some special methods such as reaction with a second "reagent plate" or "sandwich plate". This type of vaporization is generally seldom used, and is therefore not further discussed here.

☞ **Practical Tips** for vapor treatment of TLC plates:

- A method of detecting many substances with iodine vapor is nonspecific but usually does not cause decomposition. It is therefore recommended for laboratories in which TLC is often used and where old or slightly damaged 20 × 20 cm flat-bottomed chambers are always kept ready for use for this type of vapor treatment. These contain two small porcelain dishes containing elementary iodine. After introducing one or two 20 × 20 cm TLC plates or more plates of smaller format up to 10 × 10 cm, hot-air blowers are used to heat the outer wall of the chamber until violet vapors appear.
- For 5 × 5 cm TLC plates, for teaching purposes or in laboratories in which TLC is seldom used, a screw-top glass jar or preserving jar with a well-fitting lid can also be used as a vaporization chamber.
- Dwell times for plates in a vaporization chamber can range from a few minutes for detection reactions to ca. 20 h for purity tests. They should be individually determined for each task and specified in a testing procedure.
- Aluminum foil is only suitable in exceptional cases for vaporization reactions for precoated layers. With long dwell times, e.g. in the iodine chamber, the aluminum is attacked to such an extent that it becomes impossible to evaluate the chromatogram.
- TLC plates, after treatment with iodine vapor and removal from the chamber, should be covered with a clean glass plate until they are documented or evaluated.
- At the conclusion of all the documentation and/or evaluation work, it is generally advisable to keep the plate in an efficient fume cupboard until the reagent has been completely vaporized.

Examples of derivatization by vapor treatment: tests of the purity of piracetam and metoprolol tartrate

In the purity test on piracetam, the solvent mixture is first removed from the chromatograms, and these are then sprayed with 2,4-dinitrophenylhydrazine reagent. The plate can then be placed along with the untreated plate with the metoprolol tartrate chromatograms in the iodine chamber which is ready for use in a fume cupboard, with the coating facing the middle of the chamber. Warm air (type 2 hot-air drier) is blown onto the outside of the chamber until it becomes saturated with violet vapors. With both purity tests, it is necessary to keep the plates in the iodine chamber for a period of ca. 16 h. Only then can spots or zones with concentrations at the limit of detection be evaluated. Figure 41b (Section 3.7) shows the chromatogram of a piracetam purity test after 20 h, and Fig. 73b that of metoprolol tartrate next to metoclopramide hydrochloride (see Section 5.1.2) after a dwell time of 16 h in an iodine chamber.

6.3.4 Coating TLC Plates

Coating TLC plates has so far mainly been used for the detection and determination of saponins with the blood-gelatin reagent. Here, the word saponins has its usual meaning of glycosidic plant components which have the properties of hemolytic action, foam formation in aqueous solution and toxicity to fish. By hemolysis we mean the destruction of the red blood corpuscles with liberation of the hemoglobin and other components of the erythrocytes from the cell interiors and into the surrounding medium. To enable the hemolytic substances to be detected directly on the chromatograms, a blood-gelatin solution is poured onto the TLC plate, and the resulting hemolytic action is observed after gelation of the reagent. In contrast to the usual cloudy, red blood-gelatin layer, these zones are transparent and almost colorless because of the diffusion of the saponin out of the chromatogram and into the coating [104]. These hemolytic properties are still used today for the detection of saponins [105] and were formerly also used for their quantitative determination [106]. The assay has a high standard deviation (±5 % according to [106]), the analysis takes a long time, and the difficulties in obtaining the blood from cattle mean that the test is rather too laborious for routine analysis, especially in view of the fact that the experience of the user is also of the greatest importance. (For methods of coating TLC plates, see also 6.4.4 “Biological-Physiological Methods of Detection”.)

☞ **Practical Tips** for coating TLC plates with blood-gelatin reagent:

- First and foremost, it is essential to have **fresh** ox blood, as specified in all literature references. For this, you should find out from the nearest slaughterhouse on what day and at what time the cattle are usually slaughtered. At the appointed time, you should arrange to be driven to the slaughterhouse by a second person and should be equipped as follows: old, flat-soled shoes or rubber boots, dirty overall, plastic bag containing paper towels, disposable gloves, a small plastic bag containing money for two bottles of beer and, most importantly, a screw-top glass jar of ca. 500 ml capacity containing ca. 15–20 glass spheres ca. 5 mm in diameter.
  The “beer money” is a tip for the slaughterhouse worker whose job will have to be interrupted for your benefit. He will fill the glass jar with fresh blood, close it,

shake it vigorously and quickly wash the outside of the jar with water. After you have received the jar, you must shake the blood for ca. 15–20 min to prevent it from coagulating and to enable the fibrinogen to coat the glass spheres. If a second person is not available to drive the car, you must stay at the slaughterhouse until the shaking operation is completed. It must be emphasized that this visit is not for the squeamish!

- On returning to the laboratory, the blood should be transferred to smaller screw-top jars, filling them to the brim. If these are tightly closed and kept in a refrigerator, the blood can be used for ca. 4 weeks. The residue of glass spheres and coagulated blood components stays in the large glass jar and is disposed of with other waste materials.

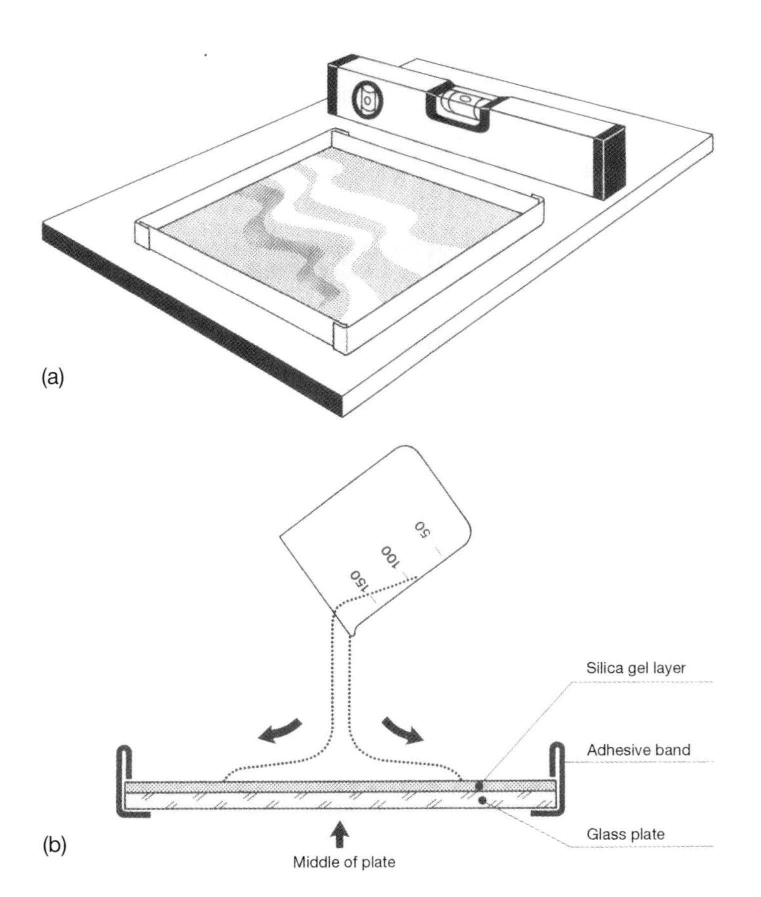

Figure 88. Schematic diagram of the process of coating a TLC plate with the blood-gelatin reagent

(a) The TLC plate, surrounded by an adhesive textile band, is leveled using a water leveling device

(b) Section through a TLC plate during pouring of the blood gelatin

- After the chromatography, the plate, from which the solvent has been removed, is surrounded on all four sides by a 3-cm wide adhesive textile band so that a border ca. 1 cm high is formed which will prevent the blood-gelatin reagent applied later from running off (Fig. 88a,b). Using a spirit level, the plate is placed on a horizontal substrate (Fig. 88a).
- If available, a cooling block forms a particularly good base for the plate which is to be coated, as this enables the reaction time to be considerably shortened.
- Further tips can be found in the description of the example.

Example of a derivatization by coating: saponin-containing drugs

The chromatography of the saponins from material of plant origin is performed on TLC silica gel 60 using the solvent system of Wagner [47, p. 266] consisting of 64 volumes of chloroform, 50 volumes of methanol and 10 volumes of water, with chamber saturation. For a migration distance of 10 cm, the development time is ca. 40 min. After removing the solvent residues in a stream of hot air, an adhesive textile band is fixed around the plate, as described in the practical tips, and this is made level. The preparation and application of the reagent is in several steps:

- **Physiological saline solution:**
  9 g sodium chloride dissolved in 1000 ml water.
- **Erythrocyte solution:**
  ca. 4 ml defibrinated ox blood with the same amount of physiological saline solution is shaken and centrifuged at 2000 rpm for 5 min. The supernatant solution is drawn off by a plastic pipette, and this "washing process" is repeated until the supernatant solution no longer has a pink color (only applies if fresh blood was not used). Approximately the same amount of physiological saline solution is added to the remaining erythrocytes, gently shaken and stirred into the cooled gelatin solution.
- **Gelatin solution:**
  Approximately 100 ml physiological saline solution is added to 4.5 g gelatin powder, and this is allowed to swell for at least 15 min. Water is heated to boiling point in a somewhat larger glass beaker, the glass containing the swollen gelatin is placed in it and heated to 80 °C (stirring thermometer!) with occasional stirring. The striation-free, pale yellow solution is then allowed to cool to 30–33 °C. The washed erythrocytes are then carefully stirred in. Any air bubbles on the surface are removed with a Kleenex cloth.
- **Pouring the blood-gelatin:**
  After adding the erythrocytes to the gelatin, the reagent should be immediately applied (risk of gelation in the glass beaker!). The reagent is initially carefully poured onto the line of the front, either directly or with the aid of a glass rod. Without interrupting the pouring process, the glass beaker can then be moved to the bottom of the plate so as to coat this area rapidly.
- **Evaluation of the plate:**
  The gelation of the reagent and the reaction of the saponins with the erythrocytes should preferably be allowed to take place overnight. The band of adhesive textile material can then be removed, the plate can be documented photographically and the hRf values calculated. Plates coated in this way can then be stored over a period of years without any change in their appearance.

Figure 89 shows a plate, coated and reacted with a blood-gelatin reagent to reveal the chromatograms of saponin, early golden-rod herb, soapwort, horsetail herb and primula root. The colored strip also included shows the accuracy of the color reproduction of the plate (see also Section 8.3.2 "Photography using 35-mm Cameras").

▶ **Figure 89:** see Photograph Section.

6.4 Special Cases of Derivatization

The postchromatographic reactions described above in Sections 6.1–6.3 are mainly suitable for the detection of the separated substances and only secondarily for the characterization of the compounds. For completeness, other important types of derivatization are described in the following Section, but biological-physiological methods of detection are only briefly discussed as we have no experience of this area, apart from the technique of coating with blood-gelatin reagent (see Section 6.3.4 "Coating TLC Plates").

6.4.1 Prechromatographic Derivatization

In contrast to postchromatographic derivatization, reactions performed **before** the chromatography are used

- to increase the selectivity of the separation,
- to increase the sensitivity of the detection,
- to improve the linearity.

6.4.1.1 Reaction With Reagents

In Section 3.1 "Prechromatographic in situ Derivatizations" of Vol. 1a of their reagent books, Jork et al. include 14 Tables of examples of the following reactions (with references):

- Oxidation
- Reduction
- Acid hydrolysis
- Alkaline hydrolysis
- Enzymatic decomposition
- Chlorination
- Bromination
- Iodination
- Nitration

- Diazotization and coupling
- Esterification and ether formation
- Hydrazone formation
- Dansylation
- Other prechromatographic derivatizations.

The above reactions can be performed either during the sample preparation or directly on the layer. In the latter case, the reagent is usually first applied pointwise or bandwise in the starting zone, and the sample solution is then applied, usually in the reagent zone while it is still wet. Here, care must be taken that the solvent used for the sample does not give rise to any undesired outwards-directed chromatographic effect on the reagent. The reagent solution can if necessary be re-applied later so that it is present in excess.

In quantitative investigations, it should be verified that the reaction proceeds to completion on the TLC plate or is at least stoichiometric and reproducible. In all cases, it is advisable also to apply the solutions of the reagent and sample separately to neighboring lanes in order to determine where the starting products are located in the chromatogram under the reaction conditions. This technique will also show whether additional by-products are formed.

6.4.1.2 Incorporation of Radionuclides

Radioactive isotopes have been used since the mid-1940s for the elucidation of metabolic irregularities. Today, radionuclides – principally ^{3}H, ^{14}C and ^{125}I – are used as tracers in all areas of pharmaceutical and biological research and in genetic technology [107]. In classical radio TLC, from the early 1960s, labeled prepared samples were applied to the TLC plate, which was then developed, and the β radiation was then measured with a TLC scanner (raytest).

Today, in radio TLC the radioactive-labeled molecules are transported by the migrating solvent along the thin-layer trace and are finally stay at a certain point along the trace on the TLC. The radioactive fraction is distributed equally in the depth of the thin-layer material (silica gel) between the surface and bottom.

Different types of detection and counting are used [108]:

- Open window gas detection
- Closed window gas counting
- Phosphor image
- Positron-Geiger-Müller scanner
- γ-Scintillation scanner

Further information on radio TLC can be obtained from the company raytest (see Section 12.5).

6.4.2 Simultaneous Derivatization and Development

To apply a reagent "homogeneously" to a layer of sorbent, an elegant approach is to add it to the solvent system. In this method, the reagent must flow with the solvent front and distribute itself uniformly over the layer. Of course, the physical processes of TLC still take place in the usual way, so that the word "homogeneously" must be placed in quotation marks as the reagent usually becomes distributed in the form of a gradient in the direction of development. Jork et al., on p. 88 of Vol. 1a of the reagent books give some examples of the addition of a reagent to solvent systems, citing the original literature references.

Example of a derivatization performed simultaneously with the development: quantitative determination of alanine from a dry extract of marshmallow root in the bronchial herbal tea Solubifix®

Only the parameters that relate to the theme of "derivatization" are listed here. A complete description of this TLC method, including validation, has been published [109], and the original data can be requested from the present author.

Alanine is an amino acid, which in this example is in the form of an aqueous solution in hydrochloric acid as a result of the sample preparation method. For this reason, a TLC silica gel 60 WF_{254s} precoated plate (Merck 16485) was used for the chromatography. After development in a ninhydrin-containing solvent, alanine is converted into a compound visible in daylight. The conditions for the assay are given below:

Reagents:	1-Butanol	Merck 101990
	Acetic acid (100 %)	Merck 100063
	Ninhydrin	Merck 1006762
	Water	HPLC quality

Reagent solution: Ninhydrin (450 mg) is added to 30 ml water and treated in an ultrasound bath for 15 min. Crystals remaining undissolved are filtered off.

Solvent system: 1-Butanol (75 ml) + acetic acid (19 ml) + reagent solution (19 ml) (for a 20 × 20 × 8.7 cm developing chamber).

Chromatography: After saturating the developing chamber (with a styrofoam hood) for at least 30 min.

Development time: Exactly 90 min (digital alarm!).

Migration distance: Approximately 9 cm.

Derivatization: After the chromatography, the back of the plate is dried with a paper towel and immediately placed on a cold plate heater. This is switched on, with a set temperature of 120 °C, until the intensity of the zones reaches a maximum (ca. 10 min).

Before the measurement: The plate is then allowed to cool in a protected environment for exactly 10 min, the Y position of the alanine is then checked and the measurement is immediately started (see Section 7.2.2.1 "Absorption Measurement").

6.4.3 Reaction Sequences

A combination of different or similar derivatization methods performed consecutively on the same TLC plate is known as a "reaction sequence". Combinations of this sort are often used in clinical or forensic chemistry, especially if a rapid and reliable method of detecting causative substances in cases of poisoning, drug abuse etc. is required. Thus, for example, the MTSS (Merck Tox Screening System) is a collection of identity tests that enable the unknown compound to be identified by means of reference substances and data in a data bank when various chromatographic systems followed by special reaction sequences are used [110]. Vol. 1b of the reagent books of Jork et al. gives 30 proven examples of frequently used "reagent sequences" (the term used in the book).

The use of reaction sequences is also common in the analysis of pharmaceuticals. For example, for the purity test for hydrochlorothiazide (HCT) the DAB specifies a nitration reaction followed by reaction with naphthylethylenediamine dihydrochloride, in which the reaction sequence is performed as in the example given below. HCT is a diuretic which is often used in combination with beta-blockers for the treatment of hypertension (high blood pressure). Diuretics of this type are also abused in sport (doping) and appear on the list of banned substances prepared by the IOC (International Olympic Committee). In the quantitative determination of HCT in urine, the red azo dyes formed in the derivatization (parent substance and metabolites) is regarded as a "doping-positive" result [111]. The complete specification for this analytical method, known as Application A-43.2, can be obtained from CAMAG.

The following example describes an unusual "trick" (which to our knowledge has not been reported elsewhere) for documenting these TLC plates.

Example of a reaction sequence for derivatization: detection limit test for known and unknown compounds in the purity testing of ambroxol hydrochloride

Reagents:	*N, N*-dimethylformamide	Merck 103034
	N-(1-naphthyl)ethylenediammonium dichloride	Merck 106237
	Sodium nitrite	Merck 106549
	Hydrochloric acid, 32 %	Merck 100319

4 N hydrochloric acid: The 32 % hydrochloric acid (41.2 g or 34.3 ml) is diluted with water to 100.0 ml.

Spray reagent: *N*-(1-naphthyl)ethylenediammonium dichloride (200 mg) is dissolved in 20 ml of a 1:1 mixture of *N,N*-dimethylformamide and 4 N hydrochloric acid.

Method: After removal of solvent residues from the TLC plate, it is detected, documented and evaluated in 254-nm and 366-nm UV light.

Practical Tip: Before further treatment, sorbent is scraped off the glass plate in the form of a path at least 5 mm wide in the shape of a 90° arc of a circle of radius ca. 3 cm with its center at the top right hand corner of the plate. This procedure prevents the plate from darkening at a later stage, as the starting point of the almost black coloration that appears is usually at the point where the plate is held in order to carry it. Like a

firebreak in a forest, the sorbent-free path usually prevents the dark area, which spreads like an outbreak of fire, from crossing over it [112].
After this has been done, the TLC plate is placed in a closed development chamber for nitration. This contains a 50-ml glass beaker containing ca. 10 g sodium nitrite. By means of a pipette, ca. 5 ml concentrated hydrochloric acid is then added to the sodium nitrite, and the plate is allowed to stand for ca. 5 min in the development chamber. The nitrous gases (highly toxic!) are then removed by blowing with a cold current of air, and the plate is placed in a refrigerator at ca. 5 °C for ca. 15 min. It is then immediately sprayed to test for unknown impurities. The testing procedure states that the plate should not be touched again after the spraying. However, as this is unavoidable for photographic documentation, the above technique of scraping off some sorbent in the shape of an arc of a circle is useful.

Figure 90a shows a derivatized plate with samples of ambroxol, which has darkened (in only a few seconds) from the right hand side. Figure 90b shows another plate in which it has been possible to stop the darkening by quickly scraping off the sorbent layer underneath the arc of a circle that can be seen on the right.

▶ **Figure 90:** see Photograph Section.

6.4.4 Biological-Physiological Methods of Detection

A very interesting technique is to detect biologically active substances by making use of their biological action. This fully exploits the fact that with TLC the individual components of a developed chromatogram are in fixed positions as if in a data storage system. Plenty of time is available for any necessary additional operations; for example, biological media can be very simply brought into close contact with the substance spots with which they are to be reacted. Spraying or coating techniques with an agar (similar to coating with blood-gelatin, see Section 6.3.4) can be used, thereby making use of the fungicidal, antibacterial or antioxidative effects of the substances to be determined. Using these screening methods, plant components from all over the world can be tested for their biological or medical effects [113].

Even quantitative determinations can be performed using biological-physiological methods. These are discussed by Jork and Weins [114] using the example of enzymatic in situ detection of pollutants.

Toxicity screening of complex mixtures following separation on a HPTLC plate is a new technique invented by Bayer. Now ChromaDex™ (see Section 12.5 for the address) has incorporated this technology into a commercial kit (Bioluminex™) utilizing the bioluminescent bacteria, *Vibrio fischeri*. CAMAG has marketed the BioLuminizer™ for the detection and documentation (see Section 8.8).

The published work on this subject encourages us to look forward to even more publications on these highly specific and sensitive methods.

6.5 Further Treatment of Derivatized Chromatograms

Examples of various methods of treating chromatograms have already been discussed in the foregoing Sections, e.g. heating, described in Section 6.4.2, and intensification of fluorescence by treatment with Macrogol 400, described in Section 6.3. Whereas in the latter case the brilliance of the fluorescence colors and hence the detection sensitivity of the flavonoids could be increased, i.e. there was an improvement in qualitative information, the minutely described technique of causing a reaction to occur by heating the plate obtained with bronchial herbal tea is followed by a quantitative evaluation, i.e., special attention is given to the aftertreatment in this case. In Vol. 1a of their reagent books, Jork et al. describe in detail aftertreatments by changing the pH, applying heat, and stabilizing colored and fluorescent zones, both from the point of view of the theoretical background and also with the aid of examples given in Tables. In the present book, therefore, only the two most commonly used methods of aftertreatment of derivatized chromatograms are described.

6.5.1 Effect of Heat

To enable the desired chemical reactions in the layer of sorbent to proceed as rapidly as possible or to enable them to proceed at all, it is advisable to heat the TLC plates. Formerly, infrared heaters, drying ovens and early forms of hotplates were used, but these are now largely replaced by modern plate heaters with a Ceran® ceramic heating surface (CAMAG) as shown in Fig. 91. These are not only stable towards all commonly used reagents but are also easy to clean. The temperature range, 25–200 °C, is sufficient for most applications. According to the manufacturers, the set temperature is maintained over the whole area of the plate. Heat treatment at 100–120 °C for 5–10 min, as is often specified in the literature, will often give complete reaction, but experiment sometimes shows that temperatures of 160 °C and above are necessary.

The use of microwave equipment to promote reactions on TLC plates has also been reported [115]. However, this is not yet widely used in the laboratory, as, apart from the fact that equipment with direct fume extraction is not yet available, aluminum foil has to be ruled out as a support for sorbents!

Figure 91. Plate heater with Ceran® ceramic heating surface (CAMAG)

☞ **Practical Tips** for heating:

- The TLC plate heater must always be operated in an efficient fume cupboard, as the vapors from the reagents must always be assumed to be highly toxic.
- For every job, and especially for quantitative work, it should be determined whether better results are obtained with a cold or a preheated plate.
- The set temperature and heating time of the equipment should be exactly specified in the testing procedures.
- When using a plate heater with an aluminum surface (e.g. the Thermoplate S of DESAGA), it is advisable to place a clean glass plate in direct contact with the aluminum block and to place the TLC plate on this. If this is not done and the equipment is preheated, the TLC plate, because of the difference between the thermal expansion coefficients of aluminum and glass, becomes warped, and the outer edge lifts slightly. This leads to poorer heat transfer and hence nonuniform heating. Quantitative work is then impossible! A further reason for having the protective plate is to avoid contaminating the surface of the aluminum with reagent.
- If chromatograms on aluminum foil supports are derivatized by the action of heat, an additional glass plate or a metal frame ca. 1 cm wide will prevent the foil from curling up. Of course, if a metal frame is used the reaction vapors will be able to escape more easily.
- Because of the high rate of heat loss in a fume cupboard, it is better to set the temperature of the plate heater at 120 °C rather than 100 °C.

Example of the use of heat for derivatization: stress tests of dexpanthenol ointment

The solvent system specified in the DAC monograph for dexpanthenol was used. After the sample treatment process the ointment is in the form of an aqueous solution, and a water-resistant silica gel layer (DC-KG 60 WF_{254s}, Merck 16485) was therefore chosen. Because of the low sensitivity of the method, the derivatization was not performed by ninhydrin reagent (as specified in the DAC), but as a reaction sequence with the 2,5-dimethoxytetrahydrofuran/4-(dimethylamino)-benzaldehyde reagent (Jork et al, Vol. 1a, p. 265). It should be noted here that the traces of ammonia that come from the solvent mixture must be completely removed from the chromatograms before using the reagent, as the background colors can otherwise by too intense.

Reagents:	2,5-Dimethoxytetrahydrofuran	Merck 802961
	4-(Dimethylamino)-benzaldehyde	Merck 103058
	Acetic acid, 100 %	Merck 100063
	Ethyl acetate	Merck 109623
	Hydrochloric acid, 32 %	Merck 100319

Reagent solution A: 2,5-Dimethoxytetrahydrofuran (0.3 ml) with ethyl acetate (24 ml) and acetic acid (6 ml).

Reagent solution B: 4-(Dimethylamino)benzaldehyde (600 mg) is dissolved in a mixture of 25.5 ml acetic acid and 4.5 ml hydrochloric acid.

Storage: Solution A can be kept for a few days and solution B for a few weeks at room temperature.

Procedure: After complete removal of the ammonia (ca. 30 min heating by a hot air blower), the plate is sprayed with solution A and then heated at 160 °C for 10 min. After cooling the plate to room temperature, it is then sprayed with solution B.

Safety precautions: When applying and handling these reagents, appropriate safety precautions must be taken, including personal protection by use of disposal gloves, protective goggles and breathing masks!

Figure 92 shows the chromatogram of stressed samples of dexpanthenol ointment after derivatization with the reagent known internally as “JFFW 1a/265”.

▶ **Figure 92:** see Photograph Section.

6.5.2 Stabilization of Colored and Fluorescent Zones

The colored and/or fluorescent chromatogram zones obtained on derivatization should be stable for at least 30 min to enable documentation and/or quantitative evaluation to be performed.

With fluorescent substances, this is usually very satisfactorily achieved by the use of fluorescence intensifiers (FITs). Vol. 1a of the reagent books of Jork et al. gives tables of lipophilic FITs (mainly mineral oils) and hydrophilic FITs such as polyethylene glycols, triethylamine, triethanolamine and especially Triton X-100, including areas of use. Treatment by spraying and dipping in various solvents is recommended. In general, this increases the fluorescence yields in comparison with those of untreated chromatograms by a factor of 5–10. However, the physicochemical processes that lead to these improvements in stability and intensity are not yet fully understood.

Example of an intensification of fluorescence: caffeic acid derivatives

See Section 6.3.1, Fig. 84: various species of Echinacea derivatized by the flavone reagent according to Neu and stabilized with a 2 % ethanolic solution of Magrocol 400.

There are no general recommendations for stabilization of the color of colored chromatogram zones. Jork et al. recommend storage in a nitrogen atmosphere in the absence of light until the plates are measured quantitatively, and they suggest suitable methods of stabilization in the case of some reagents, e.g. by the addition of salt to the reagent or by treatment of the plates with ammonia vapor.

☞ **Practical Tip** for stabilizing colored zones: On their way from the spray cabinet or plate heater to a documentation system, colored chromatograms should be covered with a clean glass plate to minimize reactions with oxygen.

7 Evaluation After Derivatization

It would have been possible to combine Chapter 5 "Evaluation without Derivatization" with the present Chapter 7 to form a single Chapter "Evaluation", as the earlier Chapter is mainly of general application. However, evaluation *before* derivatization in the course of the TLC process needed to be clearly emphasized. As the use of a TLC scanner was described in detail in Section 5.2, it would be a good idea to reread this Section at this point in the book.

Evaluation after derivatization does not only involve assessment of the chromatogram in daylight or measurement in light of wavelength >400 nm; there are also a large number of reactions whose products are detected in 254-nm or 365-nm UV light.

7.1 Visual Evaluation

A rapid assessment of colored chromatograms immediately after completion of the derivatization reaction is always necessary if single spots or all the zones undergo very rapid color change or if the whole plate loses its color.

7.1.1 Visual Qualitative Evaluation

A visual evaluation is only adequate for identity tests. However, in accordance with GMP guidelines, in the absence of any further documentation a second person should also inspect the TLC plate immediately after derivatization and the result should be entered in a database. The same applies to results to be submitted as evidence in court.

7.1.2 Visual Semiquantitative Evaluation

All limit value tests performed in accordance with the pharmacopoeias include a semiquantitative assessment of the chromatogram. Here, too, in accordance with GMP, a second person should inspect the plate if no pictorial documents are prepared.

Applied Thin-Layer Chromatography: Best Practice and Avoidance of Mistakes, 2nd Edition
Edited by Elke Hahn-Deinstrop

ISBN: 978-3-527-31553-6

7.2 Evaluation Using a TLC Scanner

The TLC scanner was developed principally as a highly effective analytical tool to provide quantitative measurement results. However, its field of use will most likely increase.

7.2.1 Qualitative Evaluation

If there is no photographic documentation system in the workplace but a TLC scanner with an analog printer is available, all chromatograms can be plotted and compared with each other. Today, this "fingerprint" method is mainly used with samples of complex composition, e.g. phytopharmaceuticals.

Example of a qualitative evaluation using a TLC scanner: identification of dry extract of marshmallow root in the bronchial herbal tea Solubifix®

In the analysis of bronchial herbal tea, dry extracts of primula root, liquorice root and marshmallow root were prepared, chromatographed and derivatized using a validated method, and this procedure was repeated with the bronchial herbal tea and a matrix mixture containing all the components of the instant tea apart from the dry extract of marshmallow (see also Section 6.4.2 "Simultaneous Derivatization and Development"). The red-violet chromatograms obtained were scanned at a wavelength of 496 nm, and the curves produced were then placed one above the other to form a 3-D graphic (angle = 0°). The chromatogram curves of marshmallow root extract and bronchial herbal tea give the same fingerprint. This confirms the identification of this extract in the herbal tea (Fig. 93).

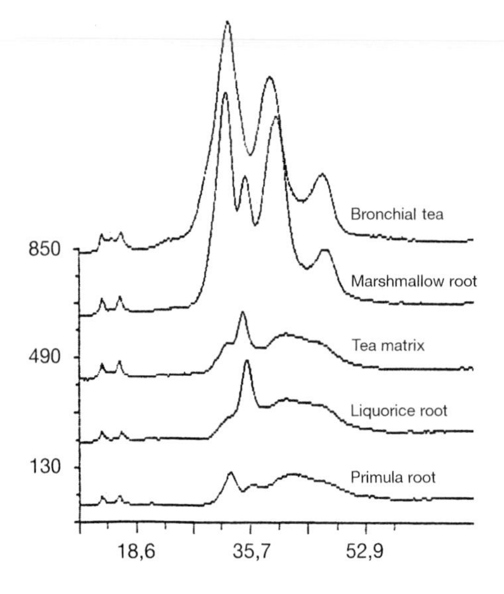

Figure 93. Qualitative evaluation with the TLC scanner: identification of marshmallow root DE in bronchial herbal tea Solubifix® Fingerprints of preparations from the individual extracts, the tea matrix and the tea, measured with the TLC Scanner CD 60 (DESAGA) at 496 nm

Modern TLC scanners (e.g. by CAMAG or DESAGA) can produce a spectrum from each spot of a TLC plate. This can be compared with the data in a library of spectra, enabling the identity of a substance to be established [116].

7.2.2 Quantitative Evaluation

The best way to obtain a thorough grounding in the practical operation of a TLC scanner is to attend a seminar presented by the manufacturer. Only in this way can one really understand the scope and limitations of the equipment and be able to use one's own scanner in an assured and reliable fashion. The two following Sections give an example of an absorption measurement and a fluorescence measurement respectively, and these are followed by a comparison of a "parallel" with a "transverse" measurement. But first we give a few more

☞ **Practical Tips** for quantitative evaluation with TLC scanners:

- Before placing a plate in a scanner, the chromatographic lanes should be visually checked for intactness.
- The hRf value of a substance is in general different from the Y position of the measurement. For example, in the case of alanine (see Section 7.2.2.1 for the measurement data), the hRf value of the substance is 28–30 and the Y position is ca. 42 mm.
- Our own experience has shown that with quantitative measurements a better standard deviation is obtained if two additional samples are chromatographed, one on each side of the sample being analyzed. These are **not** included in the calculation.
- Especially in the case of samples of complex composition, a better result can be obtained by using an aliquot from a sample applied in the form of a band rather than a spot applied at one point only.
- Software designed for "spot optimization" should enable the difference between aliquot measurement of a band (without this optimization) and measurement of a spot to be determined experimentally.
- In general, the question whether better evaluation results are obtained using peak areas or peak heights should be determined by experiment.
- To determine limits of detection and/or determination, a "blank lane" should always be used for comparison. This is scanned outside the chromatogram lanes.
- In TLC, linearity sometimes presents problems. The chosen concentration bandwidth of a substance should therefore not exceed a few orders of magnitude.
- If a calibration line is linear but does not pass through the origin, a calibration with at least three standard concentrations should be performed in order to avoid systematic errors.
- If a calibration line is not linear, the method of calibration should be changed, e.g. by using the Michaelis-Menten (or other) function [117].
- All the **Practical Tips** for working with TLC scanners given in Section 5.2 apply here also.
- The following examples in Sections 7.2.2.1 to 7.2.2.3 are taken from the 1st edition of this book. Today the output of the documents is dependant on the current software. Contact the manufacturer regarding questions about the current software versions.

7.2.2.1 Absorption Measurement

Example: quantitative determination of alanine in dry extract of marshmallow root

This example of an absorption measurement is taken from the raw data in [109]. The assay is performed by comparing the peak height of the alanine in the calibration solutions 1–3 with that in the sample solution. The densitometric measurements were performed with the TLC scanner CD 60 (DESAGA) at a measurement wavelength of 496 nm. The Version 3.2 software used calculates the mean value from the three values obtained (peak heights of alanine) for the three calibration solutions, and the linear regression is calculated from this. The calibration function is also given on the graph (see the example in Fig. 94). The program then calculates the mean value from the three values of the sample solution and uses this as the Y value in the calibration function. The calibration and test solutions are treated identically with respect to preparation, the taking of aliquots for application etc., and the mean value (mg) of marshmallow root extract, expressed as a fraction of the initial weighed amount, is taken as the analytical result.

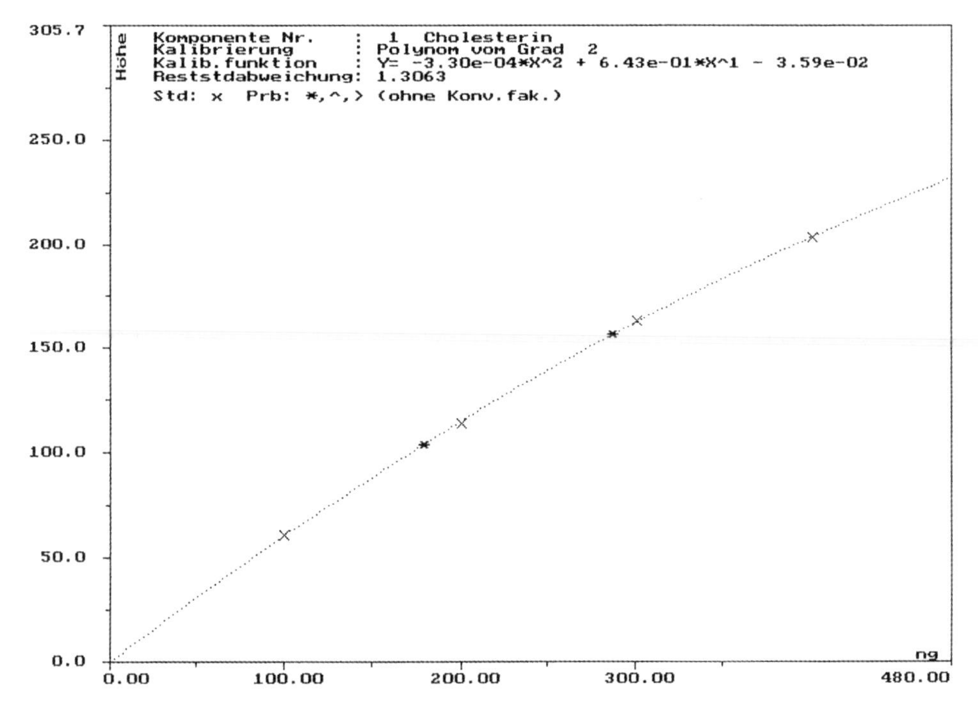

Figure 94. Calibration curve of a TLC plate with measurement data for the evaluation of peak heights for cholesterol in butter and margarine
The measurement conditions are given in Document 2 (Section 7.2.2.2)

Document 1 contains the following printed-out pages:

1. Method. This page is identified by the name of the method (in this case AS 15). All the parameters for the measurement have previously been entered in the corresponding computer file.

2. Peak list (long). This page is identified by the name of the subject of the analysis (in this case Eib 91.1). In this list, all the data for the peaks are printed out, and alanine is assigned. For reasons of space, only the first 7 out of 16 lanes are shown.

3. Peak list (short). In this list, only the substances with the automatic peak assignments are used. (For reasons of space, this list is not given here.)

4. Results (short). Printouts are given as for "Results (long)", but without listing the calibration solutions. (Again, this list is not given here for reasons of space.)

5. Results (long). In the "Standard values" column, the concentrations specified in the method are given. In the "Sample result" column, the listed values are in the units specified in the method (in this case mg). Any additional calculations, e.g. to obtain the final result "% of target", must be performed externally if this software is used.

```
                D E S A G A   C D 6 0   M E T H O D E

Name der Methode : AS 15
Kommentar        :
Quantitative Bestimmung von Alanin in Eibischwurzel TE
90, 100, 110 % Standards
Benutzer         : Hahn-Deinstrop                 Datum : 05/07/1993
                                                  Zeit  : 20:26

Startkoordinate X    :  23.0 [mm]   Spaltbreite           : 0.4  [mm]
Startkoordinate Y    :  10.0 [mm]   Spalthöhe             : 2.0  [mm]
Endkoordinate   Y    :  70.0 [mm]   Filterposition        : offen
Mäanderbreite        :   0.0 [mm]
Zahl    der Bahnen   :  15          Meßwellenlänge        :  496 [nm]
Abstand der Bahnen   :  11.0 [mm]   Referenzwellenlänge :      0 [nm]
Meßart  :  Remission   Extinktion   Signal positiv    Lampe D und W

Auflösung bei der Datenaufnahme   :     0.100   [mm]
Zahl der Messungen pro Punkt      :     8
Glättungfaktor                    :     0
Linearisierungsparameter          :     0
Skalierung der Graphik            :    Signal-Faktor   20
Untergrundkor. (nur b. Mäander)   :    nein
Automatischer Nullabgleich        :      ja
Ausdruck                          :    Ergebnisse (lang)
Datenspeicherung                  :    speichere Chromatogramme

Auswertung bei der Datenaufnahme  :      ja
Fenstergröße                 (mm) :     0.200
Schwellwert für Peakerkennung     :     1.000
Maximale Steigung Basislinie      :     2.000
Minimale Peakhöhe                 :    50.000
Minimale Peakfläche               :   300.000
Auswerteintervall    von  bis (mm) :   25.000     60.000

Zahl der Komponenten              :     1
Zahl der verschiedenen Standards  :     3
Zahl der verschiedenen Proben     :     2
Einheit des Standards             :    mg
Einheit des Ergebnisses           :    mg
Konversionsfaktor                 :     1.00000
Nachkommastellen d. Standards     :     2
Nachkommastellen d. Ergebnisses   :     2
Art der Kalibrierung              :    Gerade y = a*x + b ohne (0,0)
Kalibrieren nach Peakhöhe/fläche  :    Höhe

Zuordnung der Bahnen              :
Standard  1 :   2  6 12
Standard  2 :   3  8 13
Standard  3 :   5 10 15
Probe     1 :   1  7 11
Probe     2 :   4  9 14

Konzentration der Standards :
Standard  1 :  K1 :          58.50
Standard  2 :  K1 :          65.00
Standard  3 :  K1 :          71.50

Automatische Peakzuordnung :
Komponente    Name d. Komponente    Position [mm]   Toleranz [mm]
     1        Alanin                    42.0             4.0
```

Document 1 / page 1

```
        D E S A G A   C D 6 0   P E A K L I S T E   ( L A N G )

Name des Objekts : Eib 91.1
Kommentar        :
Wiederfindung des Extraktes
Aufarbeitung, Chrom + Scan an 1 Tag
Proben 5 + 6
Name der Methode : AS 15
Benutzer         : Hahn-Deinstrop                    Datum : 03/02/1993
                                                     Zeit  : 20:43


Peak                    Name Y-Pos.  Fläche Fläche   Höhe PM     Std.Konz.
 Nr.                         [mm]           [%]                        mg

     Bahn Nummer : 1  Extrakt Probe 5     X-Koordinate :  23.0 [mm]
  1                         33.4    844.9   20.9  249.39 f
  2                         38.0    742.7   18.4  248.42 f
  3                 Alanin  42.8   1927.5   47.8  429.74 b
  4                         51.3    518.1   12.8  146.74 b

     Bahn Nummer : 2  Extrakt 90 %        X-Koordinate :  34.0 [mm]
  1                         33.1    655.8   20.6  204.81 f
  2                         37.8    611.4   19.2  207.61 f
  3                 Alanin  42.7   1542.3   48.4  369.00 b          58.900
  4                         51.2    378.5   11.9  118.80 b

     Bahn Nummer : 3  Extrakt 100 %       X-Koordinate :  45.0 [mm]
  1                         33.4    780.1   22.3  237.87 f
  2                         38.0    624.7   17.9  225.84 f
  3                 Alanin  42.8   1694.5   48.5  403.49 b          66.000
  4                         51.1    394.5   11.3  128.31 b

     Bahn Nummer : 4  Extrakt Probe 6     X-Koordinate :  56.0 [mm]
  1                         33.2    804.2   22.3  253.78 f
  2                         37.9    701.5   19.5  244.20 f
  3                 Alanin  42.6   1729.1   48.0  401.54 b
  4                         50.9    369.2   10.2  125.16 b

     Bahn Nummer : 5  Extrakt 110 %       X-Koordinate :  67.0 [mm]
  1                         33.2    859.1   22.3  260.78 f
  2                         37.7    797.3   20.7  254.70 b
  3                 Alanin  42.3   1775.2   46.1  429.09 b          71.200
  4                         50.9    419.9   10.9  136.80 b

     Bahn Nummer : 6  Extrakt 90 %        X-Koordinate :  78.0 [mm]
  1                         33.0    724.0   22.2  225.52 f
  2                         37.8    631.8   19.4  223.84 f
  3                 Alanin  42.5   1565.8   48.0  361.40 b          58.900
  4                         50.7    337.3   10.4  112.99 b

     Bahn Nummer : 7  Extrakt Probe 5     X-Koordinate :  89.0 [mm]
  1                         33.2    811.9   23.2  249.65 f
  2                         37.8    671.5   19.2  240.60 f
  3                 Alanin  42.5   1678.5   48.0  388.18 b
  4                         50.7    338.3    9.7  113.57 b
```

Document 1 / page 2

```
      D E S A G A   C D 6 0   E R G E B N I S S E   ( L A N G )

Name des Objekts : Eib 91.1
Kommentar        :
Wiederfindung des Extraktes
Aufarbeitung, Chrom + Scan an 1 Tag
Proben 5 + 6
Name der Methode : AS 15
Benutzer         : Hahn-Deinstrop
                                                  Datum : 13/10/1993
                                                  Zeit  : 04:28

Kalibrierung     : Gerade y = a*x + b ohne (0,0)

Komponente Nr.   :  1   Alanin
Kalib.funktion   :  Y=  5.69e+00*X^1 + 3.57e+01
Stand.abweichung :  0.5664

             Y-Position       Höhe    Standard     Standard   Wert
                   [mm]               Abw. [%]                  mg
Extrakt 90 %     : 42.5      371.0         2.9                58.90
Extrakt 100 %    : 42.4      410.7         1.5                66.00
Extrakt 110 %    : 41.9      441.0         3.1                71.20
                                                    Proben Ergebnis
                                                                 mg
Extrakt Probe 5  : 42.6      402.9         5.8                64.55
Extrakt Probe 6  : 42.3      415.3         3.4                66.73
```

Document 1 / page 3

7.2.2.2 Fluorescence Measurement

Example: cholesterol in butter and margarine

Preparation of the TLC plate and design of all the documents was carried out by Maria Müller (DESAGA). After derivatization of the lipids with the manganese chloride-sulfuric acid reagent and heating the plate to 120 °C, it was then measured by the CD 60 scanner using a mercury vapor lamp with a measurement wavelength of 366 nm. A cut-off filter at 420 nm was used. Figure 95 shows all 12 chromatograms of the cholesterol. The assignment of the lanes is shown on page 1 (Method). The evaluation was performed using the DESAGA software Version 5.3. For reasons of space, only a part of the printout is shown, as Document 2, on the following pages, size-reducing the originals and omitting further commentary.

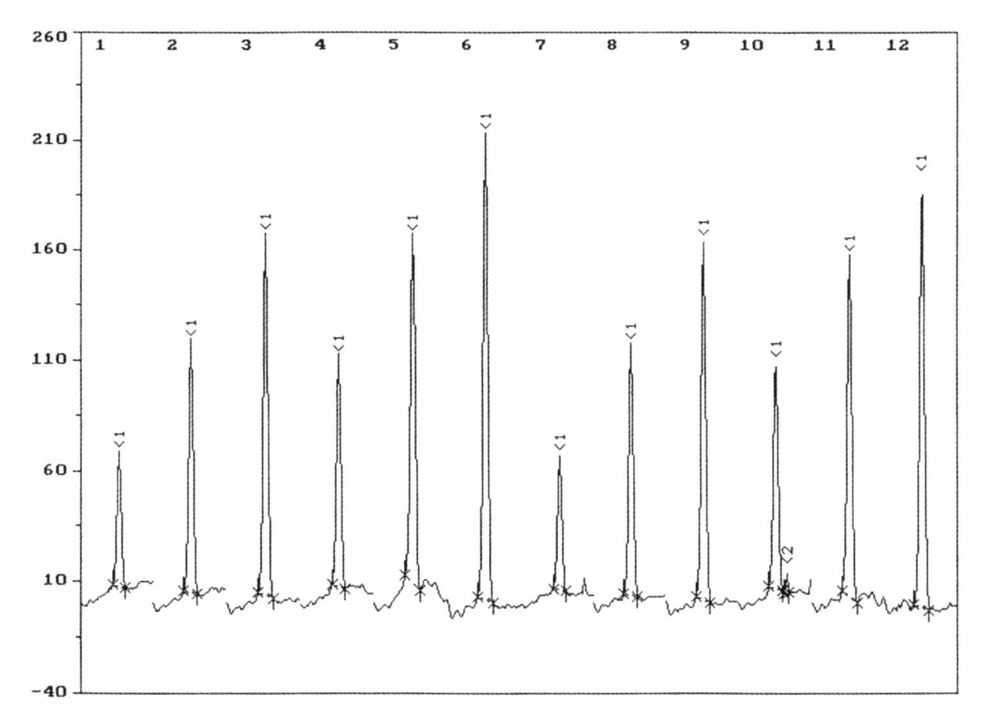

Figure 95. Direct optical evaluation of cholesterol in butter and margarine. Fluorescence measurement at 366 nm (cut-off filter 420 nm) using the TLC Scanner CD 60 (DESAGA)
The measurement conditions are given in Document 2 (Section 7.2.2.2)

D E S A G A   C D 6 0   M E T H O D E

```
This line is printed on each document. To be edited by the user.
Chromatogramm Nr. : 0000136 gedruckt    02/10/1996 13:29
erstellt von  mr                        18/05/1995 09:35
Name der Methode : DC von Lipiden

Startkoordinate X     :  16.1 [mm]   Spaltbreite            : 0.1  [mm]
Startkoordinate Y     :  12.0 [mm]   Spalthöhe              : 3.0  [mm]
Endkoordinate   Y     :  32.0 [mm]   Filterposition         : 420  [nm]
Mäanderbreite         :   0.0 [mm]
Zahl    der Bahnen    :  12          Meßwellenlänge         :  366 [nm]
Abstand der Bahnen    :   7.0 [mm]   Referenzwellenlänge :       0 [nm]
Meßart  :  Remission   Fluoreszenz    Signal positiv   Lampe Hg

Auflösung bei der Datenaufnahme     :    0.100   [mm]
Zahl der Messungen pro Punkt        :   16
Glättungsfaktor                     :    0
Linearisierung nach Kubelka/Munk    :    0
Skalenbereich des Signals           :    200
Untergrundkor. (nur b. Mäander)     :   nein
Automatischer Nullabgleich          :     ja
Ausdruck                            :   kein Ausdruck
Datenspeicherung                    :   nein
Rufe Benutzerprogramm auf           :  Starte kein Benutzer Programm

Fleckoptimierung                    : Keine Optimierung
Auflösung bei Fleckoptimierung      :    0.050   [mm]
Zahl der Messungen pro Punkt        :   16

Auswertung bei der Datenaufnahme    :     ja
Fenstergröße                  (mm)  :    0.800
Schwellwert für Peakerkennung       :    1.800
Maximale Steigung Basislinie        :    2.000
Minimale Peakhöhe                   :    0.000
Minimale Peakfläche                 :    0.000
Auswerteintervall   von  bis (mm)   :   12.000    32.000

Zahl der Komponenten                :    1
Zahl der Standardkonzentrationen    :    4
Zahl der verschiedenen Proben       :    2
Einheit des Standards               :   ng
Einheit des Ergebnisses             :   ng
Konversionsfaktor                   :    1.00000
Nachkommastellen d. Standards       :    2
Nachkommastellen d. Ergebnisses     :    2
Art der Kalibrierung                :   Polynom vom Grad  2
Kalibrieren nach Peakhöhe/fläche    :   Höhe

Zuordnung der Bahnen                :
Standard   1 :    1  7
Standard   2 :    2  8
Standard   3 :    3  9
Standard   4 :    6 12
Probe      1 :    4 10
Probe      2 :    5 11

Konzentration der Standards :
Standard   1 :   K1 :        100.00
Standard   2 :   K1 :        200.00
Standard   3 :   K1 :        300.00
Standard   4 :   K1 :        400.00

Automatische Peakzuordnung :
Komponente    Name d. Komponente    Position [mm]   Toleranz [mm]
    1         Cholesterin              23.0             2.0
```

Document 2 / page 1

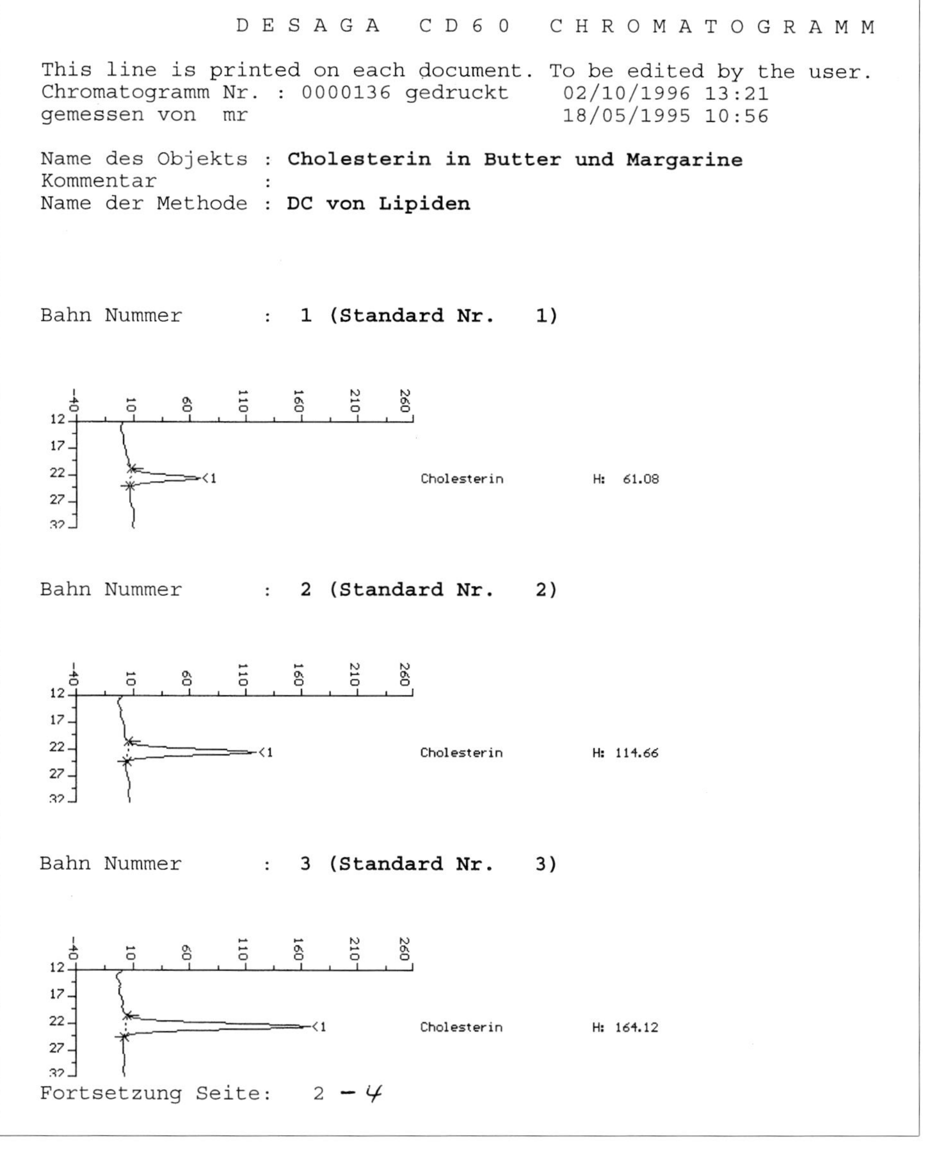

DESAGA CD60 CHROMATOGRAMM

This line is printed on each document. To be edited by the user.
Chromatogramm Nr. : 0000136 gedruckt 02/10/1996 13:21
gemessen von mr 18/05/1995 10:56

Name des Objekts : **Cholesterin in Butter und Margarine**
Kommentar :
Name der Methode : **DC von Lipiden**

Bahn Nummer : **1 (Standard Nr. 1)**

Bahn Nummer : **2 (Standard Nr. 2)**

Bahn Nummer : **3 (Standard Nr. 3)**

Fortsetzung Seite: 2 – 4

Document 2 / page 2

```
                    D E S A G A   C D 6 0   P E A K L I S T E

This line is printed on each document. To be edited by the user.
Chromatogramm Nr. : 0000136 gedruckt     02/10/1996 13:11
gemessen von  mr                         18/05/1995 11:00

Name des Objekts : Cholesterin in Butter und Margarine
Kommentar        :
Name der Methode : DC von Lipiden


Peak Komp      Name           Y-Pos.   Fläche Fläche     Höhe PM  Std.Konz.
 Nr. Nr.                       [mm]              [%]                     ng

     Bahn Nummer : 1 (Standard Nr.  1)     X-Koordinate :  16.1 [mm]
  1   1 Cholesterin             22.6      82.9 100.0    61.08 b   100.000

     Bahn Nummer : 2 (Standard Nr.  2)     X-Koordinate :  23.1 [mm]
  1   1 Cholesterin             22.6     162.0 100.0   114.66 b   200.000

     Bahn Nummer : 3 (Standard Nr.  3)     X-Koordinate :  30.1 [mm]
  1   1 Cholesterin             22.6     234.5 100.0   164.12 b   300.000

     Bahn Nummer : 4 (Proben   Nr.  1)     X-Koordinate :  37.1 [mm]
  1   1 Cholesterin             22.6     143.9 100.0   104.61 b

     Bahn Nummer : 5 (Proben   Nr.  2)     X-Koordinate :  44.1 [mm]
  1   1 Cholesterin             22.7     245.4 100.0   158.25 b

     Bahn Nummer : 6 (Standard Nr.  4)     X-Koordinate :  51.1 [mm]
  1   1 Cholesterin             22.6     305.3 100.0   211.57 b   400.000

     Bahn Nummer : 7 (Standard Nr.  1)     X-Koordinate :  58.1 [mm]
  1   1 Cholesterin             22.5      83.1 100.0    61.33 b   100.000

     Bahn Nummer : 8 (Standard Nr.  2)     X-Koordinate :  65.1 [mm]
  1   1 Cholesterin             22.6     164.4 100.0   114.11 b   200.000

     Bahn Nummer : 9 (Standard Nr.  3)     X-Koordinate :  72.1 [mm]
  1   1 Cholesterin             22.5     238.6 100.0   163.61 b   300.000

     Bahn Nummer :10 (Proben   Nr.  1)     X-Koordinate :  79.1 [mm]
  1   1 Cholesterin             22.5     151.1  96.6   103.48 b
  2                             25.5       5.3   3.4    11.79 b

     Bahn Nummer :11 (Proben   Nr.  2)     X-Koordinate :  86.1 [mm]
  1   1 Cholesterin             22.5     235.5 100.0   155.39 b

     Bahn Nummer :12 (Standard Nr.  4)     X-Koordinate :  93.1 [mm]
  1   1 Cholesterin             22.4     297.4 100.0   196.08 b   400.000
```

Document 2 / page 3

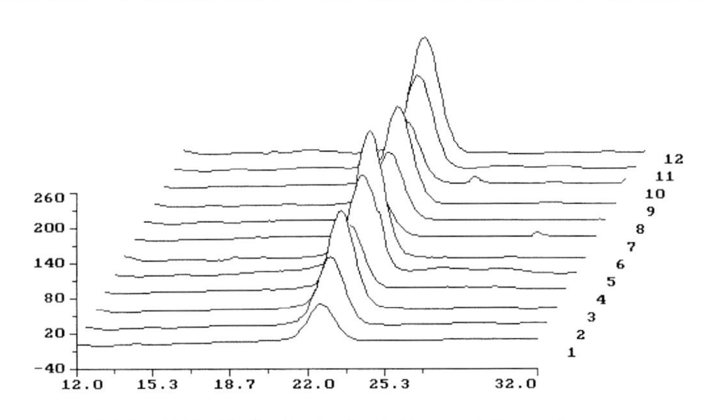

D E S A G A   C D 6 0   E R G E B N I S

This line is printed on each document. To be edited by the user.
Chromatogramm Nr. : 0000136 gedruckt 02/10/1996 13:11
gemessen von mr 18/05/1995 11:00

Name des Objekts : **Cholesterin in Butter und Margarine**
Kommentar :
Name der Methode : **DC von Lipiden**

Kalibrierung : **Polynom vom Grad 2**

Komponente Nr. : **1 Cholesterin**
Kalib.funktion : **Y= -3.30e-04*X^2 + 6.43e-01*X^1 - 3.59e-02**
Reststdabweichung: **1.3063**

	Y-Position [mm]	Höhe	VK [%]	Standard Wert ng
Std. 1	: **22.6**	**61.2**	**0.3**	**100.00**
Std. 2	: **22.6**	**114.4**	**0.3**	**200.00**
Std. 3	: **22.6**	**163.9**	**0.2**	**300.00**
Std. 4	: **22.5**	**203.8**	**5.4**	**400.00**
				Proben Ergebnis ng
Prb. 1	: **22.6**	**104.0**	**0.8**	**178.34**
Prb. 2	: **22.6**	**156.8**	**1.3**	**286.24**

Document 2 / page 4

7.2.2.3 Comparison of "Parallel" With "Transverse" Measurement

Example: sulfonamides

This measurement has already been mentioned in Section 5.2.3 "Direct Optical Evaluation below 400 nm", as the evaluation is performed without derivatization. However, the importance of this example lies in the quantitative results of the measurements of trimethoprim and sulfamethoxazole (both being pharmaceutically active), so that it seems more appropriate to discuss it here.

The preparation and measurements on the TLC plate and the design of the documents were carried out by Gerda Morlock. Especially for this comparative measurement, 16 samples (1-µl each) of the same solution were applied to an HPTLC plate and chromatographed. The plate was then scanned once in the usual way along each lane from left to right (Fig. 96a), and it was then measured a second time in a tranverse direction, i.e. scanning all the trimethoprim samples in the first pass and all the sulfamethoxazole samples in the second (Fig. 96b). The coefficients of variation of the calculated mean values of each of the 16 peak values and area values of both series of measurements are given in Table 20. The printouts of the parallel measurements are shown here in size-reduced form as Document 3.

Professor Ebel (Würzburg) often introduces his lectures on quantitative TLC by presenting the outcome of this comparison as follows:

Wer quer mißt, mißt Mist!

This may be roughly translated as "If you measure transversely, you measure rubbish!"

Table 20: Comparison of parallel measurement with transverse

Mean value height	**Parallel measurement**	**Transverse measurement**
TMP	1,1 %	4,6 %
SMZ	1,4 %	2,6 %
Mean value area	**Parallel measurement**	**Transverse measurement**
TMP	1,6 %	10,0 %
SMZ	1,4 %	3,8 %

Coefficients of variation of the measured values obtained [%]
TMP = Trimethoprim
SMZ = Sulfamethoxazole

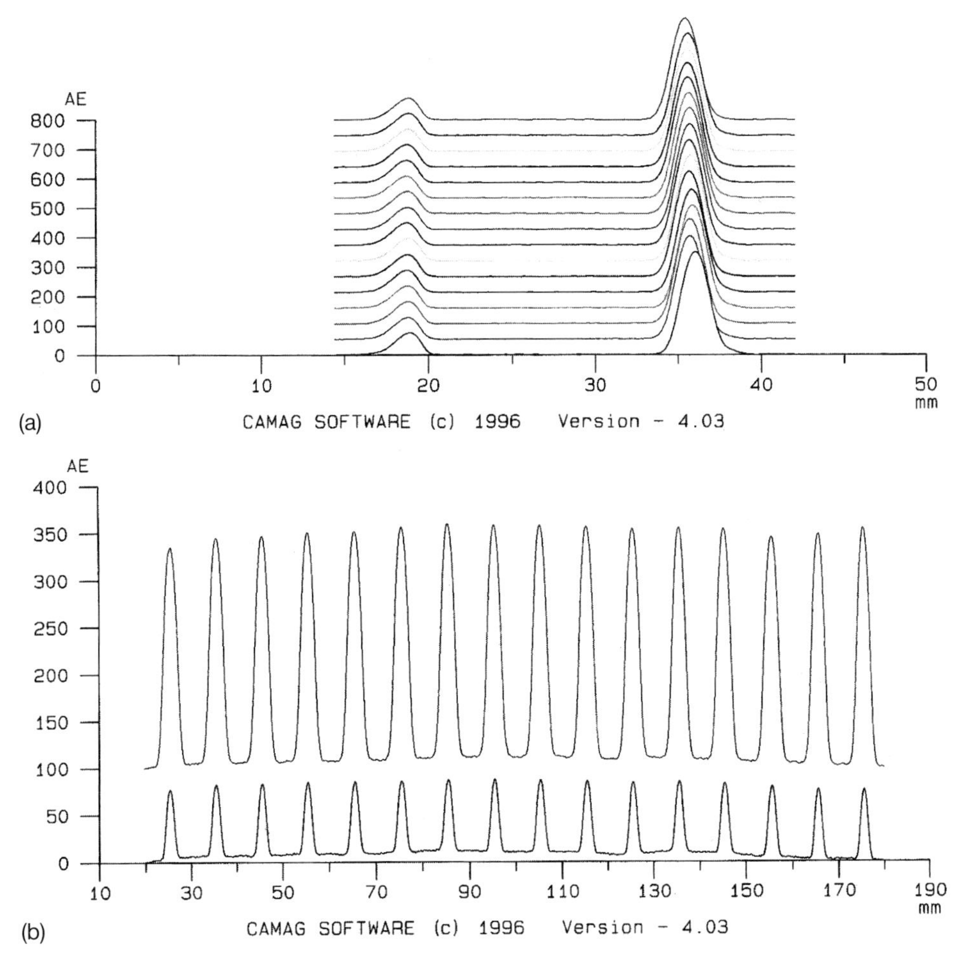

Figure 96. Comparison of the direct optical evaluation of sulfamethoxazole and trimethoprim using parallel and transverse measurement at 285 nm using the TLC Scanner III (CAMAG)
(a) Parallel measurement
(b) Transverse measurement
The measurement conditions are given in Document 3 and the results in Table 20

C A M A G TLC Auswerteprogramm
****
Test Version 4.03

TLC/HPTLC-Auswertung (CATS V4.03, S/N:6311A005 / Sc3 V1.12, S/N:020303)

"Parallelmessung"

Kalidaten	Kalibrationseingaben von : **zw**		
	Dateiname : **EHDLR**	**16/APR/96**	**13:29:10**
Messen	Messung der Rohdaten von : **zw**		
	Dateiname : **EHDLR**	**16/APR/96**	**13:47:04**
Integrieren	Integration ausgeführt von : **zw**		
	Dateiname : **EHDLR**	**16/APR/96**	**13:47:24**
Berechnen	Benutzer : **zw**		
	Datum und Zeit :	**16/APR/96**	**13:47:36**

Analysen- und Chromatographiebedingungen:

Analyse	:Bestimmung von Sulfonamiden
Schichtmaterial	:HPTLC Kieselgel MERCK 60F 254 20 x 10 cm
Fliessmittel	:Dichlormethan - Methanol 80;20
Auftragemethode	:ATS3 punktförmig 16 x 1ul
Entwicklungsmethode	:Doppeltrog gesättigt

Scannereinstellung:

Plattengrösse (B x H)...			20 x 10 cm
Auftrageposition	Y	:	8.0 mm
Frontposition	Y	:	50.0 mm
Bahnanfang	Y	:	5.0 mm
Bahnende	Y	:	45.0 mm
Startposition	X	:	25.0 mm
Bahnabstand	X	:	10.0 mm
Anzahl Bahnen		:	16
Lampe...			Deuterium
Monochromator-Bandbreite...			20 nm
Wellenlänge		:	285 nm
Spalt (Länge x Breite)...			5.0 x 0.20 mm
Messwertauflösung...			100 µm
Max. Messwert		:	1000 AE
Messmodus...			Absorption / Reflexion...
Messgeschwindigkeit		:	20 mm/s
Optisches Filter		:	Oberwellen-Filter
Abgleichmodus		:	Automatik
0-Abgleich bei		:	5.0mm, Bahn: 1
Suchscan von		:	5.0mm bis 45.0mm, alle Bahnen
Offset		:	10%
Empfindlichkeit		:	Automatik (35)
PM-Hochspannung		:	339 V

Bahnoptimierung...

Bahnoptimierung	:	AUS
Optimierungsart...		7 Bahnen x 0.4 mm
Peakschwelle, Höhe	:	100 AE
Peakschwelle, Steigung	:	5

Benutzer : zw **16/APR/1996 ID:7CC04100D2F2447** **Seite 1**
Visum : Geprüft:

Document 3 (Parallel measurement) / page 1

```
Integrationsparameter:
Video-Integration            : Nein
Basislinienkorrektur         : Ja
Peakschwelle, Höhe           :   20 AE
Peakschwelle, Fläche         :  100
Peakschwelle, Steigung       :    5
Filterfaktor                 :    3
Datenreduktionsfaktor        :    1
Position Bahnanfang          : 14.4 mm
Position Bahnende            : 42.0 mm

Kalibrationsparameter:
Kalibrationsmethode...         Reproduzierbarkeit
Interner Standard            : Nein
Statistik...                   VK
Ergebnisberechnung über...     Höhe und Fläche
Spotcheck-Methode...           Automatisch
Peakfenstergrösse (auto)     : 2000 µm

Kalibrierte Substanzen:
Substanz             Einwaage    Dimension    Typ       rResponse
TMP                  ----.----      --        Normal      1.000
SMZ                  ----.----      --        Normal      1.000


Substanzen gefunden auf Bahnen:
Automatisch (*) / per Video-Integration (v) festgelegter / (.) fehlender Peak.
                           1
Fraktion LS      1234567890123456
  1       18.8 : ****************
  2       35.7 : ****************
Bahnverwendung: aaaaaaaaaaaaaaaa

AK       : (Analysenkennzeichen) 'a' bis 'z' Unbekannte, 1-10 Standardbahnen
LS/Rf    : Laufhöhe relativ Plattenunterkante/Rf-Wert
Kali     : Einwaage, berechnet als Menge pro Fraktion (nur auf Standardbahnen)
X (ber)  : Berechnete Menge  (nur auf Analysenbahnen)
VK (H/F) : Variationskoeffizient [%] (Höhe/Fläche)
CI (H/F) : Vertrauensbereich [%] (bei 95% Sicherheit)
'<' '>'  : ACHTUNG: Dieser Wert ist ausserhalb des Kalibrierbereichs und
           wird deshalb bei der Mittelung nicht berücksichtigt !!











Benutzer    : zw                  16/APR/1996 ID:7CC04100D2F2447     Seite 2
Visum       :                            Geprüft:
```

Document 3 / page 2

Substanz: TMP

Bahn	AK	LS	Kali	Höhe	X(ber)	Fläche	X(ber)
1	a	18.9		74.8	74.77	1406.1	1406.07
2	a	18.8		75.0	75.00	1388.0	1388.02
3	a	18.8		75.7	75.67	1394.6	1394.59
4	a	18.7		75.5	75.51	1419.3	1419.25
5	a	18.7		74.9	74.90	1410.9	1410.92
6	a	18.8		74.4	74.40	1400.2	1400.17
7	a	18.8		76.8	76.81	1452.0	1452.04
8	a	18.7		76.2	76.15	1420.3	1420.28
9	a	18.8		74.8	74.77	1400.8	1400.76
10	a	18.7		75.5	75.47	1429.4	1429.41
11	a	18.7		76.2	76.22	1440.1	1440.08
12	a	18.7		75.8	75.79	1410.9	1410.94
13	a	18.7		76.0	75.98	1417.5	1417.47
14	a	18.8		74.8	74.80	1382.8	1382.82
15	a	18.8		75.1	75.14	1411.5	1411.50
16	a	18.8		73.3	73.29	1362.5	1362.50

Substanz: SMZ

Bahn	AK	LS	Kali	Höhe	X(ber)	Fläche	X(ber)
1	a	36.0		350.4	350.39	7300.5	7300.46
2	a	35.7		351.3	351.29	7243.7	7243.73
3	a	35.7		356.0	356.03	7268.6	7268.60
4	a	35.9		348.7	348.71	7170.6	7170.62
5	a	35.8		348.5	348.47	7174.9	7174.93
6	a	35.7		358.6	358.64	7339.5	7339.54
7	a	35.8		358.8	358.82	7310.4	7310.39
8	a	35.7		358.3	358.31	7262.2	7262.23
9	a	35.7		359.4	359.40	7314.8	7314.82
10	a	35.7		360.9	360.92	7337.6	7337.62
11	a	35.7		358.5	358.51	7283.1	7283.11
12	a	35.6		358.8	358.78	7286.4	7286.38
13	a	35.6		354.9	354.86	7252.4	7252.38
14	a	35.6		348.7	348.70	7127.0	7127.02
15	a	35.6		348.9	348.91	7108.1	7108.10
16	a	35.5		346.5	346.47	6984.2	6984.24

Benutzer : zw **16/APR/1996 ID:7CC04100D2F2447** **Seite 3**
Visum : Geprüft:

Document 3 / page 3

C A M A G TLC Auswerteprogramm
****
Test Version 4.03

TLC/HPTLC-Auswertung (CATS V4.03, S/N:6311A005 / Sc3 V1.12, S/N:020303)

Kalidaten	Kalibrationseingaben von : **zw**
	Dateiname : **EHDLR** **16/APR/96 13:29:10**
Messen	Messung der Rohdaten von : **zw**
	Dateiname : **EHDLR** **16/APR/96 13:47:04**
Integrieren	Integration ausgeführt von : **zw**
	Dateiname : **EHDLR** **16/APR/96 13:47:24**
Berechnen	Benutzer : **zw**
	Datum und Zeit : **16/APR/96 13:47:38**

Kalibrationsmethode... **Reproduzierbarkeit**

Analyse a: Sulfonamide

Substanz	**LS**	**Höhe: Mittel**	**VK [%]**	**n**
TMP	18.8	75.292	1.1	16
SMZ	35.7	354.201	1.4	16

Substanz	**LS**	**Fläche: Mittel**	**VK [%]**	**n**
TMP	18.8	1409.176	1.6	16
SMZ	35.7	7235.260	1.4	16

Benutzer : zw **16/APR/1996 ID:7CC04100D2F2447 Seite 4**
Visum : Geprüft:

Document 3 / page 4

C A M A G TLC Auswerteprogramm
****
Test Version 4.03

TLC/HPTLC-Integration (CATS V4.03, S/N:6311A005 / Sc3 V1.12, S/N:020303)

"Quermessung"

Kalidaten	Kalibrationseingaben von : **zw**		
	Dateiname : **EHDQUER1**	**16/APR/96**	**13:50:40**
Messen	Messung der Rohdaten von : **zw**		
	Dateiname : **EHDQUER1**	**16/APR/96**	**13:53:12**
Integrieren	Integration ausgeführt von : **zw**		
	Dateiname : **EHDQUER1**	**16/APR/96**	**13:54:16**

Analysen- und Chromatographiebedingungen:
Analyse :Bestimmung von Sulfonamiden
Schichtmaterial :HPTLC Kieselgel MERCK 60F 254 20 x 10 cm
Fliessmittel :Dichlormethan - Methanol 80;20
Auftragemethode :ATS3 punktförmig 16 x 1ul
Entwicklungsmethode:Doppeltrog gesättigt

Scannereinstellung:

Plattengrösse (B x H)...		10 x 20 cm
Auftrageposition	Y	: 8.0 mm
Frontposition	Y	: 190.0 mm
Bahnanfang	Y	: 10.0 mm
Bahnende	Y	: 185.0 mm
Startposition	X	: 17.8 mm
Bahnabstand	X	: 17.0 mm
Anzahl Bahnen		: 2
Lampe...		Deuterium
Monochromator-Bandbreite...		20 nm
Wellenlänge		: 285 nm
Spalt (Länge x Breite)...		5.0 x 0.20 mm
Messwertauflösung...		100 μm
Max. Messwert		: 1000 AE
Messmodus...		Absorption / Reflexion...
Messgeschwindigkeit		: 20 mm/s
Optisches Filter		: Oberwellen-Filter
Abgleichmodus		: Automatik
0-Abgleich bei		: 10.0mm, Bahn: 1
Suchscan von		: 10.0mm bis 185.0mm, alle Bahnen
Offset		: 10%
Empfindlichkeit		: Automatik (36)
PM-Hochspannung		: 339 V

Bahnoptimierung...

Bahnoptimierung	: AUS
Optimierungsart...	7 Bahnen x 0.4 mm
Peakschwelle, Höhe	: 100 AE
Peakschwelle, Steigung	: 5

Benutzer : zw **16/APR/1996 ID:7CC04100D361251** **Seite 1**
Visum : Geprüft:

Document 3 (Transverse measurement) / page 5

Integrationsparameter:

Video-Integration	: Nein
Basislinienkorrektur	: Ja
Peakschwelle, Höhe	: 20 AE
Peakschwelle, Fläche	: 100
Peakschwelle, Steigung	: 5
Filterfaktor	: 3
Datenreduktionsfaktor	: 1
Position Bahnanfang	: 19.7 mm
Position Bahnende	:179.8 mm

Bahn 1

Peak	Anfang		Max			Ende		Fläche	
#	**mm**	**H**	**mm**	**H**	**[%]**	**mm**	**H**	**F**	**[%]**
1	22.0	2.0	25.4	76.8	5.78	27.5	5.7	1578.7	5.58
2	32.9	6.6	35.4	82.2	6.18	37.5	6.6	1742.3	6.15
3	43.1	6.4	45.5	83.1	6.25	47.5	7.5	1738.6	6.14
4	53.4	8.7	55.4	84.6	6.37	57.8	8.0	1802.7	6.37
5	63.0	8.1	65.4	85.4	6.42	69.5	8.5	2003.1	7.07
6	72.9	10.2	75.4	85.7	6.44	77.5	11.0	1907.8	6.74
7	83.3	12.7	85.5	87.3	6.57	87.6	11.5	1946.9	6.88
8	93.3	11.3	95.4	87.7	6.60	97.3	11.2	1864.1	6.58
9	103.5	11.4	105.3	86.6	6.51	107.4	10.4	1825.9	6.45
10	113.0	9.6	115.4	85.4	6.43	117.4	8.6	1854.9	6.55
11	123.1	7.1	125.5	83.8	6.30	127.9	8.6	1809.8	6.39
12	132.4	9.9	135.5	85.1	6.40	137.7	10.3	1969.7	6.96
13	143.4	9.3	145.4	83.1	6.25	148.7	6.9	1827.8	6.45
14	153.6	5.8	155.5	79.4	5.98	157.7	4.1	1571.8	5.55
15	163.4	2.1	165.6	76.8	5.77	167.9	1.8	1451.8	5.13
16	173.2	0.8	175.6	76.2	5.73	177.5	0.9	1420.8	5.02

Höhe total = 1329.2 Fläche total = 28316.8

11,57% 100,00%

Bahn 2

Peak	Anfang		Max			Ende		Fläche	
#	**mm**	**H**	**mm**	**H**	**[%]**	**mm**	**H**	**F**	**[%]**
1	22.0	1.9	25.6	234.7	5.83	28.8	4.7	7646.2	5.84
2	32.1	5.8	35.5	245.0	6.08	39.2	6.0	8010.9	6.12
3	42.0	6.6	45.5	246.8	6.13	49.3	6.0	8049.7	6.15
4	51.6	7.5	55.5	250.6	6.22	58.7	8.0	8207.2	6.27
5	61.8	7.4	65.5	251.7	6.25	69.4	8.2	8344.5	6.37
6	71.9	8.5	75.5	256.2	6.36	78.9	10.0	8403.6	6.42
7	81.9	10.8	85.4	259.5	6.44	88.8	12.4	8578.5	6.55
8	91.8	12.2	95.3	258.1	6.41	98.5	13.5	8502.7	6.49
9	101.0	10.5	105.3	258.1	6.41	108.9	12.1	8611.3	6.58
10	111.5	10.1	115.4	256.1	6.36	118.7	9.5	8455.7	6.46
11	122.1	9.3	125.5	254.0	6.31	129.2	10.3	8339.5	6.37
12	132.3	11.4	135.6	255.1	6.33	139.4	8.4	8215.0	6.27
13	141.7	8.2	145.4	254.1	6.31	149.3	6.6	8147.2	6.22
14	152.0	6.0	155.5	245.0	6.08	159.5	4.1	7996.9	6.11
15	162.1	4.1	165.8	248.1	6.16	169.2	1.9	7682.7	5.87
16	172.0	2.3	175.5	254.7	6.32	178.9	0.0	7743.2	5.91

Höhe total = 4027.6 Fläche total = 130935.0 3,78%

2,58%

Benutzer : zw **16/APR/1996 ID:7CC04100D361251** **Seite 2**

Visum : Geprüft:

Document 3 / page 6

7.3 Evaluation Using a Video System

The first publications on the evaluation of thin-layer chromatograms by image processing appeared as early as the mid-1980s [118]. These methods, also known as "digital evaluation", were presented to the scientific world in the lectures of Prošek, who developed his own software for the purpose in 1991 [119], although only the enormous developments in the hardware (video cameras, computers, printers etc.) have made video evaluation of thin-layer chromatograms economically justifiable. Prošek presented his paper on the validation of quantitative TLC by video camera in 1997 [120].

What is "digital photography"?
Digital images consist of information about the intensity and color of light at thousands of separate points on the image. These data consist of numbers, encoded as a series of zeros and ones, that are transferred in the form of electrical impulses to or from the microprocessor of a computer and are then translated back into an image on the computer screen.

No digital image is as good as the human eye's view of an actual object. The smaller the number of stored points of the image (known in computer terminology as pixels) the poorer is its quality.

The second criterion of quality is the depth of color of a digital photograph. This depends on the number of colors encoded: for 256 colors, eight digital characters (bits) per pixel are necessary. A color intensity of 16 bits enables 65 536 colors to be represented, which is known as "high color". An almost photographically realistic representation ("true color") is only possible with 16.7 million colors, and this requires 24 bits per pixel [121].

Modular systems of the latest type for the production of images and documentation of thin-layer chromatograms consist of four components.

- **Illumination system** with the following kinds of light:
  - 254-nm short-wave UV incident light
  - 366-nm long-wave UV incident light
  - 400-nm to 750-nm visible incident light and/or transmitted light

The tubular lamps are driven by 25-kHz to 30-kHz high-frequency voltage. This not only enables the light efficiency to be optimized, but also eliminates synchronization problems with the CCD camera. Furthermore, a light-proof hood and camera bellows can be used to exclude all extraneous light from the system, enabling photographs to be taken even in a room that is not completely darkened.

- **Camera system,** consisting of:
  - 1-chip or, better, 3-chip video camera[1] with zoom objective
  - macro-objective with converter to give a full-format image of a 5 × 5 cm TLC plate
  - supplementary lenses (various focal lengths)
  - UV filter combination
  - adjustable camera stand (e.g. tripod)

- **Data processing equipment.** In order to run the software, the minimum operating system is a PC with a Pentium II processor and a 64-MB memory and Windows 95/98/NT™.

- The **image output system** should consist of:
  - video printer with 256 shades of gray, or
  - color printer, i.e. Windows-driven laser or ink-jet printer. The images which this produces on special paper (720 dpi) have a more limited gray range compared with video prints, but do not require heat-sensitive paper and can be stored almost indefinitely

The items listed under the above four headings can currently be purchased individually without difficulty from specialist suppliers, but it is becoming somewhat more difficult to obtain functional software. We would therefore recommend the purchase of a complete package from a company such as CAMAG (VideoStore) or DESAGA (Densitometer) in which the software has been developed in accordance with GMP guidelines and excludes the possibility of tampering with images. Professional image processing systems automatically "flag up" any images that have been altered. The processed images are referred to as "derivatives" (the same word as that used to describe the results of chemical reactions) [123].

7.3.1 Qualitative Video Evaluation

Many of the illustrations in this book were produced by a video system. Not only can qualitative results be obtained using such systems, but semiquantitative evaluations are also possible from a good video shot of the chromatogram (see Fig. 72 in Section 5.1.2). Practical tips for the preparation of video shots are closely linked to the documentation and given described in detail in Section 8.4.

---

[1] Digital cameras do not contain a film. Instead, a so-called CCD (charge-coupled device) chip takes the picture. This chip consists of separate sensors with red, green and blue filters so that they are sensitive to various colors. The image is not stored by the chip but in a separate memory. The image is then immediately available to be printed out or displayed on a computer screen.

7.3.2 Quantitative Video Evaluation

Both of the manufacturers of video systems mentioned in Section 7.3 are putting a great deal of effort into the development of their own software for the quantitative evaluation of thin-layer chromatograms. The first examples of this type of product were presented at conferences in early 1997. In a paper by Maria Müller, the individual zones of the chromatogram in Fig. 72 were quantified using the DESAGA system. This was performed in two series of measurements, as the individual zones showed concentration differences that were too great. The results obtained from this quantitative video evaluation are given in Table 21, enabling a visual evaluation to be performed. It can be seen that the visual evaluation of zone 1 (known impurity) is in good agreement with the result of the video evaluation. This should correspond with the calibration of this substance on the plate (lanes 1–3). The three unknown substances (Z 2, 3 and 5) were calculated with the aid of the measured values of the limit value concentrations on lanes 4 and 5. The discrepancies between the visual and the video evaluations are sometimes appreciable here.

Table 21: Comparison of semiquantitative visual and quantitative video evaluation of Fig. 72

No. of the zone in Fig. 72	Visual evaluation Lane 6	Lane 7	Video evaluation Lane 6	Lane 7
		declaration in % of 400 µg		
Z 1 = known related compound	< lane 7	< 0,5	0,38	0,44
Z 2 = unknown compound	≈ lane 7	< 0,5	0,12	0,09
Z 3 = unknown compound	trace	< 0,5	–	0,20
Z 4		Zone of the active ingredient		
Z 5 = unknown compound	≈ 0,5	< 0,5	0,63[a)]	0,23
Secondary spots total	< 1,25	< 1,75	1,13	0,96
Unknown compounds total	< 0,75	< 1,25	0,75	0,52

[a)] Limit test acc. to DAB exceeded
Visual evaluation is described in Section 5.1.2 "Detection with 254 nm UV Light".
Comments on video evaluation can be found in Section 7.3.2.

The book accompanying InCom 1996 includes an example of the "Video-Densitometric Evaluation of Thin-Layer Chromatograms" [124]. Thomas Mall concluded that, compared with the usual TLC evaluation methods, the video-densitometric method was

1. quicker,
2. linear over a wider range, and
3. represented an ideal combination of evaluation with documentation.

The evaluation software used was the BASys 1D-PAGE program available from BIOTEC-FISCHER, Reiskirchen (Germany). This program, which was developed for gel electrophoresis, was adapted to the requirements of TLC. The software gives the possibility of performing a so-called "background correction" by subtracting the curve that represents the background, thus compensating for optical effects such as nonuniform illumination of the TLC plate. Figure 97a gives an example of the calibration curve of the substance BM 96.0102 over the range 0.1–1.0 μg **without** background correction, and the same TLC plate **with** background correction is shown for comparison in Fig. 97b. These documents were made available by Thomas Mall.

7.3.3 Comparison of the TLC Scanner With Video Evaluation

A comparison between quantitative evaluation by TLC scanner and by a video system was made by the author. For this, the foil carrying the coffee chromatograms (Fig. 12) was scanned with the CD60, and a video shot, obtained with the VD 40, was evaluated with the ProResult software (all from DESAGA). The results show good agreement between similar samples, but there were differences when the matrices were different. This environmentally friendly TLC was developed especially for teaching in schools (sorbent, TLC aluminum foil RP-18 F_{254s}, solvent system: methanol + water, 60 + 40 v/v). The results are given in Table 22, and a detailed description of the experiments can be found in [124a].

Table 22: Assay of caffeine [%] in various samples of coffee and tea

	Video system	TLC scanner
Tchibo “best beans”	1,18	1,08
Jacobs “light”	0,68	0,71
Jacobs “free”	0,06	0,03
Nestlé “our best”	2,02	1,50
Darjeeling tea	1,68	1,60

Comparison of determinations using a video system and a TLC scanner

Comparative measurements have also been made on a phytopharmaceutical product. In the investigation of ginseng roots and of the medicaments prepared from this, the ginsenoside Rg_1 was determined by both 1-dimensional and 2-dimensional development. The results were presented in a poster in Vienna in September 1998 [124b].

In its house journal (CBS), CAMAG has published many comparisons between densitometric and video evaluation [124c], and has given a clear account of the current scope and limitations of the two systems:

- Classical densitometry utilizes the whole spectral range between 190 nm and 800 nm with high spectral selectivity. Absorption spectra can be produced covering the whole range.
- The video technology in general use today operates exclusively in the visible range. Therefore, substances can only be detected if they are either colored or can be converted into colored derivatives or if they fluoresce in the visible range on excitation by long-wave UV light (365 nm). Indirect detection is also possible on plates with

fluorescence indicators by excitation by UV light (254 nm or 365 nm). It is not possible to produce spectra using video technology.

The comparisons of the various methods of evaluation made in this Chapter are valid up to mid-1999, but the very rapid developments in the multimedia market will lead to further improvements in the future. A summarized comparison of quantitative evaluation methods for TLC up to the present time can be found in [124d].

7.4 Evaluation by Flat-Bed Scanner

More recently, electronic equipment used in other areas of science has been used in TLC for the evaluation of chromatograms. This includes the flat-bed scanner, which is often used in conjunction with special software for the evaluation of DNA electrophoresis results. However, the direct use of flat-bed scanners in TLC is limited to the evaluation of colored chromatograms, and the author is not aware that any work on this theme has yet been published.

7.5 Evaluation Using a Digital Camera

In today's industry, digital cameras have replaced most video systems. All information about the illumination system, data processing equipment and image output system given in Section 7.3 also apply here. Only the current software determines how to achieve a qualitative and/or quantitative evaluation of chromatograms. For the topic "Digital Camera" see "Documentation", Section 8.5.

BIOTEC-FISCHER 1D ANALYSIS REPORT

Linearitaet von BM 96.0102

```
user                 : Hahn
object               : BM 96.0102
comment              : Linearitaet 0.1-1.0ug
report name          : L0051.1DB
station.phase        : Kieselgel 60 F254nm MERCK 5715
mobil phase          : Toluol-Xylol-Dioxan-Isoprop.-NH3 1+1+3+3+2
method of detection: UV 254 nm
method of measure    : ohne Untergrundkorrektur
path length          : 10 cm

scalefactor (multiply) : -5    actual scanlength   (mm): 200
min.peak-height    (%) : 2    min. peak-width  (pixel): 3
aut.baselinecorrection :       y          CV = 0.9448
```

number [ ]	position [mm]	area [pixel]	height [%]	%area [%]	absolute val. [ ]
5	43.9024	662	14.1	6.9	302.74
7	58.5366	701	14.9	7.3	330.98
8	74.6055	645	14.9	6.7	290.43
9	88.9527	1076	20.8	11.1	602.49
10	103.8737	1057	22.4	10.9	588.74
11	119.3687	1034	22.4	10.7	572.08
12	134.5768	1467	28.2	15.2	885.59
13	150.0717	1497	30.2	15.5	907.32
14	165.5667	1514	32.9	15.7	919.62
		9653		100.0	5400.01

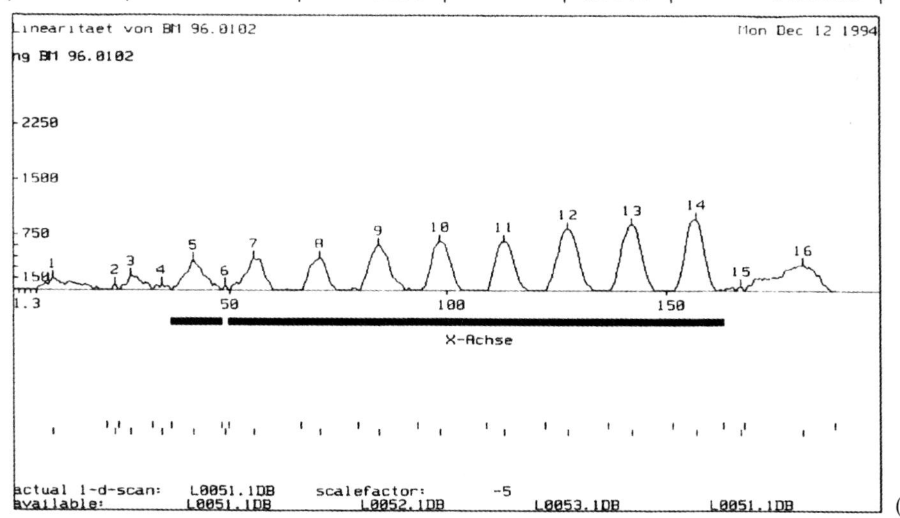

(a)

Figure 97. Direct optical evaluation with a video system. BASys software: 1D-page program (BIOTEC-FISCHER)
(a) Calibration curve of the substance BM 96.0102 (0.1–1.0 μg) **without** background correction
(b) The same plate as that in (a), but **with** background correction

BIOTEC-FISCHER 1D ANALYSIS REPORT

Linearitaet BM 96.0102

user : Hahn
object : BM 96.0102
comment : Linearitaet0.1-1.0ug
report name : L0054.1DB
station.phase : Kieselgel 60 F254nm MERCK 5715
mobil phase : Toluol-Xylol-Dioxan-Isoprop.-NH3 1+1+3+3+2
method of detection: UV 254 nm
method of measure : mit Untergrundkorrektur
path length : 10 cm

scalefactor (multiply) : 10 actual scanlength (mm): 200
min.peak-height (%) : 2 min. peak-width (pixel): 3
aut.baselinecorrection : y CV = 0.9974

number [ ]	position [mm]	area [pixel]	height [%]	%area [%]	absolute val. [ ]
3	28.6944	204	6.7	1.6	116.10
4	43.9024	443	14.5	3.5	212.75
5	58.5366	660	20.4	5.2	300.51
6	74.6055	804	24.3	6.3	358.75
7	88.9527	1138	32.9	8.9	493.83
8	104.4476	1439	37.6	11.3	615.56
9	119.3687	1602	42.0	12.5	681.48
10	134.5768	1944	47.1	15.2	819.79
11	149.7848	2201	52.9	17.2	923.73
12	165.2798	2334	55.7	18.3	977.51
		12769		100.0	5500.01

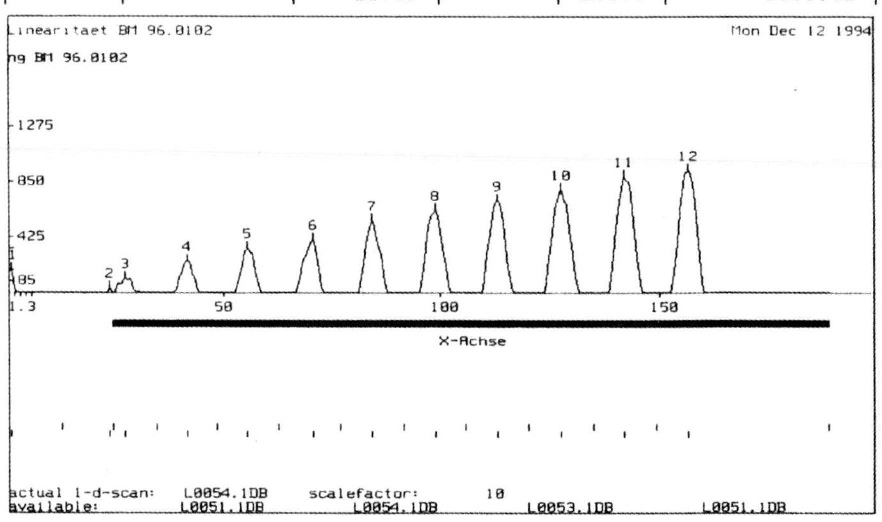

(b)

Figure 97 b

8 Documentation

The main theme of this Chapter is the documentation of a TLC plate, i.e. the "capture" and storage of the result with the object of being able to reconstruct it. Methods used can range from simple descriptions of chromatograms to modern recording systems based on electronic data processing.

8.1 Description of a Thin-Layer Chromatogram

In accordance with methods prescribed in pharmacopoeias, evaluations of thin-layer chromatograms are often performed by simple descriptions even today. As the "hRf value" is not used in the pharmacopoeias, circumlocutions which can hardly claim to be scientific are often used. An example of this was found in the data given with a commercial sample of expressed juice of red coneflowers (*Echinaceae purpureae*). The description of this chromatogram, after derivatization with the natural products reagent and observation under long-wave UV light, was as follows:
"The strongly fluorescent chlorogenic acid spot is at the top end of the bottom third of the Rf range, and the bright spot due to caffeic acid is in the top third. A fluorescent spot appears at the height of the caffeic acid. Between the caffeic acid and the chlorogenic acid are two broad strongly fluorescent spots. Just below the chlorogenic acid there is also a spot with a yellow-green fluorescence. Several weak zones can be present between the starting line and the chlorogenic acid."

As an alternative possibility, the chromatogram obtained by the author in the course of investigations of the sample has been described with use of the hRf values, and this description is given below. A small discrepancy in the chlorogenic acid region was found (Fig. 98).

▶ **Figure 98:** see Photograph Section.

The light blue fluorescent zones of chlorogenic acid (hRf value 18–23) and caffeic acid (hRf value 76–81) can be seen on the reference lane in long-wave UV light. On the sample lanes at various points between the starting line and the solvent front are a large number of zones with a bright blue fluorescence. The five strongest zones have the following hRf values (in decreasing order of intensity):

Applied Thin-Layer Chromatography: Best Practice and Avoidance of Mistakes, 2nd Edition
Edited by Elke Hahn-Deinstrop

ISBN: 978-3-527-31553-6

Zone intensity 1: 24–30
Zone intensity 2: 52–56
Zone intensity 3: 60–64
Zone intensity 4: 77–81 corresponding to caffeic acid
Zone intensity 5: 84–87

8.2 Documentation by Drawing, Tracing and Photocopying

Even today, thin-layer chromatograms are often drawn or traced on tracing paper, and the zones are then colored with colored pencils (Fig. 99). These methods are time consuming, and the fine color nuances can seldom be accurately represented. Nevertheless, these methods of documentation are perfectly acceptable for legal purposes [125].

▶ **Figure 99:** see Photograph Section.

To illustrate this theme, the cartoon below shows a young man somewhat laboriously tracing thin-layer chromatograms, while the girl is using modern video equipment for the documentation of a TLC plate.

Photocopiers are also used for documenting TLC results. However, as color copiers are rarely available, the colored chromatograms are reproduced in black and white only. It is also a disadvantage that documentation directly under UV light is not possible.

8.3 Photographic Documentation

Photography has been successfully used for many years for TLC documentation. For all photographs of thin-layer chromatograms it is essential to secure the camera firmly to a tripod or stand, as only by keeping the camera still and in the correct alignment with the TLC plate can a precise and sharp picture be ensured. The equipment should be set up in a dark room if it is necessary to have the greatest possible flexibility with respect to accessibility to the object being photographed. Systems with a modular construction for photographic documentation are supplied by CAMAG and DESAGA.

8.3.1 Photography Using the Polaroid Camera MP-4

Although instant films have been improved in quality during recent years, they do not completely fulfill the essential requirements for the documentation of thin-layer chromatograms. Fine zones on the pictures obtained cannot be seen in UV light, and in daylight photographs the reproduced colors do not correspond to the original. In long-wave UV light, red fluorescence cannot be reproduced even though various filter combinations are used. A more important disadvantage of instant films is that only one print is obtained. Extra prints, e.g. of testing procedures or licensing documents, cannot be obtained retrospectively. For this reason no additional tips for photographing with Polaroid cameras are given here.

8.3.2 Photography Using 35-mm Cameras

The best results for the documentation of thin-layer chromatograms are obtained using "normal" photography. The most suitable cameras are all single-lens reflex cameras with suitable macro-objectives and provided with by a proven filter combination for the UV range [126]. To obtain optimum results, three camera housings ("bodies") are required, as different types of film must be used to obtain the best results for each detection region. By using a standard method of photography developed by the author [127], even TLC novices can produce good photo-documents (*NB: provided there is a film in the camera* – see the cartoon below), without the necessity for any great previous knowledge of the technique of photography. The parameters needed for photography with single-lens reflex cameras are given in Table 23.

Table 23: Photographic documentation using single-lens reflex cameras

	Copying stand Light source Aperture	1 UV 254 nm 5,6	2 UV 366 nm 5,6	3 VIS 5000 K 8
		Filter and exposure time (s)		
Camera I	**Monochrome film (b/w)**	+ 2 UV filter		
	Greenish fluorescence	0,5; 1; 2	–	–
	Bluish fluorescence	0,25; 0,5; 1	–	–
Camera II	**Color transparency film 200 ISO**		+ 2 UV filter	
	Weak fluorescence	–	8–16	–
	Strong fluorescence	–	2; 4; 8	–
Camera III	**Color negative film 125 ISO**	–	–	No UV filter
	Color transparency film 125 ISO	–	–	All shots 1/8

For all shots: lens ML Macro/2.8, 55 mm

8.3.2.1 Photography in 254-nm UV Light

When detections are performed in short-wave UV light using precoated layers containing a fluorescence indicator, the substances appear as dark zones on a bright yellow-green or pale blue background. We are thus not concerned with the formation of different colors, but only with the accurate representation of small differences in brightness. For this purpose, the “hard-working” document films are candidates. Using the Agfa-Ortho 25 Professional (15 DIN), for example, it is even possible to represent traces of substances that can hardly be seen with the naked eye so that they are clearly visible. The following photographs in this book were taken with this film: 16a, 16b, 51a, 51b, 57a, 57b, 61, 73b etc.

8.3.2.2 Photographs in 365-nm UV Light

On illumination with long-wave UV light, the fluorescent zones appear bright on a dark background. Here, there can be a broad palette of colors. Very great demands with respect to true representation of color are therefore made on the film material. The representation of differences in brightness is here of only minor importance. All the professional color negative films (e.g. Agfacolor XRS 100, 21 DIN) are suitable because of their brilliant color reproduction. However, extremely good images are also obtained using diapositive films, which can be printed as "digiprints" on photographic paper for inclusion in testing procedures or other documents. The following photographs in this book were produced with the diapositive film Agfa CTX 200 (24 DIN): 38, 45a–d, 56b, 68, 73a, 73c, 92 etc.

8.3.2.3 Photographs in White Light

Photographs taken in natural daylight can give optimum results, but do not give reproducible photographs, as the spectral composition of this light changes with the time of day. It is therefore not possible to work in this way continuously. Flashlight equipment is unsuitable because of the reflections on the images. However, if white light fluorescent tubes with a light intensity of 5000 K are used, accurate documents representing thin-layer chromatograms are obtained with commercial diapositive or negative films. The following photographs in this book were produced using the RB 5000 reproduction equipment (Kaiser Fototechnik): 6, 46a, 46b, 47a–c, 62, 83, 86, 94 etc.

☞ **Practical Tips** for photography using 35-mm cameras:

- First find a qualified photographic business in the neighborhood of the workplace (or your home) where black-and-white processing is performed. You can then call in on your daily journey home. For very urgent work, it should be possible for a black-and-white film delivered at about 6 p.m. to be developed and printed and then collected by 10 a.m. the following day. Color films are developed and printed overnight in a professional laboratory and should also be ready by ca. 10 a.m. If extra copies are required later, a little more time should be allowed.
- In both UV ranges, it is advisable to take a series of exposures with increasing exposure times, e.g., 0.5 s, 1 s and 2 s in short-wave UV light, using the same part of the plate. The photographic laboratory will then produce a black-and-white print on matt paper from the negative with the highest contrast.
- For photographs in 254-nm UV light, the plates should be marked with a soft pencil if this is necessary (see Figs. 11a and 11b, which are marked with the code numbers of a TLC plate and an HPTLC plate. In short-wave UV light, the dark marking on a bright background can easily be seen).
- For very weak zones (e.g. limit of detection), at least two sections of a plate should be photographed at various distances of the camera from the plate (e.g. a photograph of the whole plate and then just that area containing the zones at the limit of detection).

- After loading a new film for the 365-nm UV light, this should first be exposed using visible light so that the automatic developing machine in the professional laboratory can detect the start of the film and will not cut off the first and subsequent negatives.
- For photographs in long-wave UV light, the plates can be marked with a bright green fluorescent marker (e.g. Color-Lites of Faber-Castell, see Fig. 84).
- When GLP-coded plates are photographed in long-wave UV light, a narrow strip of white paper should be placed underneath. The markings then appear bright on a dark background (see Fig. 23b).
- Any small fluorescent particles should be removed from the plate using a soft brush.
- Stray light on photographs should be avoided (white laboratory coats!).
- For photographs in long-wave UV light, the exposure time depends on the intensity of the fluorescences. Here also, it is advisable to take a series of photographs. Either a normal print or a "digiprint" (Agfa system) is produced from the best diapositive of the series. The latter is more brilliant than a paper print.
- It is often necessary to document TLC plates that have been kept in an iodine chamber for a long time (up to 20 h). Limit value concentrations of a wide range of substances are often visualized in this way. When handling these plates, disposable gloves should always be worn, the plates must be covered with a clean glass plate for transportation, and the document table of the camera is protected by a paper towel. If the plate has a dark brown color, a preliminary photograph is first taken (only removing the protective glass plate to take the photograph!), the plate is placed in a fume cupboard for a few minutes and a second photograph is taken. If the result is still unsatisfactory, this procedure can be repeated several times.
- Prints of color negative films produced in a professional laboratory can vary in appearance, especially if there are only a small number of colored zones visible on an almost white background. In this case, a home-made colored strip can be very useful. This is placed at the edge of the plate and is photographed with the plate. In the case of paper prints, the colored strips can be cut off at a later stage if they are not required. A colored strip is also useful for TLC plates derivatized with iodine or coated with blood gelatin, as in these cases a similar color strip can be placed with the negatives if copies are ordered later, so giving prints of similar quality.

To give a "finishing touch" to the photographs, the following tips can be useful:

- To avoid documentation of dirty marks on the photographs as far as possible, prewashed prepared coatings should be used.
- For important documentations the zone width should be >15 mm for optical reasons.
- Also for optical reasons, 40–50 % of the total area should be covered with chromatogram zones.

- Special care should be used in the production of prints for business brochures, posters, etc. It is worth mentioning here that this kind of work is not possible in a routine laboratory, as it requires peace and quiet, time and the complete absence of interruptions by third parties! The photograph illustrated in Fig. 100 "Identification of dried extract of primula root and liquorice root in Heumann bronchial herbal tea Solubifix®" was produced as follows. The reproduction equipment RB 5000 was placed in a fume cupboard and the hotplate was placed on the table. After loading the color film, the camera was mounted and correctly adjusted with the aid of a plate that had already been heat treated and a (switched on) light source. The TLC plate, from which all solvent residues had been removed, was then sprayed homogeneously with the vanillin-sulfuric acid reagent and placed on the plate heater, which was still cold. This was then switched on (the set temperature being 120 °C). After a few seconds, when the first colored zones became visible, the series of photographs was started. A "shot" was taken every 10 seconds. At the end of the series, all zones showed signs of "overheating". When the colored paper prints were obtained, the best negative was used for the artwork.

▶ **Figure 100:** see Photograph Section.

8.3.3 Archiving of 35-mm Films

Negatives and diapositives must be retrievable quickly and reliably, e.g., for the production of extra prints for licensing documents, testing procedures etc. The films must therefore be carefully documented and archived. For this, it is useful to have forms with the following information:

- a unique archive number for each film
- film type
- date of the first and subsequent references to the film at the laboratory
- image number
- light source
- subject (e.g.: "Marshmallow, plate 91")
- derivatization
- name of user
- any other information

As an example, an extract from the archive sheet containing the entry for the photograph "bronchial herbal tea" (see above and Fig. 101) is given below.

Film Archive No.: 161 ☒ Color negative film Date: 16.4.87
O Color transparency film

Shot No.	Light source	Subject	Plate No/s	Test	Remark	User
0	—	—				
1	white	Bronchialtee Solubifix	50/10	Identity	for cataloge	HD
2	"	(medical herb tea)	15	↓	↓	↓
3	"	"	20	↓	↓	↓

%

Shot No.	Light source	Subject	Plate No/s	Test	Remark	User
30	white	Solubifix	50/290	Identity	for cataloge	HD
31	"	"	300	↓	↓	↓
* 32	"	"	310	↓	↓	↓
33	"	"	320	↓	↓	↓

/s = time on the plate heater
* = best shot, 310 s

Figure 101. Extract from the film archive sheet for color negative films
Documentation of the films using the example of Figure 100

8.4 Video Documentation

As with photography, the following rule applies to video shots:

A good document can only be produced from a good presentation!

Therefore, most of the practical tips from Section 8.3.2 apply here also.

So what is different about video documentation?
The TLC plate (or any other object) which is to be documented is first displayed on the monitor as a "live" picture. All the adjustments can then be made until the optimum is reached. The image is then fixed or "frozen". A frozen image can then be provided with a captions horizontally and/or vertically, and lines, arrows, circles etc. can be added. When the image is stored, a commentary can be added, although this can only

be read later on the hard copy from a color printer. Storage of the image is in a previously assigned file, so that a "photo" can be produced by the video printer or an image of the TLC plate can be printed out from the color printer together with any added commentary. The small amount of time needed for a video shot is its greatest advantage. In the ideal case, for the production, storage and printing of the shot, only ca. 2 min is necessary, so that the image is delivered by the video printer after ca. 75 s. The time required is a little longer if the adjustments of the camera and light source have to be optimized, a commentary is included and a hard copy is desired. Nevertheless, a maximum of 15 min is normally sufficient for a video shot. The time-consuming journey to the photographer, the allocation of images to negatives and the arrangement of the film archive are all eliminated.

However, video equipment also demands care. For example, the stored images must from time to time be deleted from the hard disk, as the space required for a video shot is ca. 0.85 MB, which means that a standard 3.5 inch diskette has room for one image (bmp file) only! Therefore, special diskettes are used for archiving, e.g. "magneto-optical disks" with a storage capacity of 128 MB. As many as 149 bmp files can be copied onto such an mo disk. A list of the contents of this disk can then be printed out (the so-called catalog), which then represents "the archive". The time requirement for this naturally depends on the number of images, but should not be underestimated. To format two mo disks and then to copy 2 × 149 data files can require 2.5–3 h. It is true that the copying process is automatic, but the possibility of problems must be allowed for. This means that the functioning of the equipment must be continually monitored.

In all cases, a short but easily understandable operating manual for the video equipment is essential, beginning and ending with switching the electric power on or off! A users' book (with numbered pages) should also be provided together with an SOP for the registration and archiving of video shots. An example of an SOP for the "Registration, Printing and Archiving of Video Shots" is given below (Document 4). This includes sample pages from the users' book and catalog, sample video prints and sample computer printouts (hard copy).

STANDARD OPERATING PROCEDURE

No. HD/024/95/1

Function	Date of implementation	Date of last revision	Issue date	Page
Quality Control	4th March 1999	22.11.1995	24.02.1999	1/4

Title

Registration, printing and archiving of video shots

1. SUBJECT

Registration, printing and archiving of video shots produced by DESAGA VD 40 equipment for the following applications:

a. Thin-Layer chromatography
b. Microscopy
c. Others

2. SCOPE

All departments of Quality Control, Research and Method Development

3. RESPONSIBLE PERSONS

The person who produces the video image has the responsibility for correctly recording and printing it. There is a named person in each department who is specially trained in the use of the video equipment and who acts on behalf of all his or her colleagues. Mrs. Lena Bub is responsible for the archiving, and Dr. Peter Remmert has been nominated as her deputy.

4. MATERIALS

A User Register is kept next to the VD 40 equipment in the QC darkroom. Every video shot that is to be stored must be entered in this book. The pages of the User Register are numbered consecutively, beginning with p. 000241 in Book 5, p. 000301 in Book 6 and p. 000361 in Book 7.

Applicable to: Central Documentation, Quality Control, Darkroom

Written by Quality Control	Approved by Quality Control
Lena Bub 24.2.99	Dr. Peter Remmert 26.2.99
Approved by Head of Quality Control	**Executive Approval by Quality Assurance**
Dr. D. Koch 1.3.99	Dr. Friederike Huse 3.3.99

ORIGINAL 5. März 1999

STANDARD OPERATING PROCEDURE **No. HD/024/94/3** **Page 2/4**

Title **Registration, printing and archiving of video shots**

---

5. REGISTRATION

Entries in the User Register

a. Every video shot to be stored receives a sequential number consisting of the year (last two digits) and 5 further digits

e.g. **9901806**

b. The date is then given in the German style, omitting the year

e.g. **01.09**

c. The description of the sample consists of the name of the substance or preparation, the batch number if appropriate, any other information, and the light source used

e.g. **Olibanum, various essential oils, vanillin-sulfuric acid, spot light**

d. In the space marked "stored Y/N", the letter

Y for Yes

must be entered. The entry "**N**" is only made for video shots that are not to be stored or for a repeat printout when the number of copies is included.

e. The name of the Department must be recorded for statistical reasons:

C : Quality Control Chemistry
PP : Quality Control Pharmaceutical Products
RM : Quality Control Pharmaceutical Raw Materials
MD : Method development
R : Research
L : Licence and official affairs

f. The initials of the technician are entered in the Table. (In the video system, this is stored under "User"!)

g. The CD number is only entered after copying onto a "magneto-optical" diskette

h. "Copied by/on" is entered by the person responsible for archiving, with initials and date.

Document 4 / page 2

STANDARD OPERATING PROCEDURE **No. HD/024/94/3** **Page 3/4**

Title **Registration, printing and archiving of video shots**

6. REPRINTS

If further printouts are required, e.g. for SOPs, testing procedures, drugmaster files etc., the image is again called up from the relevant CD, and the required number of prints or color images are printed out. For this, new sequential numbers are **not** given, and the number of printouts is entered under the ordering Department in the User Register. The abbreviation "Docu" is used for the Central Documentation Department.

7. PRINTOUTS

a. Videoprints from the Mitsubishi Videoprinter

"Videoprint" is clicked in the "File" menu. The printer starts to operate immediately and requires 75 seconds for each image. The finished printout contains the following legends:

In the heading (white on a dark blue background),

"DESAGA ProViDoc"

is followed by the date and time of storage and the name of the "User":

e.g. **01.09.1999/ 17:11:57 - Hahn-Deinstrop**

The text that was stored appears in the menu line in black letters on a white background. In the right-hand part of this line is the Windows/DOS filename:

e.g. **-00170B20.BMP**

The image then follows with any text, as shown below, i.e. as it was displayed on the monitor.
(*Image No. 9901806 is used as the example. The DOS filename is cut off here.*)

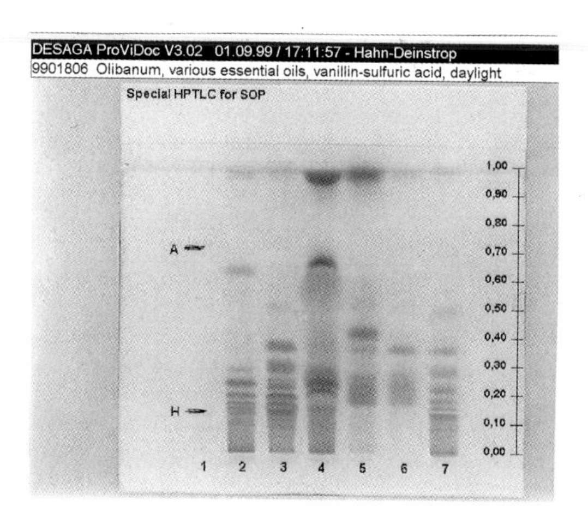

STANDARD OPERATING PROCEDURE No. HD/024/94/3 **Page 4/4**

Title **Registration, printing and archiving of video shots**

b. **Printouts produced by the Epson color printer**

Clicking on "Print" in the "Document" menu gives the screen page on which the quality of the printout (on Epson paper always 720 dpi) and number of copies are selected. Approximately 7 minutes are required for 1 printout in color. This printout is suitable as a raw data sheet. The heading is:

DESAGA Videodocumentation 'ProVoDoc'

The actual image has the fixed dimensions 16 × 11 cm. The filename appears above the image, and below it are "Date", "User", "Time" and "Path". The heading "Comments" follows. This appears every time, even if no comments have been added.
Supplement 2 shows the printout of the sample image 9901806.

8. ARCHIVING

a. **Catalog**

By clicking "Catalog" in the "Document" menu, a list of all the stored files for the plate **can** be obtained. The printout has the following headings:

Data carrier: MS_DOS_6
Path: c:\VIDEOBLD

After copying to the magneto-optical disk, a catalog printout for this diskette is prepared. The heading includes the following lines:

DESAGA Videodocumentation 'ProViDoc'
Data carrier:
Path: D:\HD-E

The sample image used here is No. 00015 on p. 1. The sequential number of the data carrier (here 5) is entered by hand on the sheet (see Supplement 3).

The catalog pages of all mo-disks are kept in a file which is also in the darkroom and should be regarded as working material.

b. **Data Backup**

Backup diskettes are produced from the mo-disks and kept in safe storage. Copies of the catalog pages are produced, placed in a file, and stored as "The Archive" by Mrs. Bub.

Document 4 / page 4

Supplement 1, SOP: HD/024/95/1

Sequential Number	Date	Subject (Substance, Lot etc.)	Saved Yes/ No	Number Videoprints/ Paperprints	Department	User	CD-No.	Copied by / on
9901806	01.09.	Olibanum, var. ess. oils, vsa, spot l.	Y	2/1	MD	HD	5	LB 3.9.

Supplement 3, SOP: HD/024/95/1

D E S A G A Videodocumentation 'ProViDoc'

Data carrier:
Path: D:\HD-E

Seq. No.	Filename/DOS-Name	Date	Time	User	Page 1 - 8
1 – 14					
00015	**9901806 Olibanum, various essential oils, vanillin-sulfuric-acid, spot light**				
	00170B20.BMP	01.09.1999	17:11:57	Hahn-Deinstrop	
16 –					

Document 4 / page 5 (Supplements 1 and 3)

Supplement 2, SOP :HD/024/95/1

DESAGA Videodocumentation 'ProViDoc'

Filename: 9901806 Olibanum, various essential oils, vanillin-sulfuric acid, spot light

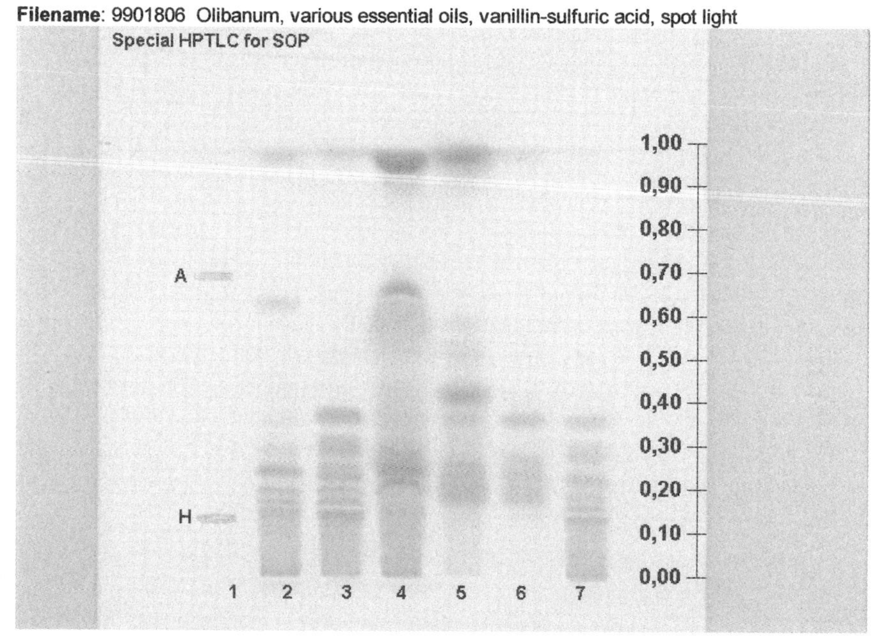

Date:	01.09.1999	**User:**	Hahn-Deinstrop
Time:	17:11:57	**Path:**	D:\HD-E\00170B20.BMP

Comments:

HPTLC of various essential oils in medicinal incense (Olibanum)

Equipment:	TLC Applicator AS 30 (DESAGA)
Bandwise:	6 mm
Sorbent:	HPTLC LiChrospher® Si 60 F_{254s} (Merck No. 1.15445) ½ plate 10 x 10 cm
Solvent system:	n-heptane + diethyl ether + formic acid (7+3+0,3 ml)
Type of chamber:	vertical tank
Chamber saturation:	no
Migration distance:	7 cm
Developing time:	12 min
Room temperature:	23 °C
Relative humidity:	58%
Derivatisation:	vanillin-sulfuric acid, heating 5 min 120 °C

Lane 1:	Anethol (A), Hydrochinon (H)
2:	Olibanum resin (Caelo), extract of Boswellia carteri
3:	Dosage form tablet: H 15
4:	Essential oil (resin from Egypt), Boswellia carteri
5:	Essential oil (resin from East Afrika)
6:	Boswellia serrata, essential oil
7:	Boswellia serrata, resin

01. Sep. 1999

HD

02.09.99 Bo

Document 4 / page 6 (Supplement 2) (Printed by Epson color printer on 720 dpi paper)

☞ **Practical Tips** for working with video equipment:

- If the video equipment has a number of users, all should be be made fully aware of how it should be used and cared for.
- If the purchase of video equipment for the documentation of TLC plates is under consideration, it is advisable only to recommend the current "best" solution, which must necessarily be expensive. Otherwise, do not become involved! At the present time, the "best" solution means at least a 3-chip camera and a PC with a Pentium II processor, with Windows XP™ as the operating system.
- If the available video equipment has a 1-chip camera with no facility for time exposures, some "tricks" are necessary to acquire video images of chromatograms of e.g. plant constituents in the 365-nm UV range. The camera requires some "residual light", i.e. it can only record the chromatogram if a certain amount of normal light is added. However, the colors then do not always match the original. However, the addition of residual light only functions if the equipment is operated **without** a closed exposure system (e.g. with an older model of a double swivel arm UV lamp, DESAGA) in a dark room. If the new cabinets of the CAB UVIS (DESAGA) or the Reprostar 3 (CAMAG), which have high-frequency lamps, are used, the results obtained with addition of residual light are no better.
- If the 1-chip camera is used, weaker fluorescences in long-wave UV light are sometimes not visible on the monitor at all. However, an attempt should be made to print out the image, as the monitor picture does not always correspond with the video print. If this only produces a black image, the documentation must be performed using a 35-mm camera with a diapositive film.
- 1-Chip cameras can always be used with short-wave UV light or daylight. For video shots in the 254-nm UV range, it is advisable to take the color out of the system in order to increase the contrast of the images.
- Even video shots in daylight require some intuitive feel to obtain a true color representation in the print-outs. After the equipment has been installed by the suppliers it is a good idea to have a multi-colored pattern always at hand which can be used as a "calibration card" over a long period of time. For this, you will not need to buy an expensive color chart (which in any case will usually be too large) from a photographic supplier. The settings of the video equipment can be checked e.g. with a perfume box (Nocturnes de Caron). This "color card" is 3.3 cm long and is always useful if a "gremlin" has caused mechanical changes to the camera.
- "Changes to the settings of the video equipment" is a particularly awkward subject, especially if several persons from various departments use the equipment. Access to these settings can certainly be controlled by means of a password, but this is only possible if there is only **one** camera. There could be second camera controlled by the PC via a second interface and linked, for example, to a microscope. Thus, to switch from the "plate camera" to the "microscope camera", the interface must be specified in the "set-up" menu (with the VD 40 equipment of DESAGA). Here, the only person who would not need to give a password would be the person responsible for servicing the equipment, as he or she must always be available for this work. To be

able to make a rapid check for any changes to the settings, it is advisable to keep a security diskette on which the optimal parameters, determined by the servicing technician, are stored.

- Video equipment can naturally be used for other purposes as well as the documentation of TLC plates. For this, a macro-objective is very useful if small objects are to be documented. For example, air inclusions in glass bottles, sharp edges on a plastic lid, black particles in a white paste and misprints on hard gelatin capsules can be recorded using a macro-objective VT B5528MA (f 2.8 / 55 mm; 2/3 inch C-Mount, Article No. 42085, Mutsubishi). For further enlargements, a Tele-Converter (VT 2x-CON, Article No. 42087, Mitsubishi) is inserted. Thus, for example, a 5-mm diameter tablet can completely fill the screen of the monitor.
- All mo disks should be numbered consecutively and stored near to the video equipment so that each user will have access to his or her stored files. However, before these are deleted from the hard disk, it is **strongly** recommended that a security diskette should be prepared and that this should be stored separately and securely (e.g. in a safe).
- The system of providing a unique number for the storage of each video shot should be repeated in the TLC raw data sheets, i.e., the archive number should also be recorded there. This is the only way to be able to produce extra printouts of a file even if the image itself is not (any longer) available on these raw data sheets. (See also Section 9.3 "GMP/GLP-Conforming Raw Data Sheets.)
- The latest (i.e. mid-1999) software for the DESAGA video documentation system for Windows 98 and NT, provided that the latest hardware is also used, is even better oriented to the need of the user. The power of the latest CPUs enables more and better functions to be used to increase the quality of the image. An example of this is the "noise suppression" function used for the digital image. Results with video shots in long-wave UV light are particularly impressive, as, if fluorescences are weak, photographic documents can be obtained simply by adding together individual images.

For example, the new software provides interfaces for several different CCD cameras (1- or 3-chip cameras with and without hardware integration). These can be directly controlled, and their status with respect to GMP/GLP-conforming documentation and improved reproducibility can be read out. A useful aid in the assessment of TLCs is the *R*f scale, which can be superimposed on the image in any desired position by specifying the positions of the start and the solvent front [127a].

8.5 Documentation With Digital Cameras

Using digital cameras is also a type of photographic documentation. Nearly all advice for practical use given in Section 8.3 apply here too. However digital cameras need a computer-controlled system, and likewise when working with video cameras. Therefore most of the advice given in Section 8.4 also applies here.

A modern computer-controlled system must fulfil the following requirements:

- An image acquisition model with the best possible resolution, high display quality and adequate speed, which can operate on all types of subjects.
- Easy to operate software that is adapted to scientific requirements and avoids all unnecessary add-on programs.
- A uniform illumination with visible and UV-light at 254 and 365 nm.
- A documentation module that reproduces even the finest nuances correctly, without loss of details and in true color.

CAMAG and DESAGA started the era of using digital photography for TLC documentation with simple point and shoot cameras, followed by non-linear 8 bit cameras. Besides snapshots they delivered useful pictures in daylight and at a 254 nm wavelength. For actively fluorescent substances, especially those with very low intensity, a scientific camera is recommended.

Both manufacturers of TLC documentation systems, CAMAG and DESAGA, have developed their own software (winCATS® rsp. ProViDoc®). They work similarly: the software supports the GxP editing and problem-free archiving of the images that have been captured. The image displayed on the monitor can be edited, labeled and marked. Each image is automatically labeled with the date, user name and an unequivocal identification number and therefore, is stored in conformity with GxP guidelines – it can even be protected by a password, if desired. The user can append a commentary of any desired length to each image. This is stored directly with the image and can be printed out with any laser or inkjet printer that is supported by Windows™.

For image capture such a system uses a high resolution 12 bit CCD-camera with a 12 mm lens for objects up to 200 × 200 mm. Alternatively for high resolution pictures of smaller objects up to 100 × 100 mm there is a 24 mm lens available. There is also a 16 mm lens for objects up to 200 × 100 mm. The photography is controlled by a PC which is connected to the camera via firewire (IEEE 1394a interface). The camera with a resolution of 3.2 million pixels produces accurately detailed images of real distinctive value with excellent color brilliance. The high light sensitivity and maximum exposure time of 67 s makes it possible to record weak fluorescence, partly invisible to the human eye.

The following photographs in this book were produced using a 12 bit Digital camera-system from CAMAG: 8, 9a and 9b.

Practical tips for working with a digital camera system:

- Users who analyze at 254 nm in black and white require software for the digital camera system which can convert to black and white.
- Sometimes only a part of the TLC plate is of interest. With the right software the image size can be altered by adding a marquee around the area of interest. Then a magnified picture is taken only from this area.
- If there are several users for one system it is advisable to use a log book with the system. Otherwise there is no way of knowing who used the system and in which order it was used after changing storage media or saving the files on different servers within a firm or institute. This is especially important if a user has found an error in the system.
- The data security must be guaranteed for the time given, according to the legislation set down for the information technology system.

8.6 TLC Scanner Documentation

After an evaluation by a TLC scanner, the complete data record, which is in the form of a number of hard-copy pages, represents the main part of the whole documentation. Different software producers emphasize different aspects of the documentation, so that the arrangement of these pages can vary. In Section 7.2 "Evaluation using a TLC Scanner", examples from various software producers are given. It is obvious that an inexperienced person can have problems in the assessment of analytical results if he or she does not know where exactly these are to be found in the data record produced by any given type of TLC scanner. The producers of evaluation software have provided comprehensive descriptions in conformity with the GMP guidelines, e.g., in the case of "CATS" by CAMAG [128]. However, I do not think that the data record is rendered any easier to follow as a result, and, moreover, it does not reflect all the operations performed in the work. In accordance with GMP, if several TLC scanners are used, the reference number of the laboratory equipment should be clearly documented, as also should the type of chamber used in the development. The record of the weighings (printed paper strips) of reference substance(s) and samples should be included with the scanner documentation. Also, it is advisable to include a summary of the results on the cover sheet of the TLC scanner documentation. An example of such a cover sheet is given in Section 9.3 (raw data sheet "TLC V").

8.7 Flat-Bed Scanner Documentation

Documentations obtained when evaluation is performed with a flat-bed scanner are always dependent on the software used and also, of course, on the scanned object. As mentioned in Section 7.4, a flat-bed scanner can be directly used for processing colored chromatograms only. The first investigations were done with the Pharmacia system, which was also used for the evaluation of DNA electrophoresis gels.

A flat-bed scanner, in conjunction with a PC and a color printer, can therefore always be used in the documentation of thin-layer chromatograms if this involves electronic storage and the inclusion of color reproductions of chromatograms in documents. For example, if the paper image obtained from a 35-mm film of a fluorescent TLC plate scanned under 365-nm UV light is transformed into an electronic data file, as many color reproductions of this plate as required can be included in documentation records [129]. The laborious process of sticking photographs into individual pages is therefore avoided.

A flat-bed scanner, with the appropriate PC software, can be a useful aid in substance identification. Colored chromatograms, such as those obtained by 2-dimensional TLC of ginseng roots, can be scanned and then inverted. This procedure gives both blue-gray-brown spots on a white background and spots in various yellow and green tones on a black background. The latter are better for differentiating than those obtained without inversion [129a].

8.8 Bioluminescence Measurements

8.8.1 Toxicity Screening Using the Bioluminescent Bacteria *Vibrio fischeri*

Toxicity screening follows the conventional TLC development and analysis process, which has to be performed in advance to separate possible toxic substances on the HPTLC plate. The developed and dried plate is dipped into a suspension of bioluminescent bacteria, *Vibrio fischeri*, thereby exposing any separated bioactive compounds to the test organisms. The luminescent activity is reduced or stopped by substances toxic to the bacteria, which may also be toxic to humans. Healthy *Vibrio fischeri*, emit very weak light of greenish color, which can not be detected with standard systems using daylight cameras. Thus, a specific bioluminescence detection system is required.

Figure 102. The CAMAG BioLuminizer™, specifically designed for bioluminescence detection

8.8.2 Detecting Bioluminescence With the BioLuminizer™

The BioLuminizer™ has been optimized for detection of bioluminescence patterns on HPTLC plates. The most important features are:

- High quantum efficiency allowing short exposure times.
- 16 bit dynamic range to detect small changes in intensity.
- Resolution optimized to resolve all details on the HPTLC plate following treatment with *Vibrio fischeri.*
- No dry-out, the bacteria are kept moist and luminescent for hours.
- Stable environment allowing long term and differential measurements.
- Predictable bioluminescence activity for reproducible results.

The CAMAG BioLuminizer™ stands out from similar system with its very smart and compact setup enabling precise and convenient positioning of the HPTLC plate, easy cleaning and compatibility with use under field conditions.

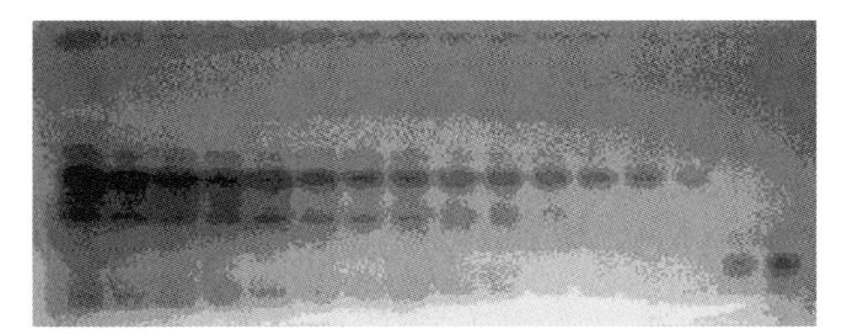

Figure 103. Comparison of different substance concentrations with two reference tracks shown on the right side. Dark spots represent reduced or stopped bioluminescence and can thus indicate the presence of toxic substances

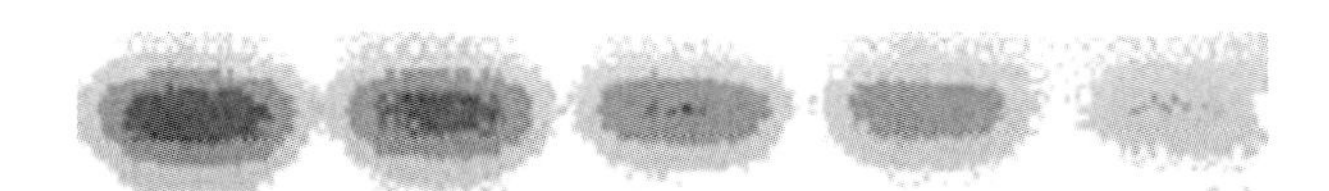

Figure 104. Bioluminescence response for different toxicity levels

9 GMP/GLP-Conforming Operations in TLC

The abbreviations "GMP" and "GLP" have already been used in various parts of this book. Both of these concepts originated in the United States. The fundamental rules for the preparation of medicaments and the guarantee of their quality lie behind the letters **GMP** (Good Manufacturing Practice). The fundamental principles are summarized and explained in "EU Guidelines on Good Manufacturing Practice for Medicaments" [130].

The ground rules of **GLP** (Good Laboratory Practice) were laid down for the orderly planning, performance and monitoring of laboratory tests, especially those aimed at detecting harmful effects of medicaments, chemicals and plant protection agents, and sometimes also those of cosmetics and food additives.

Where is work done in accordance with GMP and GLP?

The GMP rules and GLP rules are equivalent, but simply apply to different areas. It can generally be said that in pharmaceutical factories and their control laboratories the work is done in accordance with GMP principles, the so-called GQLP (Good Quality Laboratory Practice). This also applies to pharmacists who supply medicaments that they prepare themselves. All laboratories which carry out toxicological investigations of the effects of medicaments on animals, for which an application for permission from the German Institute for Medicaments and Medical Products (BfArM) must be made, require an official certificate stating that they are working in accordance with GLP. Failing this, their test results are not recognized [131]. Also for this reason, all TLC users who have to perform their work in accordance with these ground rules are recommended to read the "GLP Handbook for Laboratory Technicians" [132].

The difference between routine laboratory work as performed today and that of 30 years ago lies not only in improved analytical methods and the benefits of electronic data processing. The main difference is the enormous paper mountain that represents the comprehensive documentation demanded by GMP/GLP and the frustration that sometimes goes with it.

The parameters that have to be documented according to the GMP/GLP rules can be expressed in the following short formula:

> **Who has done what, when, where, with what, how and why?**

Applied Thin-Layer Chromatography: Best Practice and Avoidance of Mistakes, 2nd Edition
Edited by Elke Hahn-Deinstrop

ISBN: 978-3-527-31553-6

The burden of complying with these rules is sometimes relieved by witty colleagues who invent or pick up alternative meanings for the abbreviations, e.g. "GMP = gives more power, gives more paper, general minimization of productivity" etc. [133, 134]. However, humor can sometimes actually give a better understanding of these ideas. For example, Wilhelm Busch has produced cartoons which illustrate the important elements of GLP in picture form [131].

The history of GLP – whom have we to thank for all this?
The above question is often posed by young people starting their first job. It is therefore appropriate at this point to quote an apt extract from an article on Good Laboratory Practice by M. Schmidt [135]:

"In the early 1970s, the American Food and Drug Administration (**FDA**) discovered that, in animal experiments in the field of hazard identification (e.g. testing for acute and chronic toxicity, mutagenicity, teratogenicity, carcinogenicity), there was evidence of deceit and slovenly work when certification and registration were applied for. Some testing laboratories had submitted data from a two-year study within a few weeks, and some had completed 2000 studies in one year. Also, experimental animals had become muddled up, although without any intention to mislead, and other mishaps of a similar nature had also occurred, all of which placed the value of the results submitted in question.

The FDA reacted quickly by publishing the first draft of a GLP regulation in 1976. The final version came into force in mid-1979. Since then, in laboratories in the United States, the data for the assessment of possible hazards to humans and the environment, when submitted for certification, permission, registration, or reporting for information purposes, must undergo regular inspections. If irregularities come to light, severe fines or even imprisonment could be imposed.

In 1981, the OECD (Organization for Economic Cooperation and Development) adopted the American GLP rules in their concept for the mutual acceptance of test data for the evaluation of chemicals.

In 1989, the FDA/OECD-GLP rules were incorporated into **EU law,** thereby obliging the member states to include these regulations in their national law. Since 1 August 1990, this EU requirement has been fulfilled by the Federal Republic of Germany, and the fundamental principles of GLP were incorporated into the chemical law (ChemG)."

From the above discussion, the following conclusion can be drawn:

Do not falsify any analytical results!
These can come back like a boomerang with fatal consequences.

Standard Operating Procedures (SOPs). In routine laboratory work, with its GMP/GLP rules, one is governed by a large number, often hundreds, of "SOPs". In these, clear and comprehensible descriptions of **all** current routine processes are required (see, for example, the SOP "Registration, Printing and Archiving of Video Shots" in Section 8.4). M. Schmidt gives the following commentary on this [131]:

"The American origins of GLP can be clearly seen in the misconception that poor training of laboratory personnel can be compensated for by a flood of written instructions". W. Günther comments as follows: "GLP gives the laboratory legal but not technical security" [136]. However, SOPs are a valuable basis for personal training! They are especially useful for revision if "he who always sees to it" is on holiday.

Because the terms operating procedure, standard operating procedure, testing procedure, working procedure etc. are often used under various circumstances, the definition of Standard Operating Procedures (SOPs) given by the Pharmaceutical-Technical Committee of the German Association of Pharmaceuticals Manufacturers (BAH) [137] is quoted here: "Standard Operating Procedures are documents that give information and instructions for the performance of operations that do not necessarily refer to one particular product or one particular task, e.g. the cleaning of equipment and plant, clothing to be worn in the production area, or climate control. They are essentially **general** instructions of an organizational, administrative or technical nature and not specific instructions referring to specific products, manufacturing methods, cleaning instructions etc.

The **product-specific** requirements and instructions are rather to be found in documents such as manufacturing methods for the particular products, operating instructions for particular machines and plant, cleaning and disinfection instructions for particular equipment and plant etc."

Operating procedures for TLC that fulfill all requirements have not so far been published to the best of my knowledge. Therefore, all the essential components of an SOP are discussed in this book and can be referred to when drafting one's own SOPs for TLC.

Having followed these necessary expositions on the theme of "GMP" and "GLP", the question will occur to the user: "How can all these requirements be translated into practice in the area of TLC?"

☞ **Practical Tips** for GMP/GLP-conforming operations in TLC:

- **Whenever possible, use coded plates.** Laser-coded TLC and HPTLC plates precoated with silica gel 60 F_{254} considerably simplify quality assurance in the laboratory. Documentation and archiving of test results in accordance with GMP/GLP requirements is then no longer dependent on the troublesome task of writing on the plates. The name of the manufacturer, the item number, the charge number and the serial number of the plate are clearly visible and cannot be changed, being etched into the top edge of the plate (or on rotation through 90° on the left or right hand edge). These markings cannot be tampered with and are automatically included in the photographic or video documentation of the chromatographic result [138].

- **Only use validated TLC methods in routine work.** Validation in TLC is described in Section 9.1.
- **Use qualified/calibrated TLC equipment.** The qualification/calibration of the equipment used is described in Section 9.2.
- **All the necessary parameters should be documented on raw data sheets.** Suitable forms for GMP/GLP are given in Section 9.3.
- **Testing procedures**, enabling work to be performed in a truly reproducible fashion, **should be written as a result of method developments.** This subject is discussed in Section 9.4.

What will the future bring?
The rules of GMP and GLP are naturally not fixed for all time. They undergo a continual process of renewal and optimization aimed at saving time and resources and avoiding pointless investigations – especially those involving animals. The second Meeting of Experts to reexamine the principles of GLP in the OECD took place in June 1996 in Paris (80 experts from 26 nations representing industry and also official bodies responsible for supervision and assessment). The resulting recommended changes could be put into effect in late 1997 at the earliest [139]. In the second German GMP Conference in November 1996 in Frankfurt/Main an intensive dialog with representatives of the American Health Authority (the FDA) took place for the first time, and this is to be continued [140]. An important topic for the future will be the increased use of computer-supported systems. The recommendations of the project group "GLP and Electronic Data Processing" of the GLP study group of the German Chemical Industry Association have been published [141, 142]. The aim of the proposed method of working is to minimize the cost of validation of computerized systems as far as possible while at the same time fulfilling the requirements of GLP. In the field of pharmaceutical manufacturing processes, a paper on the validation of computerized systems has also been published, which points out ways of fulfilling GMP requirements while at the same time using information obtained for an effective revalidation and potential improvement of the process [143].

9.1 Validation of TLC Methods

Validation is defined as the formal determination of the suitability of a given analytical method for an intended application, achieving this by determining the reliability of results obtained [144]. The following extract from the paper "Validation of Analytical Results in Pharmaceutical Analysis" by Renger et al. serves as the basis of this Chapter. Briefly, based on differences in requirements and even in definitions of terms used in the United States and Europe, experts from these countries and from Japan, who have been attending meetings of the International Conference for Harmonization (ICH 1–ICH 3) since 1993, have prepared a scheme of uniform requirements for the validation of analytical results [145]. The framework of the present book would be exceeded if all the details and commentaries were discussed here. The interested reader is therefore referred to the literature [144].

The four analytical determinations most commonly performed are, in brief:

- Identification
- Quantitative determinations of impurities
- Limit value determinations of impurities
- Quantitative determinations of the active ingredient (also in the field of dissolution tests)

Requirements that should be fulfilled by validation data are listed in Table 24.

Table 24: Fundamental guidelines according to the ICH (from [144])

Type of analytical procedure; characteristics	Identification	Testing for impurities		Assay; dissolution (measurement only); content/potency
		Quantitation	Limit	
Accuracy	–	+	–	+
Precision:				
– repeatability	–	+	–	+
– intermediate precision	–	+ (1)	–	+ (1)
Specificity (2)	+	+	+	+
Detection limit	–	– (3)	+	–
Quantitation limit	–	+	–	–
Linearity	–	+	–	+
Range	–	+	–	+

+ signifies that this characteristic is normally evaluated
– signifies that this characteristic is not normally evaluated
(1) In cases where reproducibility has been performed, intermediate precision is not needed.
(2) Lack of specificity of one analytical procedure could be compensated by other supporting analytical procedure(s).
(3) May be needed in some cases.

The various concepts used in the assessment of analytical results are defined below and methods for determining them are recommended [146].

1. Trueness or Accuracy of the Mean

Definition: The trueness or accuracy of an analytical method is given by the extent to which the value obtained deviates from the true value. It is often expressed as the amount determined by analysis expressed as a percentage of the known amount of active ingredient added to the matrix material. It is a measure of the exactitude of the method.

Determination: The trueness or accuracy of a method can be determined by using it to analyze specially prepared samples. In these, the active ingredient is added in a similar concentration to that expected in the pharmaceutical product and also in somewhat higher and lower concentrations. The accuracy is then calculated as a % recovery.

2. Precision

Definition: The precision of an analytical method is the extent of the agreement between the measured values in a series of analyses of a homogeneous sample. The precision of a method is normally given by the standard deviation (S) or relative standard deviation (coefficient of variation or S_{rel}). The precision is a measure of the variability of the analytical method under normal laboratory conditions.

Determination: A sufficient number of analyses (usually 6–10) of a homogeneous sample are performed. The analyses must be carried out in their entirety, i.e. from weighing out the sample to the final result.

2.1 Repeatability
The repeatability or "intra-assay precision" is defined as the precision of determinations performed in the same way and within a short space of time. Its determination is of particular interest in the analytical testing for the uniformity of the content of a measured active ingredient (content uniformity test).

2.2 Laboratory Precision
The laboratory precision represents the scatter of results obtained when analyses are performed by different technicians on different days with different equipment and reagents.

3. Specificity

The ICH-2 text does not differentiate between selectivity and specificity, and defines the latter as "the ability to determine the analyte unambiguously in the presence of other components whose presence is to be expected. This includes typical impurities, decomposition products, matrix components etc."

The unrealistic requirement to demonstrate that all by-products and impurities theoretically possible have been separated is no longer maintained.

If by-products of the synthesis, impurities or decomposition products are known and accessible to chromatography, the chromatographic separation is optimized with their aid, and proof is accordingly afforded that none of these substances coincide or overlap with the analytes to be determined. If, for example in the case of new compounds and products, the potential decomposition products are unknown, stress reactions (temperature, light, acidic, alkaline or oxidizing conditions, hydrolysis etc.) are used to induce a decomposition, and the reaction mixture formed is subjected to chromatographic separation. The proof of the peak purity is in practice provided by producing in situ spectra [144].

4. Limit of Detection

Definition: The limit of detection is the lowest concentration of a substance to be analyzed that can be detected under the analytical conditions to be used, but cannot be measured quantitatively. It is normally expressed as the concentration of the substance to be analyzed in a sample (%, ppm).

Determination: The limit of detection by instrumental methods can be determined as follows: First, the magnitude and standard deviation of the background noise are determined by analysis of blank samples. Each peak with a height 2–3 times that of the background noise represents a value above the limit of detection. To check the value so determined, it is advisable to analyze a sufficient number of samples with concentrations in the region of the limit of detection.

5. Limit of Quantitation

Definition: The limit of quantitation is an important parameter in the quantitative determination of trace components in a sample, e.g. impurities in a medicament or decomposition products in a proprietary pharmaceutical product. The limit of quantitation is the lowest concentration of a substance in a sample that can be determined with acceptable precision and accuracy under the given analytical conditions. The limit of quantitation is expressed as a concentration of the substance (%, ppm) in the sample.

Determination: In instrumental methods, it is common practice to measure the magnitude of the background noise using several blank samples and to determine the standard deviation. The variation so determined is multiplied by 10, which gives an estimated value for the limit of quantitation. This limit is then confirmed by analyzing a sufficient number of samples in which the concentration of active ingredient lies in the region of the limit.

6. Linearity and Range of Measurement

Definition of linearity: The linearity of a method is its ability to provide measurement results that are directly proportional to the concentration of the substance to be measured or are directly proportional after mathematical transformation. The linearity is normally documented as the regression curve of the measured values as a function of increasing substance concentration.

Determination: The linearity is determined by evaluation of the analytical results of samples as a function of changing (increasing) concentration. The calculation is normally performed using regression curves according to the method of least squares (best straight line). However, in those cases where there is proportionality but not a linear relationship between concentrations and measured values (a curved calibration line after smoothing), the measured values of the regression analysis can be suitably transformed. A graph of the measured values as a function of the concentration is an acceptable substitute for calculation of the curve.

Definition of range of measurement: The range of measurement is the interval for which precision, accuracy and linearity have been demonstrated. Measurements outside the range are not permitted!

7. Robustness

The robustness of a method is described in the ICH-3 text as no more than part of the formal validation. It is demonstrated, for example, by the ability of the specified method to be transferred to another laboratory, where it gives consistent data. However, it is obvious that all potential external influencing factors on the analytical result should be checked during the method development, and the method parameters that must be complied with should be deduced from this.

Some of the most important parameters in HPTLC are listed below:

- The stability of the analyte in the solution being analyzed.
- The stability of the analyte on the plate before, during and after the development.
- The influence of temperature, atmospheric humidity, light, sample matrix, method of applying the sample for chromatographic separation, spot shape and evaluation [144].

☞ **Practical Tips** for the validation of TLC methods:

- The establishment of a time plan for the validation and a systematic check of the specified parameters is recommended, especially for beginners. All results are entered in a prepared table, while checking the validation for completeness.
- In the analysis of a new substance, the selectivity of the method should first be determined. The separability of all the products associated with a synthesis (precursors, by-products) and of any decomposition products (stress test) is demonstrated.
- After deciding on the chromatographic system (e.g. precoated plate, solvent system, chamber type, migration distance, detection method) and the test for robustness, the next step for quantitative determinations should be to test for the linearity of the calibration function over the required range. If this shows that the measurement results lie in a nonlinear part of this calibration curve, this working range should be changed before carrying out any further validation work.
- The limit of detection, limit of quantitation, precision and accuracy can then be determined.
- As described above under the heading "Trueness or accuracy of the mean", the recovery rate is determined by mixing various amounts of the active ingredient into the matrix (e.g. in the case of a tablet a mixture of all the auxiliary substances), and this is then quantitatively analyzed. Using this method, if the expected value is 100 %, almost 100 % of the active ingredient is usually recovered. Often, this home-made mixture of active ingredient with matrix is then used to determine the precision, and, assuming the results have a good standard deviation, these points can be used for the validation. However, it can sometimes happen that, in the analysis of the actual medicament in tablet form, low values are obtained in the assay. These can arise if there are other effects during the preparation of the tablet (e.g. interactions between the active ingredient and the matrix due to the pressure of the compression tool, causing the analyte to be incompletely detected or extracted). This example shows that the validation with the aid of mixtures of components is not always without problems [147].

9.2 Use of Qualified/Calibrated Equipment

A fundamental requirement of "Good Analytical Chemistry" is that the analytical equipment should be suitable for the intended use and should be calibrated. For this reason, equipment qualification is becoming of increasing importance in the field of analysis. Here, constructive cooperation between the manufacturer and the user plays an important role, as only an exact knowledge of the demands to be placed on a piece of analytical equipment can help manufacturers to avoid unnecessary development costs and enable them at the same time to undertake work in the area of qualification of the equipment.

Based on the lecture notes of A. Brutsche on the subject of "Calibration and Qualification of Analytical Equipment in Pharmaceutical Quality Control", the four concepts of equipment qualification in the field of analysis are listed here and elucidated in the form of headings [148]. The subject is discussed in more detail by L. Huber in the book "Validation of Computerized Analytical Systems" [149].

A. Design Qualification (DQ)

- ⇨ User requirements
- ⇨ Equipment inspection (at least two bidders)
- ⇨ Equipment evaluation/decision to purchase
- ⇨ Requirements to be met by the suppliers

B. Installation Qualification (IQ)

- ⇨ Checking of the consignment for completeness (+ documentation)
- ⇨ Installation by the suppliers (+ documentation)
- ⇨ Acceptance of the installation by the customer

C. Operational Qualification (OQ)

- ⇨ Basic calibration of the equipment (+ documentation)
- ⇨ Test run of the equipment (+ documentation)
- ⇨ Computer validation (hardware/software) if the analytical equipment is electronically controlled
- ⇨ Training of suitable operators (+ documentation)

D. Performance Qualification (PQ)

- ⇨ Test run of the equipment with three of the company's own products if possible
- ⇨ Release of the equipment for routine operation (+ final report on qualification); PQ is only necessary for complex equipment (e.g. HPLC, GC and TLC equipment used in quantitative determinations).

The user is obliged to maintain comprehensive documentation for analytical equipment. This includes

- Documentation for DQ, IQ, OQ and PQ
- SOP for equipment calibration
- SOP for cleaning and maintenance
- User book
- Log book
- Operating instructions
- Software/hardware validation certificates

The attention of any TLC user who is not yet affected by GMP/GLP and who wishes only to complete a small number of good chromatograms in the laboratory should be drawn to an SOP which is entitled "Qualification of Equipment for which Qualification is Obligatory" and which is relevant to this Chapter [150]. The object of this is to set out the procedure for the qualification of equipment that provides GMP-relevant data.

☞ **Practical Tips** for the use of qualified equipment in TLC:

- Ask (if possible) several manufacturers whether calibration and/or validation is possible at all with the equipment that they are offering for TLC.
  - Ebel [151] states that it is essential to validate automatic sample application systems if quantitative measurement results are to be obtained with their aid. Although the samples are applied before the measurement, validation of the application system presupposes a validated scanner (hen/egg problem!).
  - It has recently become possible for the user at any time to fully validate automatic samplers and scanners supplied by CAMAG. A software option "Validation" is commercially available [152].
- Before purchasing expensive analytical equipment, references should be requested from the manufacturer and the experience of other users should be used.
- For very expensive equipment, the manufacturer should be requested to provide a sample.

We conclude this subject with some very recent information. In the book accompanying the International Symposium for Planar Chromatography in April 1997 in Interlaken (Switzerland), Prošek indicates the possibility of validating quantitative TLC using a video camera system [120]. In the same Symposium, Ebel made the following points in his lecture on the theme of validation [153]:

- **The human eye cannot be validated[1)]**
- **Video scanners and flat-bed scanners can be validated under certain conditions**
- **TLC scanners can be validated completely**

These statements concerning video and flat-bed scanners are based on the present state of evaluation software. Very considerable developments in this area are likely in the future.

[1)] Example of visual evaluations: limit value tests specified by pharmacopoeas.

9.3 GMP/GLP-Conforming Raw Data Sheets

If it wasn't documented it might as well not have been done!

Adherence to this guiding principle of GMP/GLP has led to a considerable increase in the ratio of paperwork to practical work at the laboratory bench. Therefore, pre-printed forms are a very useful tool for ensuring that no essential parameters are forgotten in the hustle and bustle of the daily routine. In the area of quality control of a pharmaceutical business enterprise, these forms can be designed individually to include each testing parameter. All master copies of **authorized** forms must carry the stamp of the Quality Assurance department (usually "QA"). Copies of these are then used as working papers. The raw data sheets for TLC are thus only a part of the complete documentation for a substance (raw material) or product (proprietary pharmaceutical product) to be tested. In hardly any of the forms, therefore, is there any reference to the relevant testing procedure. The document number of the testing procedure is marked on a covering sheet (testing protocol), on which the complete scope of the testing is also recorded, classified according to parameters and pages.

Examples of raw data sheets for various TLC investigations are given below. Some of these forms were prepared specifically for the present book, and the space for the QA stamp therefore contains the author's initials. All raw data sheets, also including those for IR and pH, have the same heading design to make it easier to complete using a PC.

The following comments may help to give a better understanding of the forms:

TLC Raw Data Sheet I (Document 5)

This form was designed specifically for routine analyses to test for the identity and purity of large numbers of pharmaceutically active ingredients and intermediates. The samples are weighed out on an analytical balance with no printer. All operating procedures are given in a valid testing procedure.

The result is documented by placing a cross against the relevant parameter. This raw data sheet does not show whether the checker has actually seen the completed chromatogram or whether the formal documentation of this work has simply been authenticated!

Today the official guidelines within the pharmaceutical industry require the mandatory use of a combination of balance and printer. For this reason the form is used only rarely.

Material (description, if required) Substance X	Lot 99-05649

Thin-Layer Chromatography I

QA / stamp: HD

Balance no.: 5

Sample weights [mg]:

1	200	7	199	13		19	
2	201	8	204	14		20	
3	203	9	198	15		21	
4	201	10	203	16		22	
5	202	11		17		23	
6	203	12		18		24	

Description and weighed amounts of the reference substances [mg]:

reference substance X, USP 200
related compound A, USP 12,5
related compound C, USP 12,3

Result:

Purity: samples correspond ☒ do not correspond O not carried out O

Identity: samples correspond ☒ do not correspond O not carried out O

Deviations/ observations: —

15.06.1999 / Rö
Operator (date, sign)

16.06.99 / [signature]
Checker (date, sign)

Document: HD / TLC I Raw data sheet

Document 5

TLC Raw Data Sheet II (Document 6)

This too is a form for routine analyses. However, the instructions here can refer not only to in-house product-specific testing procedures (e.g. for proprietary medicaments) but also to monographs of valid pharmacopoeias (DAB, Pharm. Eur., USP etc.) and other sources (e.g. DAC). Samples can be weighed out on analytical balances with printout facilities. The printed paper strips should include at least the date, the code number of the reference substance and the batch numbers of the samples. The image of the chromatogram obtained should be documented and the archive number should be recorded on the raw data sheet. The result is also recorded by placing a cross in the appropriate place. Here, the checker can at least inspect and assess the images of the chromatograms.

TLC Raw Data Sheet III (Document 7)

The original version of this form (see p. 213) was produced in 1988 in accordance with the present author's instructions and was published in 1992 at a Merck Symposium [154]. Over the years, it has been modified to comply with more stringent requirements. For example, the parameters "relative humidity" (following a recommendation of Nyiredy [26]) and "other evaluations" (which could refer, e.g., to video scan or radio TLC) were included for the first time in 1997. To give the user an example of a universal type of form for TLC, the heading is given here in its original version, and the explanations below are listed.

Comments on the raw data sheet "Thin-Layer Chromatogram"
This form fulfills all the GMP/GLP requirements for the documentation of thin-layer chromatograms, and has been found to be highly satisfactory over the years, especially in method development. It is advisable to fill out this sheet before starting to apply samples to a plate.

The form is suitable for development work both before and after writing a new testing procedure, and for testing raw materials and finished products in accordance with a testing procedure that is already to hand. It can be used both for manual application and for application with the aid of the semiautomatic Linomat IV equipment. For other application equipment, appropriately designed forms should be used.

Material (description, if required)	Lot
Chlortalidone	**9924247**

Thin-Layer Chromatography II

QA / stamp

Performance according to: ⊗ (1997) Ph. Eur., DAB, BP, USP, JP, SOP No.
O Testing Procedure Pharmaceutical Product
O other*

Balance no.: 29

Plate no: 2685

Manual application: ⊗

Application by Linomat: O no.

by AS 30: O no.

weighing proceedings

Date	Time
25.06.99	09:07:16

Code 10087

10.62 mg

Chlortalidone CRS

Code 11081

10.22 mg

2-(4-Chlor-3-sulfamoylbenzoyl)-benzoe acid CRS

Code 9924247

10.32 mg

100.76 mg

Reference substance(s)

a: Chlortalidone CRS no.: 10087

b: rel. Comp. A CRS no.: 11081

c: ______ no.: ______

d: ______ no.: ______

Samples: weight [mg] + solvent [ml]:

1: 10,32 + 10,32 ≙ 1 mg/ml

2: 100,76 + 5,04 ≙ 20 mg/ml

3:

4:

5:

6:

Result: Identity: samples correspond ⊗ do not correspond O not carried out O

Purity: samples correspond ⊗ do not correspond O not carried out O

⊗ Picture no.: video 9901305 See ~~supplement~~ / back page lot no.: 9924248

* Fill out the raw data sheet TLC/HPTLC III additionally

25.06.99 HD
Operator (date, sign)

28.6.99 [signature]
Checker (date, sign)

Document: HD / TLC/HPTLC II Raw data sheet

Document 6

Hints on filling out this form

1. The serial number of the chromatogram, which depends on the subject, is entered in the heading (e.g. 44).
2. The empty space with surrounding border is intended for the subject (substance, name of preparation etc.) and any necessary additional information, e.g. "LOD" for limit of detection, "STLC" for sample chromatogram or "Stress samples" for samples subjected to "defined substance stress".
3. Characterization of the sorbent (e.g. TLC-K60 WF_{254s}), the article number and the manufacturer of the precoated layer (e.g. 16485 Merck) and the batch used (e.g. 08733902) are documented.
4. For coded precoated plates of the silica gel 60 type, the individual plate number is entered.
5. The composition of the "prewashing agent" and, if appropriate, its migration distance are noted.
6. Prewashed plates are always activated (30 min, 120 °C); amino-modified silica gel layers, for example, are activated without prewashing (10 min, 120 °C).
7. Settings of the Linomat IV: start, spaces between lanes and lane widths in mm, application rate in s/µl.
8. The chamber type is noted: N for normal chamber (or DT-N for double-trough chamber, CAMAG), H for horizontal chamber (e.g. 5 × 5 cm, DESAGA), DC-MAT for automatic development chamber (Baron), ADC for automatic developing chamber (CAMAG), etc.
9. To describe the state of chamber saturation (CS), "with" or "without" should be circled.
10. The composition of the solvent system in "ml", the migration distance in "cm", the development time in "min", the ambient temperature in "°C" and the relative humidity in "%" should be given.
11. Information on the derivatization can be in the form of either reagent names (e.g. "Dragendorff", "vanillin-sulfuric acid") or a literature reference (e.g. "JFFW 1a/265" indicates the reagent in the Jork/Funk/Fischer/Wimmer Vol. 1a, p. 265: 2,5-dimethoxytetrahydrofuran-4-(dimethylamino)-benzaldehyde reagent).
12. In the "Documentation" line a cross is placed at the appropriate place, and when the images have been obtained the archive numbers are entered in the boxes below (at a later time in the case of films!).
13. If more than 12 lanes are used, the table can be continued on the back of the form.
14. The analyst enters the date of completion of the chromatogram and also his or her signature in the spaces provided, and any additional information sheets are also dated. The checker initials this form and the last of the additional sheets on completion of all the documentation that applies to this plate (sheets with pasted-on photographs, attached scanner printouts, videoprints etc.).
15. Corrections to this form are made using a pencil of a different color so that the erroneous entry remains legible even after it has been crossed out. The correction must be dated and initialed.
16. The sample preparation method and comments can be added to the back of this form.

TLC/HPTLC- CHROMATOGRAM No. 44 **TLC/HPTLC III**

Applied: with Linomat IV No.: 2 / ~~manual~~

Dexpanthenol ointment

Stress test

Sorbent: TLC K60 WF254S	Art. no.: Merck 16485		Lot no.: 0873 9302
Plate no.:	/	Prewashed ? MeOH	Activated ? 30 min 120 °C
Start:	20	s / µl: 15	Space: 9
Band:	15	Type of chamber: vertical, N	(With) / without CS
Solvent system:	1-Butanol + methanol + NH_4-soln. 55+25+20 [acc. to DAC]		
Migration distance:	10	Developing time: approx. 70	Room temperature: 22
Derivatization:	JFFW 1a / 265		Relative humidity: 52
Documentation:	TLC scan: /	Video shot: / Film: X	Other evaluation: /

Film archive no.: 92	UV 254: /	White light: 390/28
Video archive no.:	UV 365: /	

Lane	Substance	Lot no.	mg / ml	hRf-value	Vol (µl)	µg	%
1	unexposed	PN 41/4	2,5		60	150	100
2	dry	120 °C	5,0		30	150	
3	Sun test		5,0		30	150	
4	H_2O	80 °C	5,0		30	150	
5	3% H_2O_2	80 °C	5,0		30	150	
6	0,1 N HCl	80 °C	5,0		30	150	
7	0,1 N NaOH	80 °C	5,0		30	150	
	Dexpanthenol			53 - 61			
	3-Amino-1-propanol			21 - 27			

Date: 23.3.92	Operator: [signature]	Checked by: [signature]	Date: 27.03.92

Document HD: TLC/HPTLC III raw data sheet

Document 7

TLC Raw Data Sheet IV (Document 8)

This form was designed exclusively for stability samples of preparatory pharmaceutical products. In this case it should be assumed that a valid testing procedure is available, so that it is unnecessary to record the sample preparation method and the chromatographic conditions. Above all, it is important that the developed chromatogram should be checked for the possible presence of additional zones (AZs). The number and the intensity of each individual AZ must be documented.

TLC Raw Data Sheet V (Document 9)

It is explained in Section 8.5 "TLC Scanner Documentation" why it is necessary to provide a covering sheet for the "paper mountain", giving the results in abbreviated form and documenting the number of added pages. The TLC Raw Data Sheet V (see p. 216) is the first page of this collection, and is followed by one page each on equipment parameters (type of sample application equipment, chamber type, method of evaluation and equipment with item number etc.), the weighing protocol of all reference substances (inclusive of any correction calculations to give the weight of the anhydrous substance) and the samples. The last two of these forms are not shown here for reasons of space.

Raw Data Pages and LIMS

Today in the industry Laboratory Information Management Systems (LIMS) are commonly used. When a sample requiring TLC analysis is created in LIMS, the system automatically creates the required links and prints all necessary raw data pages.

Material (description, if required)	lot
xy – Tablet	95 220871

Thin-Layer Chromatography IV

Stability Testing

QA / stamp

Balance no.: 29 Plate no.: 3312 Equipment: Linomat no.: 2 / ~~AS 30 no~~.:

Sample weights [mg] + solvent [ml]:

1 52 + 5,2 ≙ 10 mg / ml

2 49 + 4,9

3 53 + 5,3

4

Storage conditioning time of the samples: 48 month(s)

Reference substance(s): weight [mg] + solvent [ml]:

1 ref. sub. x : 50,5 + 5,05 ≙ 10 mg / ml

2 related compound A : 2,0 + 20,0 ≙ 0,1 mg / ml

Result:

Lane	Reference solution/store condition	Documentation / evaluation: print no.or shot no. taken at:				
1	ref. sub. 100 %	UV254 nm: HD – A 071		"white light": HD – A 075		
2	" 0,5 %	UV365 nm: /				
3	" 0,25 %	hRf-value major spot	Quant. of ss total intensity	hRf-value/ intensity ss 1	hRf-value/ intensity ss 2	hRf-value/ intensity ss 3
4	rel. comp. A 0,1 %					
–	Reference sample					
5	Room temperature	45 – 51	–			
6	30°C	45 – 51	2 / < 1,0	15 – 17 < 0,25	72 – 74 < 0,5	
–	40°C/75 % rel. humidity					

[ss = secondary spot] ⊗ Shot/print: see ~~back page~~ / supplement page 2

Deviations / observations: sample preparation of the 30 °C – sample has been changed, see page 3

15.07.99 Yui
Operator (date, sign)

16.07.99 HD
Checker (date, sign)

Document HD: TLC/HPTLC IV raw data sheet

Document 8

Material (description, if required)	lot
Melissae e folium aquos. sicc.	**99-07736**

Thin-Layer Chromatography V

Quantitative Determination

of Sweet balm, dry ectrakt acc. to DAB 1996

QA / stamp

☐ **Identity**: ☐ samples correspond ☐ samples do not correspond

☐ **Purity**: ______

☐ **Content uniformity test**: ☐ see supplement

☒ **Assay:**

substance	single values [%]	mean value
Rosmarinic acid	3,45; 3,51; 3,48	3,48%

Supplements:

☒	Conditions of analysis (TLC/HPTLC-sheet VI)	1 page~~s~~	☒	see lot 99-07735
☒	Weighing procedures - calculations	1 page~~s~~	☒	see lot 99-07735
☐	System suitability test	.... pages	☐	see lot ....
☒	Samples report	1 page~~s~~	☐	see lot ....
☒	Calibration report	1 page~~s~~	☐	see lot ....
☒	Reference solution chromatograms	2 page~~s~~ (korr. HD)	☐	see lot ....
☒	Blank value chromatograms	1 page~~s~~	☒	see lot 99-07735
☐	Reference solution control	.... pages	☐	see lot ....
☒	Samples chromatograms	5 pages		
☐	Content uniformity test - result	.... pages		
☐				

7.7.99 HD
Operator (date, sign)

8. Juli 1999 Ha
Checker (date, sign)

Dokument HD: TLC/HPTLC V raw data sheet

Document 9

9.4 Examples of GMP/GLP-Conforming Testing Procedures (TPs)

Testing procedures to be followed at the laboratory bench should be in the language of the country where they are to be used!

This does not mean Bavarian, Saxon or Low German, but normal written German. This is no longer obvious. In the context of the globalization of industry, operating instructions for TLC in English are no longer a rarity even in German-speaking regions. The costs of documents for the licensing authorities in the various countries of the world are thereby kept as low as possible. It is to be expected that an experienced laboratory assistant will be capable of following a testing procedure in English without making mistakes. However, for trainees and/or students and other research assistants this cannot be generally assumed. Mistakes must be foreseen in this case!

The above paragraph was originally written for the German edition of this book and mentions in the first line three typical German dialects, as spoken in the north, east and south of the Federal Republic. The direct English translation would not be generally understood, and therefore all the Spanish-speaking countries would be a good example of a case where the testing procedures would be written in the official language. Also, a cartoon could illustrate the value of an understandable testing procedure. If the probationer had used a testing procedure in his own mother tongue instead of a "recipe" in ancient Egyptian hieroglyphics, the chromatograms on the TLC plate would have turned out better.

9.4.1 Identity and Purity of a Bulk Pharmaceutical Active Ingredient and Determination of the Limit Values of Related Compounds

As an example for this chapter we have chosen the substance "chlortalidone", which has a monograph of its own in the DAB. This choice was fortuitous, as this substance was under investigation at the time of preparation of this manuscript and illustrates the problems that arise with the DAB text very well. The problems may possibly lie also with the solvents used here (see also Section 4.1 "Solvent Systems"). For example, in the Merck catalog for 1996 there are 11 different grades of 2-propanol and toluene. Each one of these is designated "Reagent DAB, Ph. Eur.", is supplied only in 500-ml bottles and is more expensive than the other grades. For reasons of cost, this grade is very seldom found in an industrial laboratory. Also, in the draft of a partial testing procedure for chlortalidone (Document 10), the grades normally available in the author's laboratory are used instead of the DAB grades. In contrast, the DAB text put together in Document 11 is to be read as only for tests by TLC. The information about the four grades of the above-mentioned solvents has been obtained from the Merck catalog of 1996 and is summarized in Table 25.

Table 25: Qualities of solvents

	Quality	**Article No.**	**Smallest quantity [l]**	**Price [€]**
2-Propanol	GR, ACS, ISO	109634	1	20,50
	Dried	100994	0.5	24,25
	Gradient grade for chromatography LiChrosolv®	101040	1	29,25
	Extra pure, Ph Eur, BP, USP	100995	1	21,75
	For spectroscopy Uvasol®	100993	1	40,25
Toluene	GR, ACS, ISO	108325	1	25,75
	Dried	108320	0.5	32,00
	For liquid chromatography LiChrosolv®	108327	1	38,75
	Extra pure	108323	1	23,50
	For spectroscopy Uvasol®	108331	1	46,00

These differ in product designations, quantities supplied and prices
Authority: VWR catalog 2005–2007, German edition

9.4.2 Identity and Purity of Various Flavonoid-Containing Plant Extracts

In the TLC analysis of dry extracts prepared from medicinal plants, the sample preparation is performed in a different way from that prescribed in the monographs for the drugs in the pharmacopoeias. Also, there is no binder in the recommended solvent system, and in these cases a validation of the new in-house method is certainly necessary.

As this not only requires adequate experience of the user, but also adds a very time-consuming operation to the laboratory routine, the development of an optimized TLC method for the dry extract is often omitted. This is often the precise reason that TLC is discredited in comparison to other analytical methods. An example of a successful optimization of the parameters of various TLC systems for birch leaves is given in Table 10, and the chromatograms for these are shown in Fig. 48a–d.

Flavonoid constituents are present in a number of drugs, and therefore a consistent sample preparation method is suggested in Document 12 together with a uniform TLC system for the dry extracts. In the sample chromatograms included with the testing procedure, there are five dry extracts and the separation of a drug. This should confirm that the chromatographic system chosen could also be used for this application.

9.4.3 Content of a Pharmaceutical Active Ingredient in a Tablet

Quantitative determinations require validated analytical methods (see Section 9.1). An example of this is the "Assay by HPTLC of theophylline in an effervescent tablet" [144], for which a proposal for a testing procedure has been prepared (Document 13). All the results of the validation, including the calibration function, chromatograms and spectra can also be found in [144]. Document 13 includes only Fig. 2, which shows the chromatograms of the standard solutions with and without the addition of auxiliary substances, a blank lane and the pure mixture of auxiliary substances.

With respect to the recommended testing procedure, it should be noted that the scope of the test for release purposes is different from that for stability testing. Whereas two applications per weighed amount of sample are sufficient for batch release purposes, the number of applications per weighed amount of sample increases to three in the case of stability tests.

For the above-described analysis for theophylline, Renger proposed a comparison (bench-marking) between the two analytical methods HPLC and HPTLC. For the three parameters investigated (material costs, operating costs, operating time), HPTLC gave the lowest figures [154a]. The results are given in Table 26.

Table 26: Benchmarking between HPLC and HPTLC for assay of theophylline tablets

	HPLC (USP 23)	HPLC optimized	HPTLC
Materials and solvents cost per test [€	55,10	19,40	8,20
Operator working time per test [h]	4.5	2.7	1.3
Working costs per test [€]	344,40	206,65	99,50

(Prices in DM from 1999 changed into €)

Special Analytical Procedure acc. to Ph.Eur.97 **PV 97/01-C**

	Current date: Datum: 03.08.1999	Previous date: Letzte Ausgabe: 16.04.1997	Page: Seite: 1/4
Substance	**Pharmaceutical active ingredient Chlortalidone**		

Topic: TLC-identification and purity testing of chlortalidone according to Ph. Eur. 97

Application: This method is suitable for testing raw material.

Published: 3.8.99 E. Hahn-Deinstrop (Elke Hahn-Deinstrop)

First version: 16.04.1997 (German edition), 03.08.1999 (English edition)

Checked: 5.8.99 A. Koch (Dr. Angelika Koch)

Approved: 6.8. 10.99 F. Huse (Dr. Friederike Huse)

Appendices:

ORIGINAL
9. Aug. 1999

HD

This document is subject to monitoring by the above named publisher. Please do not copy it yourself but request copies from the publisher. Only thus is possible to ensure that you are provided with sufficient quantities of the most up-to-date version.

The reagents and apparatus specified in this testing procedure are exemplary and can be replaced by others of comparable suitability.

Document 10 / page 1

Special Analytical Procedure acc. to Ph.Eur.97 PV 97/01-C

	Current date:	Previous date:	Page:
	Datum: 03.08.1999	Letzte Ausgabe: 16.04.1997	Seite: 2/4
Substance	**Pharmaceutical active ingredient Chlortalidone**		

Principle: Thin-layer chromatographic identification and purity testing of chlortalidone and limit test on related compounds and unknown substances.

Test sample:

Identity:	n = 1 of each container	(solution A)
Purity:	m = 1 of each lot	(solution B)

Reagents:

Acetone	Merck 100014
Ammonia solution, 26%	Merck 105428
Chlortalidone	*CRS*
CBS *	*CRS*
Dioxan	Merck 109671
2-Propanol	Merck 101040
Toluene	Merck 108325

CRS = European Pharmacopeia substance
CBS* = related compound according to Ph. Eur.: 2-(4-chlor-3-sulfamoylbenzoyl)-benzoic acid

Sorbent: Silica gel 60 F_{254} GLP
(TLC precoated plate, 20 x 20 cm, Merck 1.05566)

Solvent system: Toluene + 2-propanol + dioxan + ammonia solution
20 + 20 + 20 + 13,5 (ml)

Prepare the solvent system as follows:
First mix dioxan and the ammonia solution (solution X). Then mix 2-propanol and toluene. Give this solution on parts of about 10 ml during mixing to solution X. If turbidity will appear, add dropwise 2-propanol until the solution will be clear. Mix after each drop.

Chamber: Double trough 20 x 20 cm with chamber saturation
Weight the lid with a lead doghnut
Protect the chamber before draught

Migration distance: 13 cm (acc. to 15 cm lenght of run)

Development time: approx. 60 min, room temperature 23°C

Reference solution I: Dissolve 10 mg chlortalidone *CRS* in 10 ml acetone
(corresponding to 1 mg/ml)

Reference solution II: Dilute 1 ml of the reference solution I with 9 ml acetone
(corresponding to 0,1 mg/ml)

Reference solution III Dissolve 10 mg CBS* *CRS* in 5 ml acetone. Dilute 1 ml of this solution with 9 ml acetone (corresponding to 0,2 mg/ml)

Document 10 / page 2

Special Analytical Procedure acc. to Ph.Eur.97 **PV 97/01-C**

	Current date: Datum: 03.08.1999	Previous date: Letzte Ausgabe: 16.04.1997	Page: Seite: 3/4
Substance	**Pharmaceutical active ingredient Chlortalidone**		

Sample solution A: Pepare 10 mg of the substance in the same way as reference solution I (corresponding to 1 mg/ml)

Sample solution B: Dissolve 100 mg of the substance in acetone to give 5 ml (corrresponding to 20 mg/ml)

Applied masses: Pointwise with 5 µl-capillary

Identity: reference soln. I and sample soln. A 5 µl each (5 µg each)

Purity:

reference solution II:	10 µl	(1 µg = 0,5%)
reference solution III:	10 µl	(2 µg = 1,0%)
sample solution B:	10 µl	(200 µg = 100%)

After application allow the plate to dry about 10 min in a stream of warm air.

Detection: After evaporation of the solvent (approx. 15 min in a stream of warm air, fume cupboard!!): UV 254 nm

Storage of the plate: Video shot at UV 254 nm

Evaluation: In the short-wavelength UV light the fluorescence quenching zones of chlortalidone and the related compound CBS are detectable.

Identity
The major spot in the sample solution A matches the hRf-value, volume and intensity of the fluorescence-quenching of the major spot in the chromatogram obtained with reference solution I.

Purity
In the chromatogram obtained with the sample solution B a secondary spot at hRf-value of 16 should not be bigger and more intensive than the 1% spot of CSB. No secondary spot in the chromatogram obtained with sample solution B should be bigger or more intensive than the 0,5% spot of chlortalidone.

hRf-values:

Chlortalidone	200 µg	46 - 52
Chlortalidone	1 µg	49 - 51
CSB	2 µg	15 - 17,5

Documentation: Fill in the raw data sheet TLC II. Note the numbers of the balance used, TLC plate and reference substances. Record sample(s) weight and the quantity of the solvent. Cross the parameter's field to document the obtained result(s) and note the archive number(s) of the videoprints.
Transfer the result(s) to the testing protocol. Operator and checker have to sign the raw data sheet with date and initials. All further sheets are to be paginated and signed on the last page. Finally all TLC raw data sheets are added to the batch documentation.

Document 10 / page 3

Special Analytical Procedure acc. to Ph.Eur.97 **PV 97/01-C**

	Current date: Datum: 03.08.1999	Previous date: Letzte Ausgabe: 16.04.1997	Page: Seite: 4/4
Substance	Pharmaceutical active ingredient **Chlortalidone**		

Sample chromatogram

UV_{254} nm

Figure 69

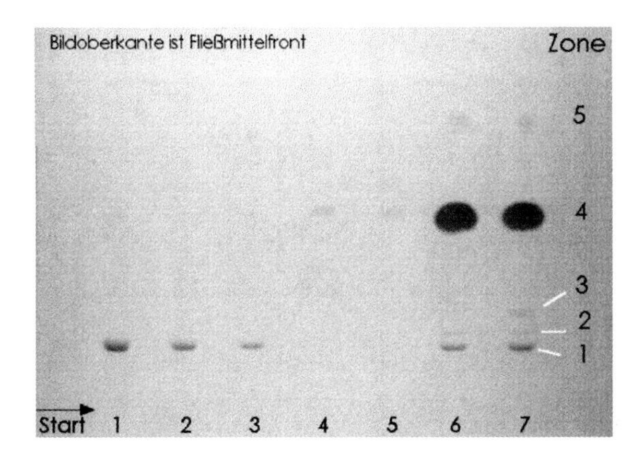

Lane 1:	Related compound CSB	4 µg = 1%	
Lane 2:	rel. comp.	2 µg = 0,5%	
Lane 3:	rel. comp.	1 µg = 0,25%	
Lane 4:	Chlortalidone	2 µg = 0,5%	(Z 4)
Lane 5:	Chlortalidone	1 µg = 0,25%	
Lane 6:	Chlortalidone (CRS)	400 µg = 100%	
Lane 7:	Chlortalidone (sample)	400 µg	

secondary spots (lane 7) at hRf:
15,7 - 17,2 (Z 1)
19,6 - 21,1 (Z 2)
23,8 - 26,7 (Z 3)
70,2 - 73,5 (Z 5)

Remark:

Figure 69 with 400 µg substance (corresponding to 100%) is here the sample chromatogram. This is double the normal amount. The visual evaluation of this plate is described in chapter 5.1.2 "Detection at 254 nm", the quantitative video determination you will find in chapter 7.3.2.

Document 10 / page 4

1997:0546

CHLORTALIDONE

Chlortalidonum

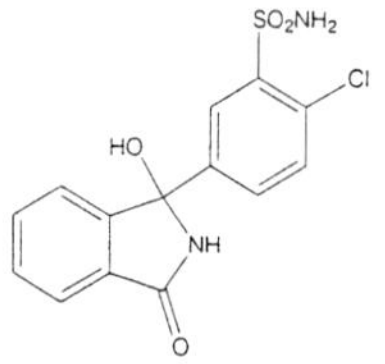

$C_{14}H_{11}ClN_2O_4S$ M_r 338.8

DEFINITION

Chlortalidone contains not less than 98.0 per cent and not more than the equivalent of 102.0 per cent of (*RS*)-2-chloro-5-(2,3-dihydro-1-hydroxy-3-oxo-1*H*-isoindol-1-yl)benzenesulphonamide, calculated with reference to the dried substance.

IDENTIFICATION

C. Examine by thin-layer chromatography (*2.2.27*), using *silica gel* GF_{254} *R* as the coating substance.

Test solution. Dissolve 10 mg of the substance to be examined in *acetone R* and dilute to 10 ml with the same solvent.

Reference solution. Dissolve 10 mg of *chlortalidone CRS* in *acetone R* and dilute to 10 ml with the same solvent.

Apply separately to the plate 5 µl of each solution. Develop over a path of 10 cm using a mixture of 1.5 volumes of *water R* and 98.5 volumes of *ethyl acetate R*. Allow the plate to dry in air and examine in ultraviolet light at 254 nm. The principal spot in the chromatogram obtained with the test solution is similar in position and size to the principal spot in the chromatogram obtained with the reference solution.

TESTS

Related substances. Examine by thin-layer chromatography (*2.2.27*), using *silica gel* GF_{254} *R* as the coating substance.

Test solution. Dissolve 0.1 g of the substance to be examined in *acetone R* and dilute to 5 ml with the same solvent.

Reference solution (a). Dissolve 10 mg of *2-(4-chloro-3-sulphamoylbenzoyl) benzoic acid CRS* in *acetone R* and dilute to 50 ml with the same solvent.

Reference solution (b). Dilute 1 ml of the test solution to 200 ml with *acetone R*.

Apply separately to the plate 10 µl of each solution. Develop over a path of 15 cm using a mixture of 20 volumes of *concentrated ammonia R*, 30 volumes of *dioxan R*, 30 volumes of *2-propanol R* and 30 volumes of *toluene R*. Allow the plate to dry in air and examine in ultraviolet light at 254 nm. In the chromatogram obtained with the test solution, any spot corresponding to 2-(4-chloro-3-sulphamoylbenzoyl)-benzoic acid is not more intense than the spot in the chromatogram obtained with reference solution (a) (1.0 per cent) and any spot, apart from the principal spot and the spot corresponding to 2-(4-chloro-3-sulphamoylbenzoyl)benzoic acid, is not more intense than the spot in the chromatogram obtained with reference solution (b) (0.5 per cent).

EUROPEAN PHARMACOPOEIA - 1997

Document 11

Special Testing Procedure **HD-052/91/5**

	Current date: 24.02.1999	Previous date: 06.12.1997	Page: 1/6
Identity, Purity (I): Dry Extracts of Flavonoid-Containing Drugs (TLC)			

Topic: Thin-layer chromatographic identification of various dry extracts and purity testing of the flavonoid components.

This special testing procedure serves as the basic for compilation of documentation, in which thin-layer chromatographic methods are described.

Application: The method is suitable for the identification and purity control of raw materials, intermediate and end products as part of their approval control.

Published: 24.2.99 E. Hahn-Deinstrop *(Elke Hahn-Deinstrop)*

First version: 09.08.1991 (German edition), 24.02.1999 (English edition)

Checked: 25.2.99 A. Koch *(Dr. Angelika Koch)*

Approved: 26.2.1999 F. Huse *(Dr. Friederike Huse)*

Appendices: 2. März 1999

The reagents and apparatus specified in this testing procedure are exemplary and can be replaced by others of comparable suitability.

Document 12 / page 1

Special Testing Procedure **HD-052/91/5**

	Current date: 24.02.1999	Previous date: 06.12.1997	Page: 2/6
Identity, Purity (I): Dry Extracts of Flavonoid-Containing Drugs (TLC)			

Method development: The chromatographic system in DAB 10 (German Pharmacopoeia) for the flavonoid drugs of birch leaves is suitable for identity testing, but the development time of approx. 105 minutes is relatively long for this test parameter. Hence, an alternative solvent system on a HPTLC layer with higher resolution and a shorter development time was employed for the identification of flavonoid-containing drugs.

In the two TLC systems mentioned above any aglycons present lie at the solvent front. For this reason an additional chromatographic system has been developed for purity testing and stability investigations. This employs prewashed and activated HPTLC plates.

The separation efficiency of the HPTLC system used here has been tested with samples of birch leaf extract, which had been exposed to extreme conditions for 4 or 20 hours. The components of the extracts were partially decomposed by dry heat, after expose to light and in warm aqueous solution and in the presence of an oxidizing agent or of acids or bases. This led to differing chromatographic fingerprints.

Related document: Purity (II) "Dry extracts of flavonoid-containing drugs (tests for aglycons)" HD-053/91/. The same dry extract was tested for decomposition products.

Principle: Thin-layer chromatographic identification of various dry extracts and purity testing of the flavonoid components.

Archive samples: Some of the samples taken were stored as archive samples in accordance with the SOP "Preparation of mixed samples and archive samples of solid materials" (SOP 064/93/).

Number of samples: identity: n = 1 per container of the lot to be tested.
purity: m = 1

Reagents:

Formic acid	Merck 100264
Chlorogenic acid*	Roth 6385
Diphenylboryloxyethylamine	Merck 159626
Acetic acid 100%	Merck 100063
Ethyl acetate	Merck 109623
Hypericin*	Roth 7929
Hyperoside	Roth 4215
Caffeic acid*	Roth 5574
Methanol	Merck 106007
Polyethylene glycol 6000	Merck 807491
Quercetin*	Roth 7417
Quercitrin*	Roth 9226
Rutoside	Roth 7176
Hydrochloric acid conc.	Merck 100317

Authentic archive samples of various dry ectracts

*)only for purposes of validation

Document 12 / page 2
The catalog numbers for chlorogenic acid, hypericin and quercitrin have changed. See Section 12.5 for Roth's address.

Special Testing Procedure **HD-052/91/5**

	Current date: 24.02.1999	Previous date: 06.12.1997	Page: 3/6
Identity, Purity (I): Dry Extracts of Flavonoid-Containing Drugs (TLC)			

Derivatisation reagents: Flavone reagent acc. to Neu
Solution A
1 % (m/V) methanolic solution of diphenylboryloxyethylamine.
The spray solution prepared freshly on each occasion.
Solution B
3 % (m/V) ethanolic solution of polyethylene glycol 6000.

Sorbent: Silica gel 60 F_{254} GLP
(precoated HPTLC plate 10 x 20 cm, Merck 1.05613)
for stability testing: prewashed with methanol + ethyl acetate (1+1), afterwards activate (120°C, 30 minutes)

Solvent system: Ethyl acetate + formic acid + methanol + water
50 + 2,5 + 2 + 4 (ml)

Typ of chamber: Normal chamber without chamber saturation

Migration distance: 7 cm

Development time: approx. 30 min

Sample solution: Treat approx. 200 mg dry extract with approx. 20 ml water and boil briefly (approx. 1 min in the microwave). After cooling to room temperature transfer the solution to a separating funnel, rinse the vessel used to dissolve the sample with approx. 5 ml water and adjust the pH of the combined solutions to pH 1 with hydrochloric acid. Then extract with approx. 30 ml ethyl acetate. Dry the organic phase over sodium sulfate, evaporate to dryness under vacuum and take up the residue in 1 ml methanol (equivalent to 200 mg dry extract/ml).

Reference solution: Work up 200 mg authentic archive sample in the same manner as the sample solution (equivalent to 200 mg/ml). A sample from an approved batch serves as authentic archive sample.

Comparison solution: Dissolve 1 mg each of rutoside and hyperoside in 1 ml methanol.

Validation solution: Dissolve 1 mg each of chlorogenic acid, quercitrin, hypericin and caffeic acid in 1 ml methanol.

Application volume: 10 µl each of sample and reference solution (equivalent to 2 mg dry extract)
5 µl each of comparison and validation solution (latter only as required)

bandwise, 15 mm

After application dry the plate 5 minutes in a stream of warm air (air heater, grade 1).

Document 12 / page 3

Special Testing Procedure **HD-052/91/5**

	Current date: 24.02.1999	Previous date: 06.12.1997	Page: 4/6

Identity, Purity (I): Dry Extracts of Flavonoid-Containing Drugs (TLC)

Detection: After evaporation of the solvent

(approx. 15 min. in a stream of warm air, grade 2, fume cupboard):

a) UV_{254} nm. Visual examination and photographic documentation.

b) Preheat layer to 60 - 80 °C, then first spray the still warm plate with flavone reagent A, then with flavone reagent B. Dry again at room temperature and inspect in UV at 365 nm. Visual evaluation and photographic documentation.

hRf values:

	hRf value	*Colour of fluorescence*
Rutoside	15 - 17	yellowish orange
Chlorogenic acid	34 - 38	pale blue
Hyperoside	38 - 41	yellowish orange
Quercitrin	63 - 65	yellowish orange
Hypericin	76 - 81	red
Caffeic acid	90 - 93	pale blue
Quercetin	93 - 95	yellowish orange

Evaluation:

Identity

The hRf values under short-wavelength UV light are determined from the fluorecence-quenching zones (flavonoids and phenolcarboxylic acids) on a pale green, fluoescent background.

Under long-wavelength UV light the colours of the intrinsically fluorescent substances are recorded in addition.

Purity (I)

Changes in the chromatogram of the sample with respect to number of zones, position of zones and fluorescence colour in comparison to the chromatogram of the corresponding reference dry extract are to be noted.

Documentation: Fill in the raw data sheet TLC II (and if necessary raw data sheet TLC III). Note the numbers of the balance used, TLC plate and reference substance(s). Record sample(s) weight and the quantity of the solvent. Cross the parameter's field to document the obtained result(s) and note the archive number(s) of the videoprints.

Transfer the result(s) to the testing protocol. Operator and checker have to sign the raw data sheet with date and initials. All further sheets are to be paginated and signed on the last page. Finally all TLC raw data sheets are added to the batch documentation.

Document 12 / page 4

Special Testing Procedure **HD-052/91/5**

	Current date: 24.02.1999	Previous date: 06.12.1997	Page: 5/6

Identity, Purity (I): Dry Extracts of Flavonoid-Containing Drugs (TLC)

Sample chromatograms

UV_{254} nm Fig. 94/2.12/19

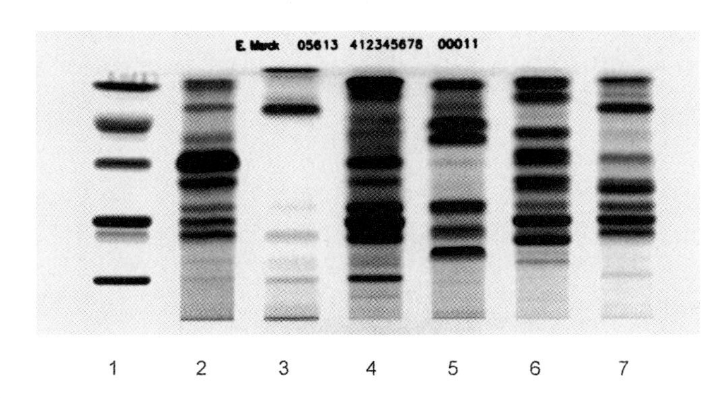

1 2 3 4 5 6 7

UV 365 nm after spraying with the flavone reagent acc. to Neu Fig. 94/Dec/29

[Insert the picture with the derivatizised chromatograms]

1 2 3 4 5 6 7

1: Comparison substances with increasing hRf values: rutoside, chlorogenic acid, hyperoside, quercitrin, hypericin, caffeic acid
2: Early golden-rod herb, dry extract
3: Red coneflower, dry herb
4: St. John's wort, dry extract
5: Horsetail herb, dry extract
6: Birch leaf, dry extract
7: Hawthorn leaf, dry extract

Special Testing Procedure **HD-052/91/5**

	Current date: 24.02.1999	Previous date: 06.12.1997	Page: 6/6
Identity, Purity (I): Dry Extracts of Flavonoid-Containing Drugs (TLC)			

Birch leaf dry extract, exposured samples

UV_{254} nm Fig. 94/16.1/5

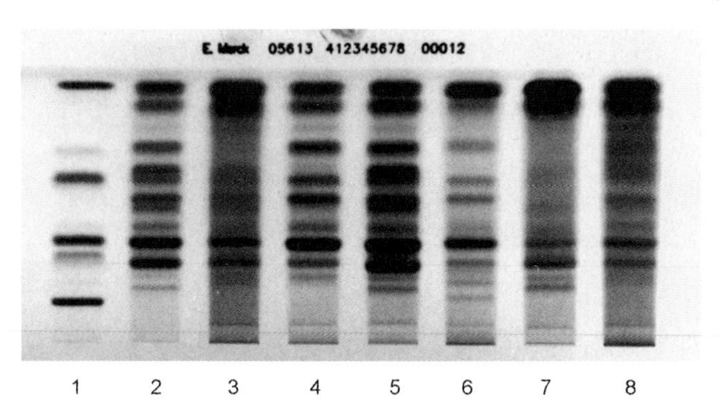

1 2 3 4 5 6 7 8

UV_{365} nm after spraying with the flavone reagent acc. to Neu Fig. 95/Jan/8

[Here figure 96 will be inserted]

1 2 3 4 5 6 7 8

1: Comparison substances with increasing hRf values: rutoside, chlorogenic acid, hyperoside, quercitrin, quercetin
2: Dry extract, freshly worked up
3: Dry heat, 120°C, 4 h
4: Exposure to light (sun test): 0,5% aqueous suspension, 20 h
5: Heat, 0,5% aqueous suspension, 80°C, 20 h
6: Action of hydrogen peroxide, 0,5% aqueous suspension, 80°C, 20 h
7: Hydrolysis: 0,5% hydrochloric acid suspension (pH 1 - 2), 80°C, 4 h
8: Hydrolysis: 0,5% alkaline suspension (pH 9 - 10), 80°C, 4 h

Document 12 / page 6

Preliminary Testing Procedure **QTP 97/01-T**

Department: Quality Control	Current Date: Datum: 12.07.1999	Previous date: letzte Ausgabe: 29.05.1997	Page: Seite: 1/3
Pharmaceutical Product **Theophylline in Effervescent Tablets**			

Topic: Assay of theophylline in effervescent tablets using high performance thin-layer chromatography and further densitometric measurement. This preliminary testing procedure was worked out acc. to literature [144]: "Validation of analytical procedures in pharmaceutical quality control" from B. Renger, H. Jehle, M. Fischer und W. Funk.

Application: This method is suitable for batch releases and stability testings.

Published: 12.7.99 E. Hahn-Deinstrop (Elke Hahn-Deinstrop)

First version: 29.05.1997 (German edition), 12.07.1999 (English edition)

Checked: 14.7.99 A. Koch (Dr. Angelika Koch)

Approved: 19.7.1999 F. Huse (Dr. Friederike Huse)

Appendices: ORIGINAL 21. Juli 1999

The reagents and apparatus specified in this testing procedure are exemplary and can be replaced by others of comparable suitability.

Document 13 / page 1

Preliminary Testing Procedure **QTP 97/01-T**

Department: Quality Control	Current Date: Datum: 12.07.1999	Previous date: letzte Ausgabe: 29.05.1997	Page: Seite: 2/3
Pharmaceutical Product **Theophylline in Effervescent Tablets**			

Principle:	HPTLC determination of theophylline.
No. of samples:	Batch release: 2 per each lot, 2 spots per each weight Stability: 2 per each lot, 3 spots per each weight
Reagents:	Acetic acid glacial..............................Merck 100063 Methanol..Merck 106007 2-Propanol...Merck 101040 Theophylline......................................CRS Toluene ....Merck 108325 Water, HPLC-quality
Sorbent:	Silica gel 60 F_{254} GLP (HPTLC precoated plate, 20 x 10 cm, Merck 105613) prewashing of the plate here is not necessary.
Solvent system:	Toluene + 2-propanol + acetic acid glacial 16 + 2 + 1 (v/v), no chamber saturation migration distance 3 cm, developing time approx. 15 min
Solvent:	Methanol + water (1+ 1 v/v)
Calibration stock solution:	Add approx. 40 ml solvent to 100,0 mg Theophylline, exactly weighed, in a 50ml volumetric flask. Treat in an ultra sonic bath for dissolution. Allow the solution cool to room temperature and fill up to the ring mark. (acc. to 2 mg/ml)
Calibration solution I:	Dilute 5,00 ml calibration stock solution with solvent to give 100,0 ml (acc. to 100 µg/ml)
Calibration solution II:	Dilute 6,00 ml calibration stock solution with solvent to give 100,0 ml (acc. to 120 µg/ml)
Calibration solution III:	Dilute 7,00 ml calibration stock solution with solvent to give 100,0 ml (acc. to 140 µg/ml)
Test solution:	Triturate 20 tablets. Bring the mass of 1 tablet, exactly weighed (acc.to 200 mg Theophylline), in a 50ml volumetric flask, add approx. 40 ml solvent and treat approx. 15 min in an ultrasonic bath. Allow the suspension cool to room temperature and fill up to the ring mark; mix and centrifuge 2 x 2 ml (Eppendorf, Safe-lock-vials) at 12000 rpm. Dilute 3,00 ml of the clear supernatant with solvent to give 100,0 ml. (acc. to 120 µg/ml)
Applied samples:	Bandwise using CAMAG Automatic TLC Sampler III Data-pair-technique 1 µl each Band 4 mm Space 6 mm Calibration solution I, II, III 3 x each Test solution batch release 2 x Test solution stability 3 x Allow the plate to dry 5 min in a stream of warm air (fan, grade I)

Document 13 / page 2

Preliminary Testing Procedure **QTP 97/01-T**

Department: Quality Control	Current Date: Datum: 12.07.1999	Previous date: letzte Ausgabe: 29.05.1997	Page: Seite: 3/3
Pharmaceutical Product **Theophylline in Effervescent Tablets**			

hRf-value: approx. 27

Evaluation: CAMAG TLC Scanner II using CATS Software,
Absorption at 274 nm,
Slit dimension 0,1 x 2 mm.

Calculation: The concentration of theophylline per tablet is calculated from the peak heights of sample spots and a calibration curve generated from three standard concentrations. Then, the percentage of the nominal value is calculated (200 mg = 100 %).

(Remark: quantitative evaluation using TLC scanner see Chapter 7.2.2.)

Sample chromatograms:

Calibration solution with and without addition of matrix (mixture of excipients), blank and matrix.

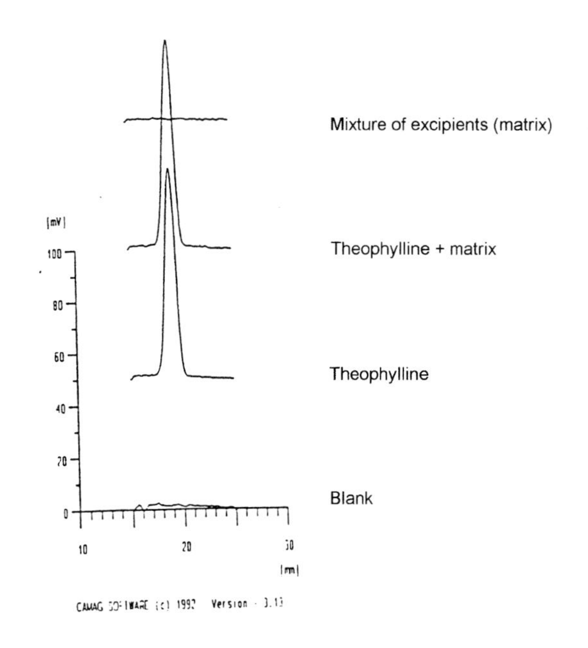

10 Effects of Stress

Two different aspects are considered under the above Chapter heading:

1. A controlled amount of stress on a substance before the analysis and subsequent TLC of the prepared samples.
2. Stress effects during the course of the TLC.

10.1 Controlled Stress on a Substance

To obtain information on the stability of a substance and to confirm the applicability of the testing method for stability tests, mainly in the field of pharmacy, controlled amounts of stress are applied to the substance. For this, the term "stress test" is generally used.

Usually, 20–50 mg of a pure active ingredient or ca. 100–500 mg of plant extracts, drugs or mixtures of active ingredients are weighed out. These are stored either as the dry substance or after addition of 10 ml solution.

Stress is applied by

- **thermolysis** of the dry substance at 120 °C
- **photolysis** by the action of light
- **oxidation** by a 3 % solution of hydrogen peroxide at 80 °C
- **hydrolysis** at 80 °C by both addition of acid (0.1 N HCl) and addition of base (0.1 N NaOH)

For comparison, a sample containing 10 ml water as well as the substance (extract etc.) is stored at 80 °C. After the predetermined storage time, the samples are worked up, i.e. undissolved substance must be brought into solution and acidic and basic solutions must be neutralized. Drugs and extracts are usually subjected to an often complex sample preparation before the preparations for TLC can be started.

Examples of the TLC investigation of stressed samples are shown in Fig. 92 (dexpanthenol) and in the testing procedure "Identity and Purity of Various Flavonoid-Containing Plant Extracts" is given in Section 9.4.2.

▶ **Figure 105:** see Photograph Section.

Applied Thin-Layer Chromatography: Best Practice and Avoidance of Mistakes, 2nd Edition
Edited by Elke Hahn-Deinstrop

ISBN: 978-3-527-31553-6

☞ **Practical Tips** for the stress test:

- Storage of the samples initially for ca. 100 h has given useful results.
- If only 50–70 % of the original substance remains, this is a clear indication of decomposition. If there is 100 % decomposition, the decomposition products could have decomposed further.
- If no detectable changes appear on the thin-layer chromatogram after 100 h, the concentrations of acid, alkali and hydrogen peroxide and also the storage time are increased.
- If 100 % decomposition occurs after the 100 h, the storage time must be reduced. This can be as little as 10 min., e.g. in the case of light-sensitive substances.

10.2 TLC-Sensitive Substances

Sensitive samples and measures necessary to preserve them have already been referred to in several parts of this book, but a discussion of all aspects of this subject will undoubtedly be helpful at this point. Further reading on some theoretical aspects of possible reactions can be found in Section 1.1 of Vol. 1b of the reagent books of Jork et al. [155].

10.2.1 Interactions with Sorbents

In Section 3.3 "Activation", it was reported that there is a considerable danger that sensitive substances will be decomposed on contact with highly active coatings. This effect, also known as "artifact formation", can be avoided in most cases by the use of a less active stationary phase (e.g. surface-modified sorbent). Interactions between sorbent and applied substance occur very frequently when the surface is also subjected to stress due to exposure to light (see also the example given in Section 10.2.3).

10.2.2 Effect of Elevated Temperature

Here we must distinguish between the effect of temperature both before and after the development (drying of the plate) and the effect of heat during chromatography. In the first case, the choice of a more suitable solvent system can prevent loss or decomposition of volatile or temperature-sensitive test substances. In the second case, development should be performed at lower temperatures, e.g. in a TLC "Thermo-Box" (see Section 4.2.3 "Effect of Temperature in Chromatography").

10.2.3 Effect of Light

In many operating instructions referring to light-sensitive samples, the precautions that have to be taken are mentioned either simply in the form of footnotes or not at all. However, especially in these circumstances it is essential that the necessary advice on light-protected operation should be given clearly and emphatically at the start of the testing procedure. This is also the case if relatively insensitive substances in solid form decompose rapidly when in solution.

Also, in the DAB monograph on the extremely light-sensitive compound nifedipine, special handling methods are prescribed:

"The substance, on exposure to daylight or artificial light of certain wavelengths, is rapidly converted into a nitrosophenylpyridine derivative. The action of UV light causes the formation of a nitrophenylpyridine derivative. Solutions should be prepared in the dark or in long-wave light (>420 nm) immediately before use and should be stored in the dark." Our own investigations have shown that ca. 85 % of dissolved nifedipine in a white graduated flask in daylight was decomposed after 10 min. Zieloff, in unpublished results which agree with our own, reports that this extremely light-sensitive compound can be tested by TLC if daylight is excluded and special lighting is used in the workplace. In his book, he shows the absorption spectrum of a nifedipine spot as a function of the duration of its exposure to light [2, p. 223].

In the case of pharmaceutical chemicals which are not discussed in a monograph of a pharmacopoeia, or for which no information on light sensitivity of the dissolved substance is provided in the monograph, these should first be subjected to a controlled amount of light stress (see Section 10.1). The result of this stress test could then lead to the conclusion that further experimental investigation of this substance in dissolved form could take place by daylight (or under ceiling lights) or that graded precautionary measures such as "subdued natural light", "semi-darkness" or even "darkroom conditions" are necessary. These phrases are defined as follows:

- **"Subdued natural light"** means winter daylight without the use of ceiling lights or, in bright summer days, at least the use of light-colored roller blinds to shade the windows.
  Pharmaceutical plants, which are submitted for investigation as pulverized drugs, should not be exposed to direct sunlight, and should therefore have some protection from the light. Our own investigations with prepared samples of *Echinaceae purpurea* (red coneflower) confirm that its constituents decompose relatively rapidly if a white graduated flask containing a test solution is left in a brightly lit window for 2 h. However, oxygen can also contribute to the decomposition of the constituents of this particular drug, and it therefore still remains to be clarified to what extent enzymatic processes make a contribution.

- **"Semi-darkness"** means that blinds should be lowered to cover ca. 80 % of the window area. Under these illumination conditions, it is just possible to work at the laboratory bench.

In the DAB, the monographs on nitrazepam and nitrofurantoin give advice on protection from light during tests. With respect to the TLC test for substances related to nitrazepam, it is stated that: "The test must be performed with exclusion of light." In the case of nitrofurantoin, similar advice applies to the UV measurement ("The test must be performed with exclusion of direct light."), but no recommendation is given about the TLC test for related substances. In order to forestall any doubts about the developed TLC plates, in both cases, after preparation of the samples, the application and development should be performed in a darkroom or in semi-darkness with the addition of a black cover over the TLC equipment.

- Work in the "**darkroom**" must be by the light of a special red lamp. It starts with opening the sample package at the analytical balance and finishes with evaluation of the developed TLC plate.
  Working in the dark is arduous, and is usually more difficult than was formerly thought. The following method of working is therefore recommended for beginners. First one should get used to working with all items of equipment in the dark, using "placebos" in place of actual samples. This includes, for example, grinding tablets, weighing samples, filling graduated flasks, using the ultrasonic bath and centrifuge, adding the test solutions to the autosampler vials when using a TLC automatic sampler and handling the spray nozzles when working with a Linomat. Before work with the actual samples is started, the development chamber, filled in readiness for development under conditions of chamber saturation, can be placed at the back of the laboratory bench. Application of samples by hand in the darkroom is not recommended, as the lighting is insufficient for this work.

Example of the influence of light: test of purity of an active ingredient

To determine the effect of light on a dissolved substance or on the substance in association with the sorbent, the following test is performed (during a short June night):

A solution of ursodeoxycholic acid (reference substance) in methanol (20 mg/ml) in a white graduated flask was allowed to stand in a window from ca. 2.0 p.m. until ca. 10.0 a.m. the following day. From this reference substance and from a new batch to be tested, test solutions of equal concentration and also a diluted reference solution were prepared. The reference substance (100 µg and 0.5 µg) was applied to two TLC silica gel 60 plates without fluorescence indicator, and both plates were dried for 3 min with a warm-air fan heater (grade I). One plate was then exposed to direct sunlight for 5 min. The development was then performed in the solvent system described in the DAC 1992. Figure 106 shows the two plates after derivatization with the phosphomolybdic acid reagent.

▶ **Figure 106:** see Photograph Section.

The following results were obtained:

a) On both plates, the sample stored overnight shows two additional zones (AZs) above the main zone (MZ).
b) Up to lane 3 (limit value concentration), all the lanes of the light-exposed plate (A) clearly show a decomposition product on the starting spot as well as the above-mentioned two AZs above the MZ.
c) Substance residues on the application points can also be seen on plate B, but are considerably weaker.

The results obtained can be interpreted as follows:

1. In the case of this substance, storage of a solution in the light in clear glass containers leads to clearly visible decomposition products after ca. 20 h. This means in practice that only freshly prepared solutions should be used, and the analyses should be performed as soon as possible.
2. The comparison also shows that exposure to direct sunlight after application leads to clearly detectable decomposition before the development.

10.2.4 Oxidative Effects

To prevent oxidation of substances on the layer, protective measures must be taken even before the samples are applied. Using the CAMAG Linomat IV with inert gas blanket, nitrogen (for example) can be applied before and during application of the sample. "To develop thin-layer chromatograms with exclusion of O_2, the chamber is purged with CO_2 if the substances are not affected by this gas, or alternatively with N_2. For this, a lid fitted with a tube with a double-valve gas supply system is useful" [20, p. 69].

11 Special Methods in TLC

This book would not be complete without a mention of a number of other variations of TLC. These have the following features in common:

- The separation makes use of various types of special equipment, the name of the method of separation usually being based on the name of the equipment required.
- All the equipment is extremely expensive (see also Section 4.2 "TLC Developing Chambers").

The equipment described in the following Sections is found in only a few laboratories. The interested user is therefore advised to examine its appearance and function personally at the manufacturers' stands in one of the large trade exhibitions (e.g. Achema, Analytica, InCom).

As the author has as yet had no practical experience with this type of equipment, Practical Tips are not included in this Chapter.

11.1 AMD – Automated Multiple Development

AMD is based on a development by Burger, who used it mainly for trace analysis of drinking water [156, 157]. This was the first method (before HPLC and GC) to be made into a German Standard (DIN 38407 Part 11) for the determination of plant protection agents in groundwater and drinking water [158].

Features of AMD:

- Preselected automatic solvent mixture gradient from polar to nonpolar
- Migration distances graded from a short to a long distance
- Focussing of the zones at each step, giving increased separating power (increased by a factor of 3 compared with normal HPTLC)
- Improvement of the selectivity by suitable selection of composition gradient
- No low-volatility solvent components (e.g. water) used
- Use of precoated plates with a very thin sorbent layer (100 μm)
- Possibility of coupling with HPLC
- Can be operated automatically outside normal working hours

Applied Thin-Layer Chromatography: Best Practice and Avoidance of Mistakes, 2nd Edition
Edited by Elke Hahn-Deinstrop

ISBN: 978-3-527-31553-6

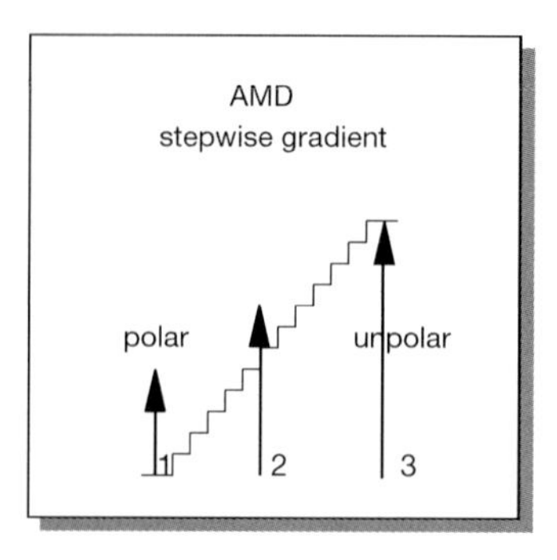

AMD is the only method with which reproducible gradient elution on silica gel is possible. In column chromatography, gradient elution can in practice only be used in the reverse phase; silica gel would be irreversibly changed, which, with a technique involving frequent re-use of the stationary phase, is not acceptable. This aspect is not relevant in TLC. (Note that the TLC plate is a disposable item!) [159].

The AMD systems 1 and 2 (without and with electronic data processing) are obtainable from CAMAG, which can provide the interested user with a series of information sheets describing various applications.

The main application areas of AMD are:

- Environmental analysis
- Analysis of natural products
- Analyses in the areas of pharmaceutically active substances and biochemistry

11.2 OPLC – Overpressured Layer Chromatography

The possibility of using pressure in TLC to achieve constant flow rates and to eliminate the effect of the gas phase on the TLC plate was first described in 1977 [160]. Here, the TLC plate is sealed with a "cushion". The pressure, which is applied to the layer by the cushion, must be higher than that causing the mobile phase to be transported through the layer. Since 1997, fully automatic equipment for OPLC has been available from CAMAG, together with the necessary plates and films sealed at the edges with a polymer [161].

This type of chromatographic development makes the following possible:

- Reduction of the development time
- Use of large migration distances without peak broadening by diffusion
- A choice of the most suitable H/u ratio [162]. (H = theoretical plate height, u = linear flow rate)
- Continuous chromatography, i.e. flow of the mobile phase beyond the end of the layer

- RP chromatography with solvent systems with a high water content on strongly lipophilic stationary phases with short development times

It should be noted that the exclusion of the gas phase can cause the OPLC chromatogram to differ considerably from one obtained with the same mobile phase in a normal chamber [161].

The following themes under the general subject heading "Working with OPLC" were presented at the International Symposium for Planar Chromatography in Interlaken in 1997:

- Food dyes [163]
- Food preservation agents [164]
- Biogenic amines in foods [165]

11.3 HPPLC – High Pressure Planar Liquid Chromatography

This variation, developed by Kaiser, is a fully instrumented circular TLC under pressure (which can be high if necessary) [166]. Here, a 10 × 10 cm HPTLC plate, with the layer underneath, is gripped between two reinforced glass disks. The sample is applied to the plate from the center of the lower reinforced glass plate via an inert capillary. The mobile phase can be "dosed" into the layer by a digitized syringe or via an HPLC pump at a constant flow rate. Of course, continuous chromatography is also possible by providing for continuous removal of the mobile phase. The mechanical pressure can be produced by an articulated lever press (see [167]).

11.4 TLC-FID/FTID – Combination of TLC and Flame-Ionization Detector or Flame-Thermionic Ionization Detector

This equipment, known as the IATROSCAN®-TLC-FID Analysator, is available from SES Analytical Systems. The company information sheet states [168]: "The analysis of high-boiling organic substances often presents problems. The IATROSCAN offers interesting possibilities by combining the advantages of GC and TLC.

Using this equipment, the separation takes place by the thin-layer method with the aid of so-called Chromarods (TLC material sintered onto fused silica rods), and detection is by the well-known FID method used in gas chromatography. The analysis is performed with the aid of various types of additional equipment. By attaching an FTID detector, not only carbon atoms but also halogen- and nitrogen-containing components can be detected with very high sensitivity.

The TLC-FID/FTID Analysator IATROSCAN TH-10 is especially suitable for the analysis of those compounds that show no UV absorption or fluorescence, are unsuitable for GC analysis and are difficult or impossible to detect by chemical reactions."

Applications:

- Fatty alcohols, fatty acids, fatty acid esters and glycerides
- Natural and synthetic fats and oils
- Lipids
- Antibiotics, pharmaceuticals, sulfonamides
- Vitamins and tocopherols
- Terpenes, steroids, hormones
- Carbohydrates
- Surfactants, emulsifiers, antioxidants
- Industrial emulsions, lubricants, parting agents, rolling oil, leather grease
- Hydrocarbons, mineral oil products, carbon hydrogenation products
- Polymers

Further information on the new IATROSCAN® MK6 can be obtained from the company SES (see Section 12.5).

11.5 TLC-NDIR

By coupling the Chromarod separation technique described in Section 11.4 with the reliable and robust NDIR (nondispersive infrared) technique for detecting the CO_2 formed in the combustion, completely new ways of analyzing hydrocarbons are opened up. This principle of detection has the advantage that it is specific for CO_2, i.e. cross-sensitivities are largely avoided.

The company ECH has developed the "Chromatherm" analytical equipment for the determination of low-volatility hydrocarbons. The extracted mixture of substances is first applied to the Chromarod rods. These are then developed in a solvent system to separate the sample into substance groups. The Chromarods are then thermally scanned by moving them forward at a constant rate into a stream of oxygen at 800 °C, where combustion of the substances occurs. This does not damage either the support material or the coating on the Chromarod rods, which can be reused many times. The determination of the proportions of the separated substance groups in the mixture of substances is performed by CO_2-selective NDIR detection [168a].

Using TLC-NDIR detection, it is possible to differentiate between important industrial products of biogenic and nonbiogenic origin. Mineral oil diesel and bio-diesel can be clearly distinguished from each other. The same applies to mineral oil-based lubricants and bio-lubricants based on rape oil, whose chromatograms are shown in Fig. 107a,b.

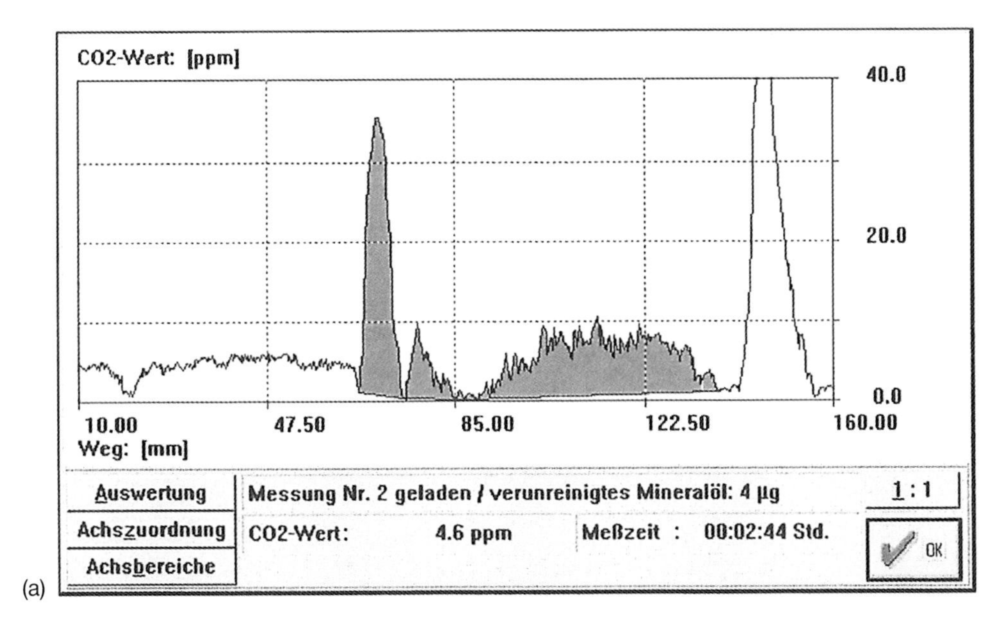

(a)

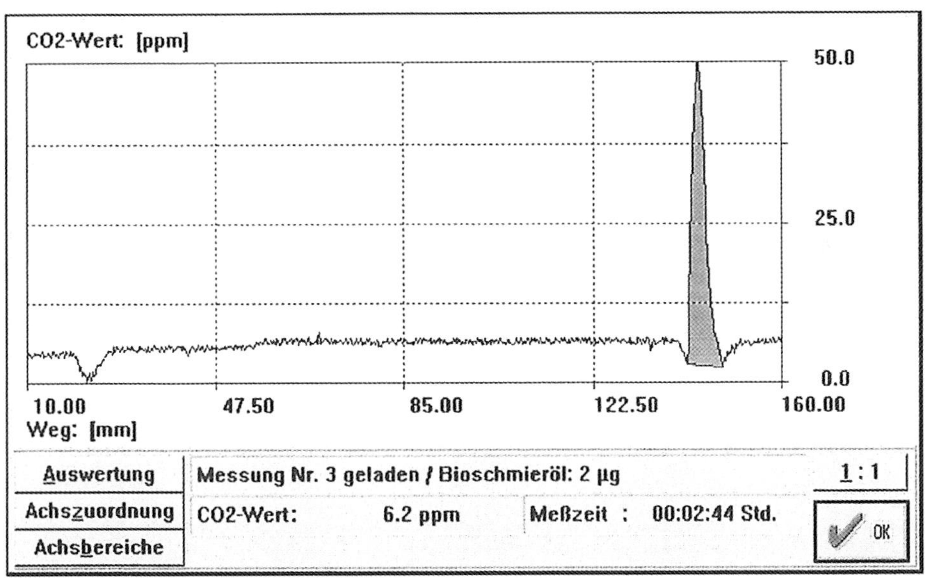

(b)

Figure 107. Comparison of (a) a mineral oil-based lubricant with (b) a rape oil-based lubricant with the aid of NDIR detection technology

11.6 RPC – Rotation Planar Chromatography

RPC, also known as centrifugal TLC, is mainly used for preparative purposes. Here, the substances are separated on a rotating disk by a solvent mixture which is fed continuously. The layers needed for this are prepared by the user by pouring the sorbent-containing suspension onto a roughened glass disk (diameter 24 cm). After the layer is bonded and dry, it is scraped down to the desired thickness. Usually, a mixture of 100 g sorbent (e.g. silica gel 60 PF_{254} containing gypsum, Merck 1.07749) with 200 ml water is used to prepare a layer 4 mm thick. A maximum of 2 g substance is then dissolved (in the solvent system if possible) and near the center of the plate. The development is started by feeding the solvent system into the layer, which is rotating at ca. 800 rpm. The maximum flow rate is ca. 10 ml/min. In most cases, the separation of the substances is observed by illuminating the disk with short-wave UV light, so that the desired fractions can be collected. In this version of the equipment, the layer is covered with a fused silica disk, as normal glass is not transparent to short-wave UV light. This enables ca. 500 mg of pure substance to be obtained from the 2 g of substance mixture originally used [169]. The new Chromatotron Model 7924T (with a Teflon funnel for solvent supply by gravity), available from Harrison (see Section 12.5 for the address), can be used to obtain substances by this method (Fig. 108).

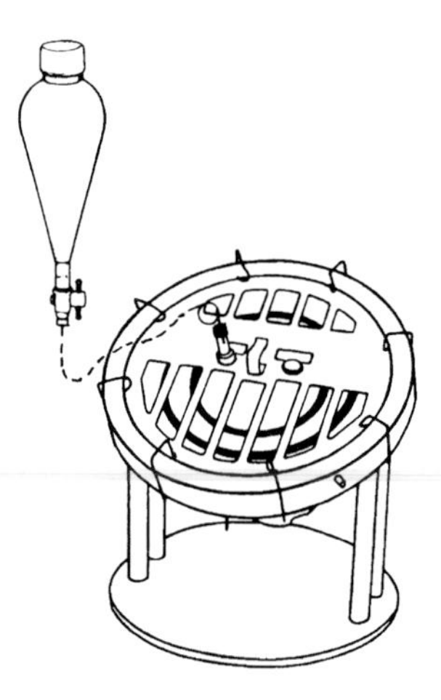

Figure 108. Model 7924T Chromatotron

12 Appendix

12.1 CHROMart

As a reward for the interested reader who has worked through to this part of the book, we include here an unusual aspect of TLC which is worthy of some study. TLC is probably the only scientific analytical method with which artistic results can also be obtained.

During our literature researches, we found some curious colored images in a book [11]. These, on closer investigation, turned out to be developed thin-layer chromatograms. This artistic technique originated from some research by the company Merck in 1981 in connection with the analysis of dyes by chromatography. Some phenomena observed in the course of this work by H. Halpaap († 1987) and H.E. Hauck led to a large number of surprising and attractive effects.

The discoverer named the result of these experiments "CHROMart", a name that expresses the idea of using a scientifically based analytical technique for the production of aesthetically pleasing images. Here, artistic creativity is complemented by a knowledge of thin-layer chromatography and hydrodynamics.

The diversity, creative possibilities and not least the beauty of the resulting compositions of colors and shapes eventually led to a patent application for the process [170].

The most important feature of the CHROMart technique is that the finished image is determined not only by the choice of the color-producing components and the preselected shape and size of the areas of application of the solvents but also by a deliberate change to the original arrangement of colors and shapes achieved by a type of "thin-layer-chromatographic development" in one or more consecutive steps. Every image produced is a genuine "one-off". The variety shown in the appearance of these images is remarkable and is clearly reminiscent of pictures by abstract artists.

The impetus for our own investigations into this technique was provided by a gift: a small bottle containing a few ml of a mixture of dyes. Carefully guarded for some months, its contents were first used at the conclusion of a fundamental course on chromatography for apprentices. In the last two hours of this course, the participants were invited to experiment with the colored solution. The results of this were so outstanding that Merck not only included them in a contribution to their house journal, but used a detail of one of the CHROMarts as a cover picture [171].

Up to this point in time, the colored "works of art" could be viewed in daylight. However, depending on the derivatization process, fluorescence colors can be produced in the analysis of some medicinal plants, so that the next step was to use substances of this type for CHROMart experiments also. Since 1991, the author, in her

Applied Thin-Layer Chromatography: Best Practice and Avoidance of Mistakes, 2nd Edition
Edited by Elke Hahn-Deinstrop

ISBN: 978-3-527-31553-6

spare time, has developed "artistic TLCs" in which the main medicinal plants used were *Ginkgo biloba, Echinacea purpurea, Hypericum perforatum, Betula folium, Solidago virgauria* and *Chelidonium majus*. The CHROMarts obtained have been shown at trade fairs (Interlaken 1995, InCom 1998) and in our own exhibitions (Ansbach 1998 [172], Eckental 1999 [173]). Experts as well as newcomers to TLC have now taken up this art, and the title pages of brochures and scientific journals are now being adorned with CHROMarts [174–176].

What is the difference between CHROMart and normal TLC?

If dyes or phytopharmaceutical preparations, for example, are used individually or in combination, the CHROMart process usually leads to an image that gives the observer the impression that the chromatogram was produced exactly as intended, just as a painter produces his picture. However, this is not the case, as the result is only partially foreseeable.

One of the possibilities for producing a CHROMart is to deliberately damage the sorbent layer after applying a "sample" solution. This can be achieved, for example, with the aid of a ruler and a sharp knife. The plate is then developed in a solvent system suitable for the "sample" and is, if necessary, derivatized. A further possibility is to use a piece of cotton wadding to apply the sample solution in a triangular shape at all four corners of the plate and to chromatograph the plate in four directions towards the middle.

"Samples", as mentioned above, can be mixtures of dyes or substances that can be converted into colored derivatives. It is not always necessary to use medicinal drugs and preparations containing them. Often, one can find suitable plants in one's own garden or on a nature walk. This search can be helped by easily obtainable books on plants that often also give information on plant constituents, e.g. Schilcher's "Kleines Heilkräuter-Lexikon" ("Small Encyclopedia of Medicinal Herbs") [177] or Ennet's encyclopedia "Heilpflanzen und Drogen" ("Medicinal Plants and Drugs") published by the Bibliographisches Institut (Duden-Verlag, Leipzig, in the former German Democratic Republic) [178].

CHROMarts need not necessarily be colored. Sometimes developed "chromatograms" photographed in black-and-white can be amazing works of art. Before derivatization, photographs have been taken under short-wave UV light using a document film.

Some idea of TLC art can be obtained from p. 298 of this book. This shows four examples from my own collection, which contains over 100 CHROMarts. The choice was not easy, as this small number of images can hardly give a true impression of the many possibilities.

12.2 References

[1] Stahl, E.: Pharmazie 11, 633 (1956)

[2] Frey, H. P., K. Zieloff: Qualitative und quantitative Dünnschichtchromatographie, Grundlagen und Praxis, VCH, Weinheim 1993, ISBN 3-527-28373-0

[3] Weins, Chr.: "Fortschritte und Entwicklungsmöglichkeiten in der Dünnschichtchromatographie", GIT Spezial, Chromatographie 1/96, 34–37 (1996)

[4] Geiss, F.: Die Parameter der Dünnschichtchromatographie, Vieweg, Braunschweig 1971

[5] Geiss, F.: Fundamentals of Thin Layer Chromatography, Hüthig, Heidelberg 1987

[6] Ennet, D.: "Dünnschichtchromatografische Prüfung", Monographie "Oleum Menthae piperitae", DAB 7 – DDR 64 in the commentary to the DAB 7- DDR, Akademie, Berlin, 1969

[7] "V.6.20.2 Chromatographie sur couche mince", Pharmeuropa 6, No. 2, June 1994, 95–97

[8] Altorfer, H. R., Chr. P. Christiansen, W. Dammertz, J. Lang, B. Renger in Readers' Tribune: "Thin-Layer Chromatography in the European Pharmacopoeia", Pharmeuropa 8, No. 1, March 1996, 172–174

[9] European Pharmacopoeia Commission, draft of method 2.2.27 "Thin-Layer Chromatography", 1997

[10] Hauck, H. E., A. Junker-Buchheit, R. Wenig: "Reproduzierbarkeit in der DC", GIT Fachz. Lab. 11/93, 973–977 (1993)

[11] Wintermeyer, U.: Die Wurzeln der Chromatographie, GIT, Darmstadt 1989 ISBN 3–921956-82-X

[12] ChromBook MERCK, German edition (1995), W 210001

[13] Hahn-Deinstrop, E.: "Dünnschicht-Chromatographie im DAB, Teil 1", Dtsch. Apoth. Ztg. 135, 2589–2594 (1995)

[13a] Fischer, W., H.-E. Hauck, G. Wieland: "Dünnschicht-Chromatographie aktuell", LABO 06/97, 56–62, Hoppenstedt, Darmstadt 1997

[13b] Koch, A., E. Hahn-Deinstrop: "Weihrauch – ein traditionelles Naturheilmittel wird wieder aktuell", BIOforum 6/98, 352–357, GIT, Darmstadt 1998

[13c] Hahn-Deinstrop, E., A. Koch, M. Müller: "Guidelines for the Assessment of the Traditional Herbal Medicine 'Olibanum' by Application of HPTLC and DESAGA ProViDoc® Video Documentation", J. Planar Chromatogr. 11, 404–410 (1998)

[13d] Koch, A., E. Hahn-Deinstrop, R. Richter, K. Jahn: "Chromatographic Methods for Determination of the Gum Resins of Olibanum", Poster at the Joint Meeting of the ASP, AFERP, GA and PSE – 2000 Years of Natural Products Research, 26–30 July, Amsterdam, The Netherlands (1999)

[14] Zerschneiden von DC-Folien, Merck-Spectrum (own publication), Darmstadt 2/1991, 24–25

[15] Funk, W.: "Derivatisierungsreaktionen im Bereich der quantitativen Dünnschicht-Chromatographie", Fresenius Z. Anal. Chem. 318, 206–219 (1984)

[16] Maxwell, R. J., S. W. Yeisley, J. Unruh: J. Liq. Chromatogr. 13, 2001–2011 (1990)

[17] Stahl, E. (ed): Dünnschicht-Chromatographie, ein Laboratoriumshandbuch, Springer, Berlin 1962

[18] Herstellung konstanter Luftfeuchtigkeit in geschlossenen Gefäßen, Merck Tabellen für das Labor (own publication), Darmstadt 1990

[19] Hauck, H. E., M. Mack: "Sorbents and Precoated Layers in Thin-Layer Chromatography", in: J. Sherma, B. Fried (Ed.), Handbook of Thin-Layer Chromatography, Chromatographic Sciences Series, Volume 71, Marcel Dekker, New York, USA 1996, ISBN 0-8247-9454-0

[20] Janssen, A., A. Neitzel, A. Lau: "Identifizierung von Zimtstoffen mittels DC und HPLC", in: CHROMATOGRAPHIE – Chronologie einer Anaylsentechnik, Merck KGaA (ed), 166–176, GIT Verlag 1996, ISBN 3-928865-21-8

[21] Kraus, Lj., A. Koch, S. Hoffstetter-Kuhn: Dünnschichtchromatographie, Springer Labor Manual, Berlin 1996, ISBN 3-540-59309-8

[22] Stahl, E. (ed): Dünnschicht-Chromatographie, ein Laboratoriumshandbuch, 2. Auflage, Springer, Berlin 1967
[22a] Oka, H., J. Patterson: "Chemical Analysis of Tetracycline Antibiotics", in: Oka, H., H. Nakazawa, K. Harada, J. D. MacNeil (eds), Chapter 10 in: "Chemical Analysis for Antibiotics Used in Agriculture"; AOAC INTERNATIONAL, Arlington, USA (1995), ISBN 0-935584-57-9
[23] Fulde, G., M. Wichtl: "Analytik von Schöllkraut", Dtsch. Apoth. Ztg. 134, 1031–1035 (1994)
[24] Fulde, G.: "Untersuchungen zur Verteilung der Alkaloide in Chelidonium majus L.", Dissertation, p. 136, Marburg 1993
[25] Hahn-Deinstrop, E., A. Koch: "Planar Chromatographic Determinations of Chelidonium Majus L.", in: Proceedings of the 9th International Symposium on Instrumental Planar Chromatography, Interlaken 9–11 April, 1997, 115–122
[26] Nyiredy, Sz.: Editorial in J. Planar Chromatogr., Vol. 9, No. 6 (1996)
[27] Sprecher, E., A. Koch: In memoriam Ljubomir Kraus, 44. Tagung der Gesellschaft für Arzneipflanzenforschung in Prag, 3.–7.9.1996, printed by the Institut für Pharmazeutische Biologie der Universität Hamburg (1996)
[28] Waters Sep-Pak® Cartridge Application Bibliography, Millipore Corporation, Bedford, MA 01730, USA 1986
[29] Merck LiChrolut® Applikationen zur Probenvorbereitung, E. Merck, Abt. LPRO CHROM 1, Darmstadt
[30] Jork, H., H. Wimmer: Bibliography "Quantitative Auswertung von Dünnschicht-Chromatogrammen, I./3–3, Auftragefehler", GIT, Darmstadt 1989
[31] Rößner, E., Heumann Pharma, Qualitätskontrolle Chemie, September 1996 (private communication)
[32] Jork, H., H. Wimmer: Bibliography "Quantitative Auswertung von Dünnschicht-Chromatogrammen, Kapitel I./3–4, Kolbenspritzen", GIT, Darmstadt 1989
[33] Junker-Buchheit, A.; H. Jork: "Prächromatographische in situ-Derivatisierung von Fettsäuren im Picomol-Bereich", Fresenius Z. Anal. Chem. 331, 387–393 (1988)
[34] Jork, Funk, Fischer, Wimmer: Dünnschicht-Chromatographie, Reagenzien und Nachweismethoden, Vols. 1a and 1b, VCH, Weinheim 1989 and 1993, ISBN 3-527-26848-0 and 3-527-28205-X
[35] CAMAG, TLC Catalog, 1996/97
[36] DESAGA, Bedienungsanleitung "DC-Auftragegerät PS 01"
[37] CAMAG, Brochure "DC-Probenautomat III"
[38] DESAGA, Brochure "TLC-Applicator AS 30"
[39] BARON, Brochure "TLS 100"
[40] Bethke, H., W. Santi, R. W. Frei: J. Chromatogr. Sci. 12, 392–403 (1974)
[41] Trappe, W.: Biochem. Z. 305, 150 (1940) and 306, 316 (1940)
[42] Halpaap, H., J. Ripphahn: "Entwicklung, Daten und Ergebnisse der Hochleistungs-Dünnschicht-Chromatographie (HPTLC)", Kontakte (Merck), 3, 16–34 (1976)
[43] Snyder, L. R.: J. Chromatogr. Sci. 16, 223–234 (1978)
[44] Kirkland, J. J., J. L. Glajch in Gertz, Ch., W. Fellmann: Fresenius Z. Anal. Chem. 323, 342–349 (1986)
[45] Nyiredy, Sz., K. Dallenbach-Toelke, O. Sticher: " The 'PRISMA' Optimization System in Planar Chromatography", J. Planar Chromatogr. 4, 336–342 (1988)
[46] Koch, A., Frohme Apotheke Hamburg, Juli 1996 (private communication)
[47] Wagner, H., S. Bladt, E. M. Zgainski: Drogenanalyse, Springer, Berlin/Heidelberg/New York 1983, ISBN
[48] Pachaly, P.: DC-Atlas, Dünnschichtchromatographie in der Apotheke, Wissenschaftliche Verlagsgesellschaft, Stuttgart 1994 (loose-leaf collection), ISBN
[49] Niesel, S.: "Untersuchungen zum Freisetzungsverhalten und zur Stabilität ausgewählter wertbestimmender Pflanzeninhaltsstoffe unter besonderer Berücksichtigung moderner phytochemischer Analyseverfahren", Dissertation, Berlin 1992, p. 163
[50] Stieneker, F.: "Horizontal-DC", Dtsch. Apoth. Ztg. 134, 4442–4448, (1994)
[51] Jork, H., W. Funk, W. Fischer, H. Wimmer: Dünnschichtchromatographie, Reagenzien und Nachweismethoden, Vol. 1b, p. 347–349, VCH, Weinheim 1993

[52] Applikationen zur DC-Aluminiumfolie RP-18 F 254s, Art. Nr. 5559, E. Merck, Darmstadt 1993
[53] Surborg, K.-H.: "Über eine Möglichkeit, die DC-Analyse preiswert(er) zu gestalten", Dtsch. Apoth. Ztg. 121, 1414–1416, (1981)
[54] Hahn U.: Abendgymnasium Oldenburg, Mai 1997 (private communication)
[55] Tschirsch, C., Lj. Kraus: "Goldregen-Alkaloid Cytisin", Dtsch. Apoth. Ztg. 131, 1876–1878, (1991)
[56] Koch, A., M. Müller: "Metabolism and Pharmacokinetics of Aloin: Optimization Procedures – Analysis and Documentation of Aloin in Human Feces", J. Planar Chromatogr. 9, 56–61 (1996)
[57] Koch, A.: "Quantitation of Methadone (D+L) and L-Polamidone in Soft Drinks" in: Proceedings of the 9th International Symposium on Instrumental Planar Chromatography, Interlaken 9–11 April, 1997, 155–159
[58] Quality Control Methods for Medicinal Plant Materials (1992), WHO/Pharm/92.559
[59] Geiss, F., H. Schlitt, A. Klose: Z. Anal. Chem. 213–331 (1965)
[60] "PAK-Bestimmung in Wasser mit Hilfe der HPTLC", Merck Spectrum 2/94, 17 (own publication), Merck Darmstadt (1994)
[61] Lichius, J. J., M. Daniel, Th. Fingerhut, K. Wärtgen: "Der Nektar von Digitalis", Dtsch. Apoth. Ztg. 130, 2191–2199 (1990)
[62] Klump, B., H.-U. Melchert: "Erhöhung von Trennqualität und Reproduzierbarkeit dünnschicht-chromatographischer Analysen mit Hilfe von Styroporkästen", Fresenius Z. Anal. Chem. 310, 252–253, (1982)
[63] Pachaly, P.: "Alternativen zu DC-Vorschriften des Arzneibuchs", Dtsch. Apoth. Ztg. 133, 3463–3469, (1993)
[64] Ropte, D., B. Volkmann: "Dünnschichtchromatographie im DAB", Dtsch. Apoth. Ztg. 135, 2956–2958, (1995)
[65] Kennedy, J. F.: Carbohydrate Analysis, A practical Approach, Oxford University Press, 2nd. edn (1994), ISBN 0-19-963449-1
[65a] Madaus, A.: "Analytische Untersuchung der wertbestimmenden hochmolekularen Schleimpolysaccharide von Althaea officinalis", Dissertation, Regensburg 1989
[66] Stahl, E., U. Kaltenbach: J. Chromatogr. 5, 351 u. 458 (1961)
[67] Ebel, S., S. Völkl: "DC-Analytik von Naturstoffen", Dtsch. Apoth. Ztg. 130, 2162–2169 (1990)
[68] Veit, M.: "Mikro-DC", Dtsch. Apoth. Ztg. 136, 2686–2693, (1996)
[69] Consden, R., A. H. Gordon, A. J. P. Martin: Biochem. J. 38, 224 (1944)
[70] Kraffczyk, F., R. Helger, H. Lang, H. J. Bremer: Clin. Chim. Acta 35, 345–351 (1971)
[71] Broschüre: "Mitteilungen zur Dünnschicht-Chromatographie X, Aminosäuren im Harn", E. Merck, Darmstadt (1974)
[71a] Koch, A., E. Hahn-Deinstrop: "Quality Control of Ginseng Roots by a Rapid & Sensitive HPTLC-Method", Poster at the 46th Annual Congress, Society for Medicinal Plant Research, Vienna, Austria, 31.8.–4.9.1998
[72] Hahn-Deinstrop, E.; "Schöllkraut, Dünnschichtchromatographische Untersuchungen", Dtsch. Apoth. Ztg. 134, 4449–4454 (1994)
[72a] Hahn-Deinstrop, E., A. Koch: "Planar Chromatographic Determinations of Chelidonium Majus L.", Poster at the 9th International Symposium on Instrumental Planar Chromatography, Interlaken, Switzerland, 9.–11. April 1997
[73] Heisig, W., M. Wichtl: "Bestimmung pflanzlicher Glycoside. Methodenkombination von zweidimensionaler DC und HPLC. I. Inhaltsstoffe aus Calendulae officinalis flos", Dtsch. Apoth. Ztg. 130, 2058–2062 (1990)
[73a] Hahn-Deinstrop, E., A. Koch: "Ginkgo biloba hilft dem Gehirn auf die Sprünge (Ginkgo biloba Trains the Brain)", BIOforum 7–8/98, 428–434, GIT, Darmstadt 1998
[74] Henning, W., L. Bertling: "Farbstoffe in Fleischerzeugnissen", GIT Supplement 2/86, Lebensmittelchemie, 27–29, GIT, Darmstadt (1986)
[75] Hahn-Deinstrop, E.: "Identification of Red Sandalwood", J. Planar Chromatogr. 6, 81–83 (1993)
[76] Jork, Funk, Fischer, Wimmer: as [34], Vol. 1b, pp. 347–348

[77] Huse, F.: "Thin-layer chromatographic comparison of branded perfumes and their imitator scents", Cosmetics and Toiletries Manufacture Worldwide, 236–238, Aston Publishing Group, Hongkong (1993)
[78] Stahl-Biskup, E., E. Wilhelm: "Die ätherischen Öle im Spiegel der europäischen Arzneibücher"; Dtsch. Apoth. Ztg. 136, 3019–3032 (1996)
[79] CAMAG Applikation A-18.2: " Quantitative Bestimmung von Glycyrrhizinsäure in radix liquiritiae (Liquorice root), (7/87)
[80] Jänchen, D., H. Jele, M. Zeller: "Standard Operating Procedure Dünnschichtchromatographie", CAMAG-Brochure, 31.5.1995
[81] Burger, K.: "AMD (Automated Multiple Development), Anwendungen und online Kopplung mit reversed phase HPLC", InCom Sonderband "Dünnschicht-Chromatographie in memoriam Prof. Dr. Hellmut Jork", 31–71 (1996)
[82] Kubeczka, K.-H.: "Analyse von Naturstoffen", Dtsch. Apoth. Ztg. 127, 2443–2450 (1987)
[83] Jork, Funk, Fischer, Wimmer: as [34], Vol. 1b, p. 31
[83a] Reimers, C.: "Screening mineralölkontaminierter Böden mit Hilfe der Dünnschichtchromatographie", GIT Spezial, Separation 1/99, 32–33, April (1999)
[84] Jänchen, D., CAMAG, CH-Muttenz, Dezember 1996, see also CAMAG Applikationsschrift A 28.6, Seite 6 (private communication)
[84a] Jork, H., Z. Anal. Chem. 221 (1966) 17–33
[84b] Spangenberg, B., K.-F. Klein, J. Chromatogr. A 898 (2000), 265–269
[84c] Spangenberg, B., K.-F. Klein, J. Planar Chrom. 14 (2001) 260–265
[85] Glauninger, G., K.-A. Kovar, V. Hoffmann: "Possibilities and limits of an on-line coupling of thin-layer Chromatography and FTIR-spectroscopy", Fresenius' Z. Anal. Chem. 338, 710–716 (1990)
[86] Kovar, K.-A., V. Hoffmann: "Möglichkeiten und Grenzen der direkten DC-FTIR-Kopplung", GIT Fachz. Lab. 11/91, 1197–1201 (1991)
[87] Kovar, K.-A., J. Dinkelacker, A. M. Pfeifer, W. Pisternick, A. Wössner: "Identifizierung von Suchtstoffen mit Hilfe der HPTLC-UV/FTIR-Kopplung", GIT Spezial, Chromatographie 1/95, 19–24 (1995)
[88] Petty, C., N. Cahon: "The analysis of thin layer chromatography plates by near-infrared FT-Raman", Spectrochim. Acta 49A, No. 5/6, 645–655 (1993)
[89] Moss, S., M. Zeller: "On-line Raman-Spektroskopie in der Dünnschichtchromatogaphie", Tagungsband InCom 1994, p. 46
[89a] Sollinger, S., J. Sawatzki: "TLC-Raman für Routine-Anwendungen", GIT Labor-Fachzeitschrift 1/99, 14–18, GIT, Darmstadt 1999
[90] Koglin, E.: "Combining HPTLC and Micro-Surface-Enhanced Raman Spectroscopy (Micro-SERS)", J. Planar Chromator. 2, 194–197, (1989)
[91] Koglin, E.: "Kombination von Dünnschicht-Chromatographie und SERS", CLB Chemie in Labor und Biotechnik, Vol. 47, No. 6, 257–261 (1996)
[92] Busch, K. L.: "Thin-Layer Chromatography coupled with Mass Spectrometry", in Handbook of Thin-Layer Chromatography, J. Sherma, B. Fried; Chromatographic Science Series, Vol. 55, 1991, 183–209, Marcel Dekker, New York
[93] Guser, A. J., A. Proctor, Y. J. Rabinovich, D. M. Hercules: "Thin-Layer Chromatography Combined with Matrix-Assisted Laser Desorption/Ionization Mass Spectrometry", Analytical Chemistry, Vol. 67, No. 11, June 1, 1805–1814 (1995)
[93a] Morlock, G., S. Häberle, U. Jautz, W. Schwack: „New HPTLC-MS method for determination of heterocyclic aromatic amines", CAMAG Bibl. Service CBS 93, p. 14–15 (2004)
[93b] Luftmann, H.: „A simple device for the extraction of TLC spots: direct coupling with an elctrospray mass spectrometer", Anal Bioanal Chem 378, p. 964–968 (2004)
[93c] Morlock, G., W. Schwack: "Quantifizierung von ITX in Lebensmitteln mittels HPTLC/FLD gekoppelt mit ESI-MS und DART-MS", CAMAG Bibl. Service 96, p. 11–13 (2006)
[94] Jork, Funk, Fischer, Wimmer: Dünnschicht-Chromatographie, Reagenzien und Nachweismethoden, Vol. 1a, p. 77, VCH, Weinheim 1989 and 1993
[95] Klaus, R., W. Fischer, H.E. Hauck: "Reagent-Free Detection of Substances on Amino-Modified Silica Gel Layers – Thin-Layer Chromatographic Separation and Detection of Fruit Acids", LC-GC INT: Volume 8 Number 3, 151–156 (March 1995)

[96] ATLAS Material Testing Technology BV, brochures and other company information on the Suntest CPS equipment (1996)
[97] Funk, W., M. Kornapp, G. Donnevert, S. Netz: "Quantitative HPTLC-Determination and Characterization of Organotin Compounds", J. Planar Chromatogr., Vol. 2, 276–281 (1989)
[98] Funk, W., M. Azarderakhsh: "Derivatisierung von β-Blockern in der quantitativen HPTLC", GIT Fachz. Lab. Supplement "Chromatographie", 31–39 (1990)
[99] E. MERCK, Brochure: "Anfärben in der Dünnschicht-Chromatographie", No. W 215 002
[100] Krebs, K. G., D. Heusser, H. Wimmer, in: Stahl, E. (ed): Dünnschicht-Chromatographie, ein Laboratoriumshandbuch, 2. Auflage, Chapter "Sprühreagenzien", p. 814 (spray scheme of D. Waldi), Springer, Berlin 1967
[101] Jork, Funk, Fischer, Wimmer: Dünnschicht-Chromatographie, Reagenzien und Nachweismethoden, Vol. 1a, p. 277: "Diphenylborsäure-2-aminoethylester-Reagenz (Naturstoffreagenz nach Neu)" , VCH, Weinheim 1989 and 1993
[102] Bauer, R., H. Wagner: "Echinacea", Handbuch für Ärzte, Apotheker und andere Naturwissenschaftler, Wissenschaftliche Verlagsgesellschaft mbH, Stuttgart (1990), ISBN 3-8047-0999-0
[102a] Hahn-Deinstrop, E., R. Bauer: "Echinacea – Chromatographischer Nachweis einer Kreuzung von E. purpurea und E. angustifolia", BIOforum 3/99, 109–114, GIT, Darmstadt 1999
[102b] DESAGA, information sheet on the ChromaJet DS 20, 1999
[102c] Koch, A., E. Hahn-Deinstrop, M. Müller: "HPTLC-Quantification of Boswellic Acids for Quality Assurance of *Olibanum*. A Comparison between DESAGA Opto-Electronic- and Video Densitometer after Derivatisation by Autospraying with the DESAGA ChromaJet CS20", Poster at the Joint Meeting of the ASP, AFERP, GA and PSE – 2000 Years of Natural Products Research, 26–30 July, Amsterdam, The Netherlands (1999)
[102d] Hahn-Deinstrop, E., A. Koch, M. Müller: "HPTLC-Measured Values of Primulasaponins in Extracts of *Primulae radix* using the DESAGA TLC-Scanner CD 60. A Comparison after Derivatisation between Handspraying, Dipping and Autospraying", Poster at the Balaton Symposium 99, 1–3 September, Hungary (1999)
[102e] Bauer, R., P. Remiger, H. Wagner: "Echinacea, Vergleichende DC- und HPLC-Analyse der Herba-Drogen von E. purpurea, E. pallida und E. angustifolia, 3. Mitt.", Dtsch. Apoth. Ztg. 128, 174–180 (1988)
[103] Jork, Funk, Fischer, Wimmer: Dünnschicht-Chromatographie, Reagenzien und Nachweismethoden, Vol. 1a, p. 82 ff, VCH, Weinheim 1989 and 1993
[104] Stahl, E. (ed): Dünnschicht-Chromatographie, ein Laboratoriumshandbuch, Springer, Berlin 1962, p. 400 ff
[105] Niesel, S.: "Untersuchungen zum Freisetzungsverhalten und zur Stabilität ausgewählter wertbestimmender Pflanzeninhaltsstoffe unter besonderer Berücksichtigung moderner phytochemischer Analyseverfahren", Dissertation, Berlin 1992, Sonderteil "Freisetzung und Stabilität hämolysierender Substanzen aus Birkenblättern, Goldrutenkraut, Orthosiphonblättern und Primelwurzel", p. 229 ff
[106] Kartnig, T., F. J. Graune, R. Herbst: "Zur Frage der Saponinverteilung bei Aesculus Hippocastanum während verschiedener Keimungs- und Wachstumsstadien", Planta Med. 12, 428–439 (1964)
[107] Maushart, R.: "100 Jahre Radioaktivität und die Folgen – der Rede wert! Teil 3", GIT Fachz. Lab. 12/96, pp. 1257–1265 (1996)
[108] Dietzel, G., R. Grugel: „Radiochromatography and Its Applications", The Column, an Advanstar Publikation, April 2006 issue
[109] Hahn-Deinstrop, E.: "Eibischwurzel, Identifizierung von Eibischwurzel-Extrakt und Gehaltsbestimmung in einem Instant-Tee", Dtsch. Apoth. Ztg. 135, 1147–1149 (1995)
[110] MTSS – Merck Tox Screening System, Instruction Manual, published by E. Merck, Darmstadt and Department of Analytical Chemistry and Toxicology, University Centre for Pharmacy, State University, 9713 AW Groningen, The Netherlands (1991)
[111] Bernhard, W., S. R. Rippstein, A. N. Jeger (Institut für Gerichtschemie, Basel), Poster at the Symposium für klinisch-toxikologische Analytik, Salzburg 1987

[112] Schmitt, J.: Heumann Pharma, Dept. Chemical Production, January 1997 (private communication)
[113] Marston, A., M. Maillard, K. Hostettmann: " The role of TLC in the investigations of medicinal plants of Africa, South America and other tropical regions", (in preparation for printing by GIT, 1997; private communication, April 1997)
[114] Jork, H., Chr. Weins: "Toxikologische Bewertung von Schadstoffen durch enzymatische in situ Detektion in der Dünnschicht-Chromatographie" , InCom special volume "Dünnschicht-Chromatographie in memoriam Prof. Dr. Hellmut Jork", pp. 225–237 (1996)
[115] Heisig, W., M. Wichtl: "Mikrowellen-Bedampfungstechnik", Ein neues Verfahren zum Nachweis von Substanzen in der Dünnschichtchromatographie, I. Nachweis von Flavonoiden, Dtsch. Apoth. Ztg. 129, 2178–2179 (1989)
[116] "CATS Spektrenbibliothek – jetzt unter MS-Windows", CAMAG literature service CBS No. 76, 9 (March 1996)
[117] Ebel, S.: "Correct Calibration and Evaluation of Analyses in HPTLC", Lecture, 8th International Symposium on Instrumental Planar Chromatography, Interlaken / Switzerland, 5–7 April, 1995
[118] Prošek, M.; A. Medja, R.E. Kaiser: "Quantitative Evaluation of TLC with a Digital Camera", Cobac, Graz 1986
[119] Prošek, M.; I. Drusany: " Image Processing – New Perspective in QTLC" in Proceedings of the Sixth International Symposium on Instrumental Planar Chromatography, Interlaken/Switzerland, 23–26 April, 1991
[120] Prošek, M.: "Validation of Quantitative TLC Using Digital Video Camera" in Proceedings of the 9th International Symposium on Instrumental Planar Chromatography, Interlaken / Switzerland, 9–11 April, 1997
[121] Cibulka, R.: "Digitale Fotografie: Retuschieren ohne Grenzen" Nürnberger Nachrichten, page 24, 1 January 1997
[122] CAMAG, Brochure "VideoStore 2", German edition 1996
[123] Herzog, R., AISYS Bildverarbeitungssysteme, D-90537 Feucht, February 1997 (private communication)
[124] Mall, Th.: "Video-Densitometrische Auswertung von Dünnschichtchromatogrammen" in the TLC special publication (in memoriam Prof. Dr. Hellmut Jork), InCom, 148–156, 1996
[124a] Hahn-Deinstrop, E., U. Hahn: "Untersuchung von Kaffee mit Hilfe der Dünnschicht-Chromatographie", Zeitschrift Naturwissenschaften im Unterricht Chemie, NiU-Chemie 9, No. 43, 23–26, Friedrich, Seelze 1998
[124b] Koch, A., M. Müller, E. Hahn-Deinstrop: "Quantification of Ginsenosides – A Comparison of Densitometry and DESAGA Video Documentation ProViDoc", Poster at the 46th Annual Congress, Society for Medicinal Plant Research, Vienna, Austria, 31.8.–4.9.1998
[124c] CAMAG, House journal CBS 78, 14–16, March 1997 and CBS 79, 2–7, September 1997
[124d] Hahn-Deinstrop, E.: "Quantitative Auswertung in der Dünnschicht-Chromatographie", CLB, 50. J., Vol. 2, 55–60, Umschau, Frankfurt 1999
[125] Machbert, Prof., Institut für Rechtsmedizin der Universität Erlangen-Nürnberg, Privatmitteilung, February 1997 (private communication)
[126] Hahn-Deinstrop, E.: "Photographie von Dünnschicht-Chromatogrammen im UV-Licht", GIT Suppl. 3 (Chromatographie), 29–31, 1989
[127] Hahn-Deinstrop, E.: "The Documentation of Thin-Layer Chromatograms by Non-Densitometric Methods", J. Planar Chromatogr., Vol. 4, 154–157, 1991
[127a] Drotleff, V.: DESAGA, Entwicklungsabteilung, April 1999 (private communication)
[128] Jänchen, P.: "Berücksichtigung von GMP/GLP bei der CAMAG TLC-Auswertesoftware 'CATS'"
[129] Abel, G., Planta Med Arzneimittel, Neumarkt, May 1997 (private communication)
[129a] Koch, A., Frohme Apotheke, Hamburg, September 1998 (private communication)
[130] Europäische Gemeinschaften: "EG-Leitfaden einer Guten Herstellungspraxis für Arzneimittel", Editio Cantor, Aulendorf (1990), ISBN 3-87193-115-2; Reprinted from "Die pharmazeutische Industrie", Pharm. Ind. 52, No. 7, pp. 853–883 (1990)

[131] Schmidt, M. (Rottenburg): "Gute Labor-Praxis (GLP) mit Wilhelm Busch", PTA heute (monthly supplement to the DAZ), Vol. 8, No. 2, pp. 121–124 (February 1994). *Remark:* Wilhelm Busch (1832–1908) was a famous German writer and cartoonist.

[132] Christ, G.A.; S.J. Harston, H.W. Hembeck: " GLP – Handbuch für Praktiker", GIT, Darmstadt (1992), ISBN 3-928865-03-X

[133] Galle M. (pseudonym): "GMP – Das magische Wort", circulated as a leaflet in the QA department, origin unknown

[134] Gessner, U., R & D Searle, Nürnberg, March 1997 (private communication)

[135] Schmidt, M. (Rottenburg): " Gute Laborpraxis – was ist das?", PTA heute (monthly supplement to the DAZ), Vol. 4, No. 12, pp. 614–615 (December 1990)

[136] Günther, W.: "GLP = Gute Laborpraxis = Richtige Analytik?", CLB Chemie in Labor und Biotechnik, Vol. 43, No. 10, pp. 536–540 (1992)

[137] Bundesfachverband der Arzneimittel-Hersteller e.V. (Hrsg.): "Standardverfahrensanweisungen (SOPs) der fiktiven Firma "Muster" für die Arzneimittel-Herstellung einschl. verwandter Produkte, Teil 1", Ubierstr. 71–73, 53173 Bonn (no ISBN number)

[138] Wieland, G.: "Lasercodierte DC und HPTLC Fertigplatten erobern die GLP-Laboratorien", circular from the LPRO/CHROM 1 department of Merck KGaA (Darmstadt) 26.03.96

[139] Hembeck, H. W.: "Revision der GLP-Grundsätze der OECD", GIT Laboratory Journal, 2, p. 220 (1997)

[140] Hartmann, L., R. Schnettler, R. Völler, K. Haberer, A. Hardtke: "2. Deutsche GMP-Konferenz", Pharm. Ind. 59, No. 2, pp. 159–164 (1997)

[141] Christ, G., S. Harston, H. Morgenthaler, R. Rauchschwalbe, G. Wagner-Youngmann: "Einsatz computergestützter Systeme bei GLP-Prüfungen", Part 1, Pharm. Ind. 59, No. 1, p. 24 (1997)

[142] Christ, G., S. Harston, H. Morgenthaler, R. Rauchschwalbe, G. Wagner-Youngmann: "Einsatz computergestützter Systeme bei GLP-Prüfungen", Part 2, Pharm. Ind. 59, No. 2, pp. 116–120 (1997)

[143] Gietz, N.: "Validierung computergestützter Systeme", Labor 2000, pp. 112–118 (1995 edition)

[144] Renger, B., H. Jehle, M. Fischer, W. Funk: "Validierung von Analysenverfahren in der pharmazeutischen Analytik, Beispiel: Gehaltsbestimmung von Theophyllin in einer Brausetablette mittels HPTLC", Pharm. Ind. 56, No. 11, pp. 993–1000 (1994)

[145] CPMP Working Party on Quality of Medicinal Products: "Tripartite ICH Text: Validation of Analytical Procedures", Draft No. 3 (II/5626/93-EN) Brussels (September 1993)

[146] According to USP 23

[147] Ammicht, R., Heumann Pharma, Quality control department Pharma, April 1997 (private communication)

[148] Brutsche, A.: "Qualifizierung analytischer Geräte", Concept Heidelberg No. 11996, Vol. 17, Pharma Technologie Journal, 3 SSW 0931–9700, current requirements of the FDA

[149] Huber, L.: "Validierung computergesteuerter Analysensteme. Ein Leitfaden für den Praktiker", Springer Berlin Heidelberg, ISBN 3-540-60839-4

[150] Renger, B., Private communication (April 1997) concerning the the draft SOP "Qualifizierung von qualifizierungspflichtigen Ausrüstungen"

[151] Ebel, S.: "Validierung von DC-Scannern", Chapter 8 of the course of lectures by S. Ebel on TLC/HPTLS, University of Würzburg

[152] Liesenfeld, O.: "Software-gesteuerte Validierung eines quantitativen DC-Auswertesystems", Lecture from an IAS Analysis Seminar on 8.10.1996 at the University of Leipzig and on 10.10.1996 in the Eurotec Moers Technology Park

[153] Ebel, S.: Lecture from the 9th International Symposium on Planar Chromatography, 9–11 April 1997, Interlaken

[154] Hahn-Deinstrop, E.: "GMP/GLP-gerechte Dokumentation von Dünnschicht-Chromatogrammen", Lecture from the Merck Forum, Heidelberg, 29. 9. 1992

[154a] Renger, B.: "DC neben HPLC in der Pharma Qualitätskontrolle", Lecture from InCom '99, Düsseldorf, 22.03.1999

[155] Jork, Funk, Fischer, Wimmer: Dünnschicht-Chromatographie, Reagenzien und Nachweismethoden, Vol. 1a and 1b, p. 13 ff., VCH, Weinheim 1989 and 1993

[156] Burger, K.: Fresenius Z. Anal. Chem 318, (1984) 228–233
[157] Burger, K.: GIT Fachz. Lab. Supplement "Chromatographie" (1984) 29–31
[158] Burger, K.: "AMD (Automated Multiple Development) – Anwendungen und online Kopplung mit reversed phase HPLC", special publication, pp. 31–37, InCom 1996
[159] CAMAG Brochure, AMD (2) System (1997)
[160] Tyihak, E., H. Kalasz, E. Mincsovics, J. Nagy: Proc. 17th Hung. Annu. Meet. Biochem., Kecskemét, 1977; C.A. 88 (1978) 15386
[161] CAMAG Brochure, OPLC – Planarchromatographie mit gesteuerter konstanter Flußrate 1997
[162] Hauck, H. E., W. Jost: "Investigations and Results Obtained With Overpressured Thin-Layer Chromatography", J. Chromatogr. 262, 113–120 (1983)
[163] Rózylo, J. K., R. Siembida: "Analysis of Food Dyes Using OPLC" in: Proceedings of the 9th International Symposium on Instrumental Planar Chromatography, Interlaken 9–11 April, 1997, 305–310
[164] Siembida, R.: "OPLC Analysis of Food Antioxidants" in: Proceedings of the 9th International Symposium on Instrumental Planar Chromatography, Interlaken 9–11 April, 1997, 321–324
[165] Simon-Sarkadi, L., Á: Kovács, E. Mincsovics: "Determination of Biogenic Amines in Food by OPLC, in: Proceedings of the 9th International Symposium on Instrumental Planar Chromatography, Interlaken 9–11 April, 1997, 325–329
[166] Kaiser, R. E.: "Grundlagen und Anwendungen der HPPLC", special publication, pp. 112–123, InCom 1996
[167] Kaiser, R. E.: "Einführung in die Hochdruck-Planar-Chromatographie", Hüthig-Heidelberg (1987), ISBN 3-7785-1563-1
[168] Brochure "IATROSCAN TH 10" of the company SES Analysensysteme
[168a] Sivers, P. v., M. Hahn: "Bestimmung schwerflüchtiger Kohlenwasserstoffe mittels DC", CLB, 49, Vol. 11, pp. 424–428, 1998
[169] Schrodt, A., Heumann Pharma, Process development, April 1997 (private communication)
[170] Hauck, H. E., H. Halpaap: "Verfahren zur Herstellung von Form- und Farbkompositionen mit Hilfe von Flachbett Chromatographie-Techniken", Patent Application, DE 3204 094 A1, 6.2.82
[171] "Chrom art"; Merck Spectrum 3/92, p. 25 and cover picture
[172] "Fließende Grenzen", Fränkische Landeszeitung Ansbach, 14.05.1998
[173] "CHROMart im Rathaus Eckental – Verschmelzung von Wissenschaft und Kunst", Wochenblatt Eckental, 13.01.1999
[174] DESAGA Brochure, UV-Analysenlampen (1995)
[175] LC·GC International, cover picture, Vol. 8, No. 8, August 1995
[176] Deutsche Apotheker Zeitung, cover picture, Vol. 51/52, 1996
[177] Schilcher, H.: "Kleines Heilkräuter-Lexikon" Walter Hädecke, Weil der Stadt, 1994, ISBN 3-7750-0252-9
[178] Ennet, D., H.D. Reuter: "Lexikon der Pflanzenheilkunde", Hippokrates, Stuttgart, 1998, ISBN 3-7773-1286-X

12.3 Abbreviations

κ	Flow constant or velocity coefficient
μg	Microgram
μl	Microliter
2D-TLC	2-Dimensional TLC
ADC	Automatic Developing Chamber (CAMAG)
AMD	Automated Multiple Development (CAMAG)
AZ	Additional zone (in purity testing)
BfArM	German Institute for Pharmaceuticals and Medicaments (Bundesinstitut für Arzneimittel und Medizinprodukte)
BP	British Pharmacopoeia
CCD	Charge Coupled Device (light-sensitive element in modern video cameras)
CHC	Chlorinated hydrocarbon
CHROMart	Art produced by TLC (formed from the words "chromatography" and "art")
CPU	Central processor unit
CRS	Reference substance from the European Pharmacopoeia
CS	Chamber Saturation
DAB	German Pharmacopoeia (Deutsches Arzneibuch)
DAC	German Pharmaceutical Codex (Deutscher Arzneimittel Codex)
DART-MS	Direct Analysis in Real Time Mass Spectrometry
DE	Dry extract
DM	Deutsche Mark (German currency up to 2001)
DQ	Design Qualification
DT-N	Double-trough chamber for normal development
DZ	Double zone
E.	*Echinacea* (coneflower)
EDTA	Ethylenediaminetetraacetic acid (Titriplex III)
ESI-MS	Electrospray-Ionization Mass Spectrometry
€	Euro (European currency since 2002)
FDA	Food and Drug Administration (USA)
FI	Fluorescence Indicator
FID	Flame Ionization Detector
FLD	Fluorescence detection
FIT	Fluorescence Intensifier
FTIR	Fourier-Transform Infrared
GC	Gas Chromatography
GF	Graduated Flask
GLP	Good Laboratory Practice
GMP	Good Manufacturing Practice
h	Hours
H	Theoretical plate height
H-chamber	Developing chamber for horizontal development
HAB	Homeopathic Pharmacopoeia (Homöopathisches Arzneibuch)
HPLC	High Pressure Liquid Chromatography
HPPLC	High Pressure Planar Liquid Chromatography (circular TLC under high pressure)
HPTLC	High Performance Thin-Layer Chromatography
hRf value	Retardation factor R_f multiplied by 100
ICH	International Conference for Harmonization (meetings of experts held at irregular intervals to produce uniform guidelines for the validation of analytical methods worldwide)
IQ	Installation Qualification (for analytical equipment)
IR	Infrared spectrometry
IS	Identity Substance

LC	Liquid Chromatography (see also HPLC)
LED	Light emitting diode
LIMS	Laboratory information management system
LOD	Limit of Detection
MB	Megabytes
min	Minutes
ml	Milliliter
mo-disk	Magneto-optical disk (128 MB storage capacity diskette)
MS	Mass Spectrometry
MTSS	Merck Tox Screening System
MWLS	Multiple Wavelength Scan
MZ	Main zone (in purity testing)
N-chamber	Normal developing chamber for vertical development
nl	Nanoliter
OECD	Organization for Economic Cooperation and Development
OPLC	Overpressured Layer Chromatography (TLC with forced flow)
OQ	Operation Qualification (for analytical equipment)
PAH	Polycyclic Aromatic Hydrocarbon
PC	Personal Computer
Pharm Eur or Ph. Eur.	European Pharmacopoeia
PKU	Phenylketonuria (congenital metabolic disease of babies that must be treated until the brain is fully developed)
PM	Photomultiplier
PQ	Performance Qualification (for complex analytical equipment)
PRS	Primary Reference Substance
PS	Pharmacopoeia substance
QA	Quality Assurance
R	Residue
Ref. Sol.	Reference Solution
Ref. Sub.	Reference Substance
rel. c.	Related compound
R_f	Retardation Factor
RPC	Rotation Planar Chromatography (circular chromatography on a rotating disc – for preparative purposes)
rpm	Revolutions per minute (e.g. in centrifuging)
RT	Room Temperature
s	Seconds
S	Standard deviation
S-chamber	Sandwich chamber (narrow developing chamber for vertical development in which the TLC plate is fixed to a second glass plate)
SERS	Surface-Enhanced Raman Spectroscopy
SOP	Standard Operating Procedure (for general repetitive operations)
S_{rel}	Variation coefficient (relative standard deviation)
SRS	Secondary Reference Substance
SS	Solvent system
STLC	Sample chromatogram (e.g. for testing procedures and approval documents)
Stress test	Definite stress on a substance as a model for the presence of decomposition products
t	Developing time [s]
TLC	Thin-Layer Chromatography
TLC scanner	Measuring equipment for the direct optical evaluation of thin-layer chromatograms
TLC-FID/FTID	Combination of TLC with Flame Ionization Detector or Flame Thermionic Ionization Detector
TLC-MAT	Automatic developing chamber (Baron; DESAGA)
TP	Testing procedure (for the analysis of a substance/product)

u	Linear flow rate
USB	Ultrasonic Bath
USP	United States Pharmacopoeia
UTLC	Ultra thin-layer chromatography
v/v	Volumes (in solvent system compositions)
WHO	World Health Organization
Z_0	Distance between solvent level and starting line [mm]
Z_F	Distance of the solvent front from the solvent level [mm]
Z_S	Distance of the substance zone from the starting line [mm]
Z_{St}	Distance of the standard substance from the starting line [mm]

12.4 Acknowledgements

I would like to thank all the people listed below without whose help and support the present book would have been rather less comprehensive. This help included the procurement of hard-to-obtain literature, precoated layers, photographs, graphics, scanned chromatograms, lecture notes, plants and other test materials. I also thank all those who corrected various parts of the text and of course those colleagues who offered practical advice, which has been gratefully received and acted upon. I give a specially hearty "thank you" to Norbert Barth, my illustrator, who translated ideas into cartoons, bringing various laboratory situations to the more comical side of life. I am also grateful for being made aware of one or two errors in the German edition and for some tips that will help with the translation into English.

Dr. Gudrun **Abel**, BIONORICA AG, 92318 Neumarkt/Opf., Germany
Lothar **Baron**, 78479 Insel Reichenau, Germany
Norbert **Barth**, 15, Rue du Barry, F-34150 Montpeyoux, France; Rubenstraße 1, 54294 Trier, Germany
Prof. Dr. Rudolf **Bauer**, Institute of Pharmacognosy, University of Graz, 8010 Graz, Austria
Dipl.-Ing. Günther **Dietzel**, raytest, 75334 Straubenhardt, Germany
Prof. Dr. Siegfried **Ebel**, formerly Institute of Pharmacy, 97074 Würzburg, Germany
Dr. Karla **Halpaap-Wood**, Houston, Texas, USA
Ian T. **Harrison**, Harrison Research Inc., Palo Alto, California 94306, USA
Dr. Heinz-E. **Hauck**, Merck KGaA, PLS/R&D, 64271 Darmstadt, Germany
Gustav A. **Heidt**, Sarstedt AG (DESAGA), 51582 Nümbrecht, Germany
Collegues of the Departments Quality Control and Chemical Development of **Heumann Pharma**, 90537 Feucht, Germany
Prof. Dr. Kurt **Hostettmann**, Institut de Pharmacognosie, 1015 Lausanne, Switzerland
Dr. Dieter **Jänchen**, CAMAG, 4132 Muttenz, Switzerland
Prof. Dr. Rudolf E. **Kaiser**, Institut für Chromatographie, 67098 Bad Dürkheim,. Germany
Dr. Angelika **Koch**, Frohme Apotheke, 22457 Hamburg, Germany
Prof. Dr. Karl-Artur **Kovar**, formerly Pharmazeutisches Institut, 72076 Tübingen, Germany
Dr. Matthias **Loppacher**, CAMAG, 4132 Muttenz, Switzerland
Dr. Thomas **Mall**, Boehringer Mannheim, Abt. TF-CAA, 68298 Mannheim, Germany
Joachim **Mannhardt**, J&M GmbH, 73431 Aalen, Germany
Dr. Klaus **Möller**, Macherey-Nagel GmbH, 52313 Düren, Germany
Dr. Gerda **Morlock**, Institute of Food Chemistry, University of Hohenheim, 70599 Stuttgart, Germany
Maria **Müller**, Sarstedt AG (DESAGA), 69153 Wiesloch, Germany
Dr. Takeshi **Omori**, Taisei Kako Co Ltd., Kanagawa-ken, 253-01 Japan
Prof. Dr. Peter **Pachaly,** formerly Pharmazeutisches Institut, 53115 Bonn, Germany
Dr. Eike **Reich**, CAMAG, 4231 Muttenz, Switzerland
Dr. Klaus **Reif**, Phytolab GmbH, 91487 Vestenbergsreuth, Germany
Dr. Bernd **Renger**, Vetter Pharma-Fertigung GmbH, 88212 Ravensburg, Germany
Pharmacist Rita **Richter**, Institute of Organic Chemistry, University of Hamburg, 20146 Hamburg, Germany
Dipl.-Ing. (FH) Michael **Schulz**, Merck KGaA, PLS/R&D, 64271 Darmstadt, Germany
Prof. Bernd **Spangenberg**, Fachhochschule Offenburg, 77625 Offenburg, Germany
Dr. Markus **Veit**, LAT GmbH Dr. Tittel, 82166 Gräfelfing, Germany
Dipl.-Ing. (FH) Margit **Werther**, CAMAG, 12169 Berlin, Germany
Dr. Klaus **Zieloff**, CAMAG, 12169 Berlin, Germany

12.5 Market Overview*

1. Companies which manufacture products for TLC

Baron
Lothar Baron Laborgeräte
Im Weiler 10
78479 Insel Reichenau (Germany)
Tel: +49–7534-7380
Fax: +49–7534-7843
e-mail: lothar.baron@t-online.de
Internet: http://www.baron-lab.de

CAMAG
Sonnenmattstrasse 11
4132 Muttenz (Switzerland)
Tel: +41-61-467 3434
Fax: +41-61-461 0702
e-mail: info@camag.ch
Internet: www.camag.ch

Camag Scientific Inc.
515 Cornelius Harnett Drive
Wilmington, NC 28401 (USA)
Tel: +1 800-334-3909
Fax: +1 910–343-1834
e-mail: tlc@camagusa.com
Internet: www.camagusa.com

ChromaDex™
2952 S. Daimler
Santa Ana, CA 92705 (USA)
Tel: +1 949.419.0288
Fax: +1 949.419.0294
Internet: www.chromadex.com

DESAGA
products via Sarstedt AG & Co
P.O. Box 1220
51582 Nümbrecht (Germany)
Tel: +49–2293-305-0
Fax: +49–2293-305-285
e-mail: info@sarstedt.com
Internet: www.sarstedt.com

Sarstedt Inc.
Peter Rumswinkel
1025, St. James Chruch Road
P.O. Box 468
Newton, NC 28658-0468 (USA)

ECH
Elektrochemie Halle GmbH
Weinbergweg 23
06120 Halle (Germany)
Tel: +49–345-5583-711
Fax: +49–345-5583-710
e-mail: info@ech.de
Internet: www.ech.de

Harrison
Research Inc., 840 Moana Court
Palo Alto, CA 94306 (USA)
Tel: +1 650 949-1565
Fax: +1 650 948-0493
e-mail: ithres@ix.netcom.com
Internet: www.chromatotron.com

J&M
Analytische Mess- und
Regeltechnik GmbH
Robert-Bosch-Strasse 83
73431 Aalen (Germany)
Tel: +49–7361-92810
Fax: +49–7361-928112
e-mail: infoj-m.de
Internet: www.j-m.de

Macherey-Nagel
GmbH & Co.KG
P.O. Box 101352
52313 Düren (Germany)
Tel: +49–2421-969-0
Fax: +49–2421-969199
e-mail: sales@mn-net.com
Internet: www.mn-net.com

* The lists are in alphabetical order

Merck KGaA
64271 Darmstadt (Germany)
Fax: +49–615172-60 80
e-mail: chromatography@merck.de
Internet: www.merck.de

EMD Chemicals Inc.
480 S. Democrat Road
Gibbstown, NJ 08027 (USA)
Tel: +1 800-222-0342
Fax: +1 856-423-4389
Internet: www.endchemicals.com

raytest
Isotopenmeßgeräte GmbH
Benzstrasse 4
75334 Straubenhardt (Germany)
Tel: +49–7082-92550
Fax: +49–7082-20813
e-mail: info@raytest.com
Internet: www.raytest.com

SES GmbH, Analysensysteme
Friedhofstrasse 76-79
55234 Bechenheim (Germany)
Tel.: +49–6736-1301
Fax: +49–6736-1305
e-mail: SES–Analysesysteme@t-online.de
Internet: www.tlc-fid.com

Whatman
Inc., 9 Bridewell Place
Clifton, NJ 07014 (USA)
Tel: +1-973-773-5800
Fax: +1-973-472-6949
e-mail: info@whatman.com
internet: www.whatman.com

Whatman International Ltd.
Whatman House
St. Leonard's Road, 20/20 Maidstone
Kent ME160LS (United Kingdom)
Tel.: +44-1622-676670

2. Companies which supply reference substances for TLC (selection)

addipharma® Referenzsubstanzen
speciality: standard natural substances with certificate of analysis
Phytolab GmbH & Co. KG
Dutendorfer Strasse 5–7, 91487 Vestenbergsreuth (Germany)
Tel: +49–9163-88-216, Fax: +49–9163-88-349
e-mail: welcome@phytolab.de, Internet: www.phytolab.com

CARL ROTH GmbH + Co. KG
speciality: standard natural substances
Schoemperlenstrasse 3, 76185 Karlsruhe (Germany)
Tel: +49–721-56060, Fax: +49721-5606-149
e-mail: info@carlroth.de, Internet: www.Carl-Roth.de

LGC Promochem Queens Road
Teddington, Middlesex TW11 OLY (United Kingdom)
e-mail: askus@lgcpromochem.com
Internet: www.lgcpromochem.com

with offices in France, Germany, Italy, India, Poland, Spain and Sweden official distributor for USP reference substances and supplier of reference substances of the European Pharmacopoeia, British Pharamcopoeia, WHO, phytochemical reference standards from ChromaDex, pharmaceutical impurity standards from Mikromol GmbH

3. Company, product of which can be used in TLC

Atlas
Material Testing Technology GmbH
Vogelbergstrasse 22, 63589 Linsengericht/Altenhaßlau
P.O. Box 1842, 63558 Gelnhausen (Germany)
Tel.: +49–6051-707–140, Fax: +49–6051-707–149
e-mail: infot@atlasmtt.com
Internet: xenotest.com
Tel. USA: +1-773-327-4520/5787

4. Supplier for TLC products worldwide

VWR
International GmbH
Hilpertstrasse 20A
64295 Darmstadt (Germany)
Tel: +49–6151-3972-500
Fax: +49–6151-3972-440
e-mail: hotline@de.vwr.com
Internet: www.vwr.com

Photograph Section

Applied Thin-Layer Chromatography: Best Practice and Avoidance of Mistakes, 2nd Edition
Edited by Elke Hahn-Deinstrop

ISBN: 978-3-527-31553-6

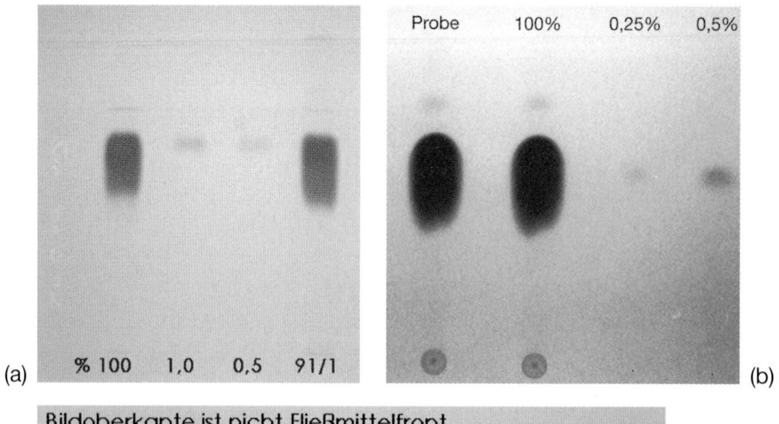

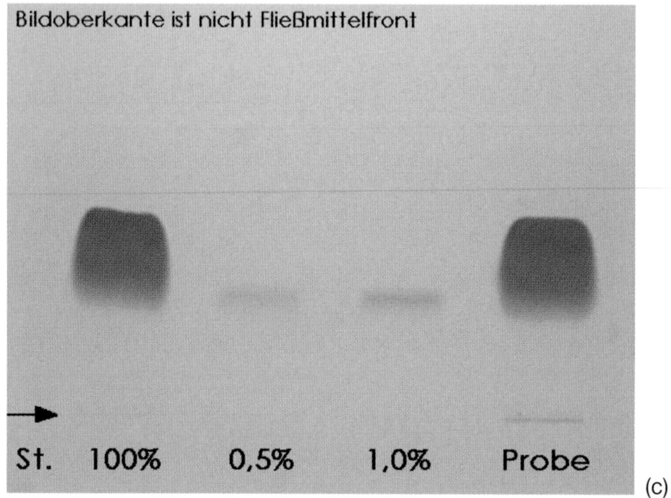

Figure 2. Purity test of pharmaceutically active ingredients, each reference substance applied at 100 % of the expected concentration, at limit test concentrations and in one sample
(a) 200 mg substance A ≈ 100 %; 1,0 %; 0,5 %; sample
(b) Sample; 500 mg substance B ≈ 100 %; 0,25 %; 0,5 %
(c) 300 mg substance C ≈ 100 %; 0,5 %; 1,0 %; sample

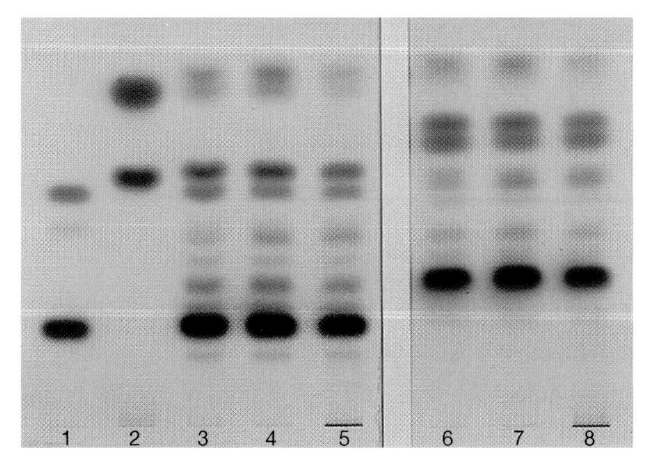

Figure 6. Identification of peppermint oils on sorbents from different manufacturers
Lanes 1–5 TLC silica gel 60 F_{254}, Merck Article No. 5715; *lanes 6–8* Durasil-25 UV254, Macherey-Nagel Article No. 812008 (solvent front is at the top)
Reference substances with ascending hRf values:
Lane 1 menthol, menthone; *lane 2* menthyl acetate, menthofuran; *lanes 3– 8* peppermint oils from different manufacturers

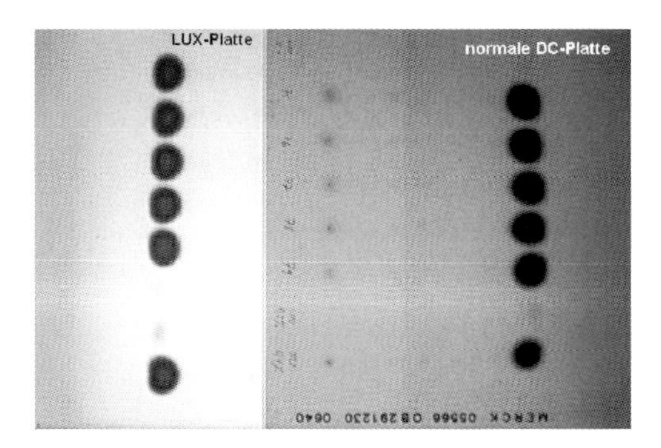

Figure 7. Comparison of brightness between Lux®-plate and Normal TLC-Plate

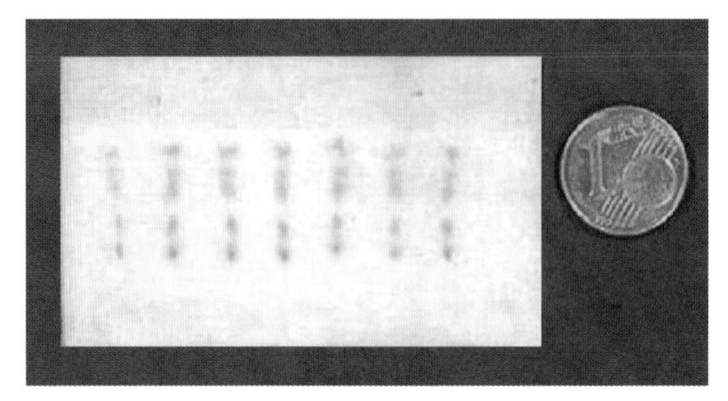

Figure 8. Separation of lipophilic dyes on UTLC-plate
(in comparison with a one Euro Cent)

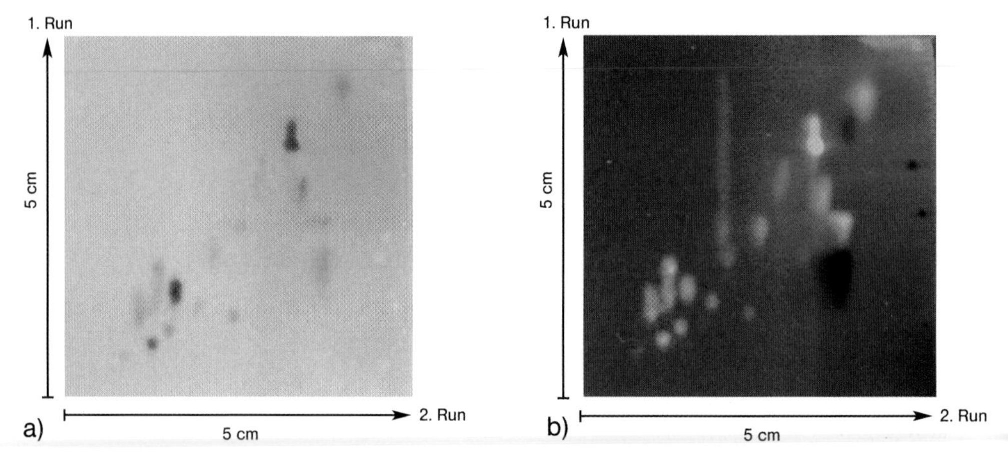

Figure 9. Peptides obtained by tryptic digest of Cytochrome C
Stationary phase:
ProteoChrom HPTLC Cellulose shetts, 10 × 10 cm
(a) 0,5 % Ninhydrin in 2-Propanol (shot taken in white light)
(b) 0,2 % Fluram® in acetone + 10 % triethylamine in dichloromethane + 10 % Triton® X100 in acetone (shot taken in UV-light 365 nm)

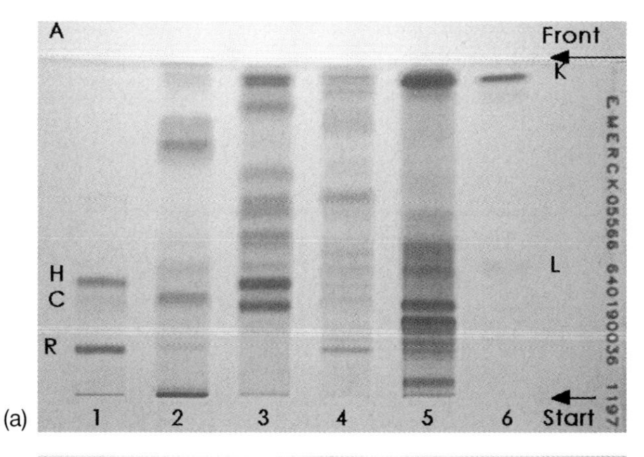

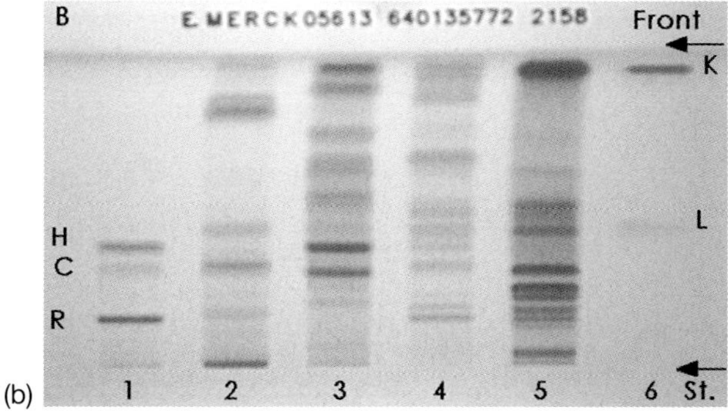

Figure 11. Comparison of separation efficiencies of TLC and HPTLC precoated plates of one manufacturer (shots taken in 254-nm UV light)

(a) TLC silica gel 60 F_{254} GLP precoated plate (Merck Article No. 1.05566)

(b) HPTLC silica gel 60 F_{254} GLP precoated plate (Merck Article No. 1.05613)

Reference substances with ascending hRf values:
Lane 1 rutoside (R), chlorogenic acid (C), hyperoside (H); *lane 6*: luteolin 7-glucoside (L), caffeic acid (K);
Lanes 2–5 preparations of pharmaceutically used parts of the plants;
Lane 2 artichoke leaves, *lane 3* birch leaves, *lane 4* elderflowers, *lane 5* leaves of the maidenhair tree (Ginkgo biloba)

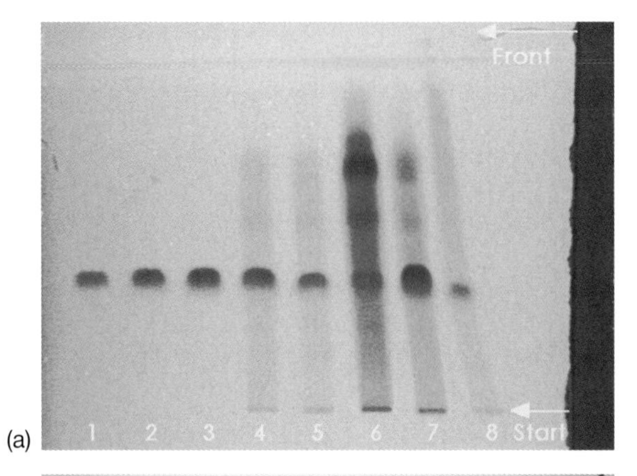

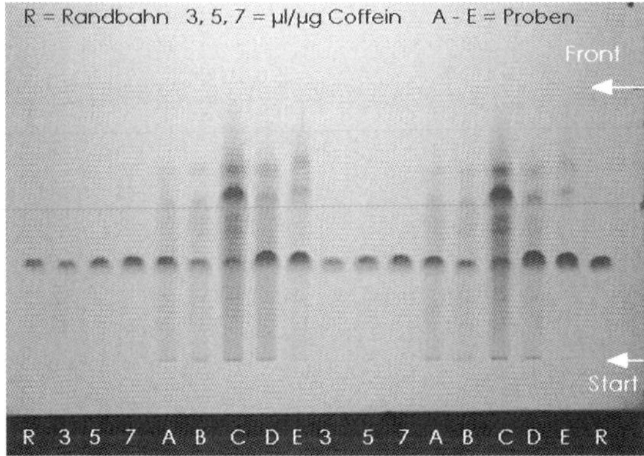

Figure 15. Influence of the cutting angle on the chromatographic result when using aluminum foil

(a) Poorly cut right-hand edge of the foil

(b) Foil correctly cut by the manufacturer

Four coffee samples and one tea sample, each containing three concentrations of caffeine, applied to a TLC aluminum foil RP 18 F_{254s} (Merck Article No. 1.05559)

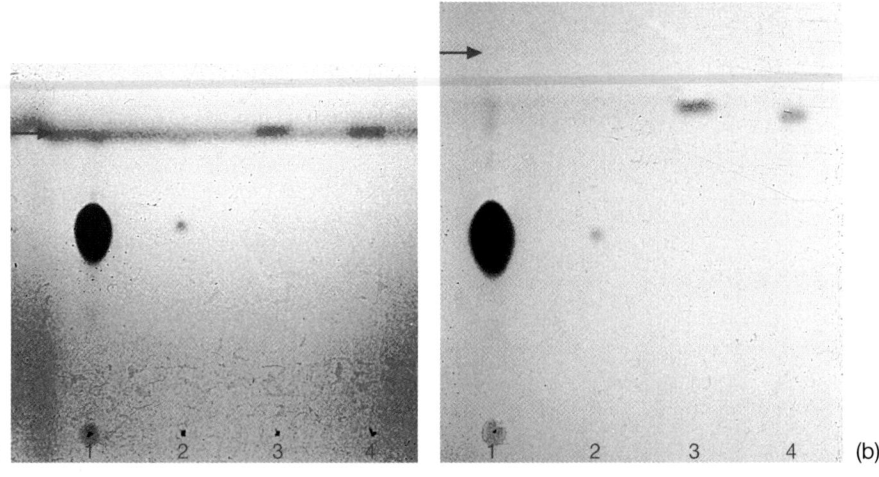

Figure 19. Influence of "prewashing" (arrows indicate the solvent front)

(a) TLC plate not prewashed: related compounds of the active ingredient metoprolol are covered over by the "dirt" front

(b) Prewashed plate

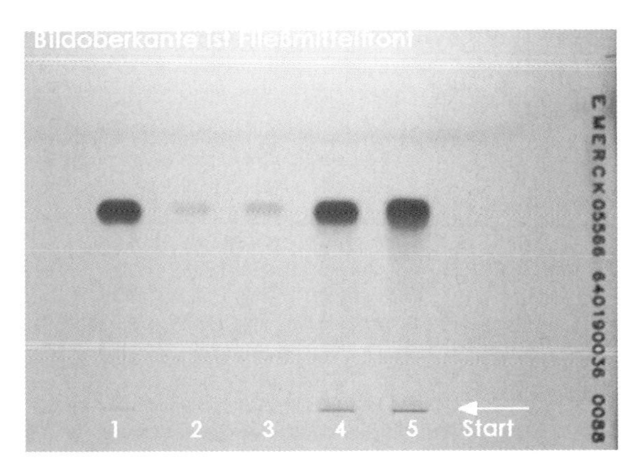

Lane 1 200 μg reference substance
Lane 2 1 μg ≈ 0.5 % reference substance
Lane 3 2 μg ≈ 1.0 % reference substance
Lanes 4 and *5* 200 μg active ingredient from a stored tablet
(Note: *lane 5* was not sufficiently dried before the development)

Figure 20. Influence of various solvent systems on prewashing and on the chromatography
The TLC silica gel 60 F_{254} precoated plate was prewashed with ethyl acetate + methanol (1 + 1 v/v), and the chromatography was performed with the solvent system toluene + 2-propanol + conc. ammonia solution (70 + 25 + 1 v/v)

(a)

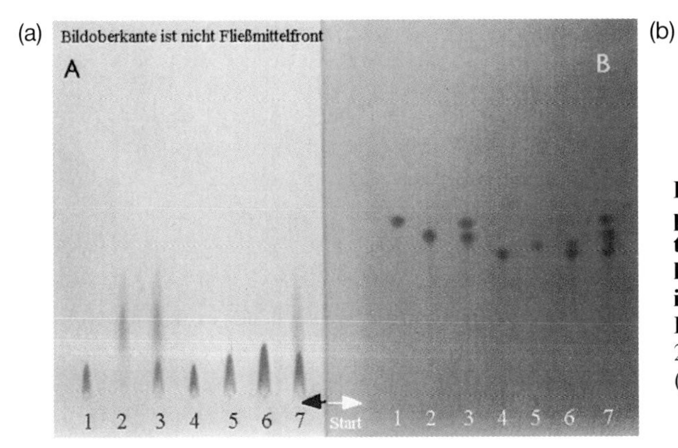

(b)

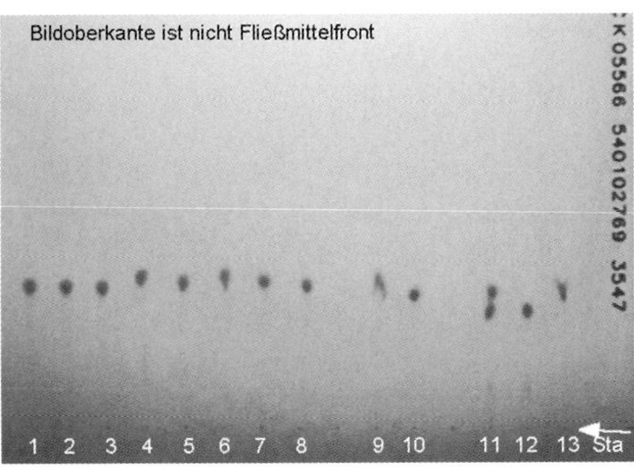

(c)

Figure 22. Influence of the impregnation with Titriplex III on the chromatography of doxycycline hyclate and related compounds in identity testing according to the DAB (all shots were taken in UV 254-nm light)

(a) TLC silica gel 60 F_{254} precoated plate (not impregnated, Merck Article No. 1.05715)
(b) Adequate EDTA impregnation of the TLC plate with the following substances:
Lane 1 doxycycline hyclate,
Lane 2 tetracycline,
Lane 3 mixture of *1* and *2*,
Lane 4 6-epidoxycycline hydrochloride,
Lane 5 metacycline hydrochloride,
Lane 6 mixture of *4* and *5*,
Lane 7 mixture of all 4 named substances
(c) Incomplete EDTA impregnation; all the substances listed in (b) above were applied

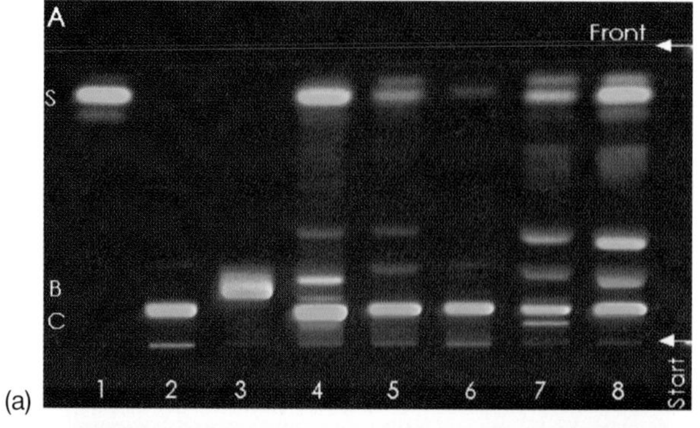

(a)

(b)

Figure 23. Influence of impregnation (by predevelopment with the solvent system) on the hRf values of greater celandine substances (shots in 365-nm UV light)
Sorbent: HPTLC silica gel 60 F_{254} precoated plate (Merck 1.05566)
(a) Chromatograms with no predeveloping
(b) Chromatograms with predeveloping

Lane 1 (with ascending hRf values) chelidonine, sanguinarine, *lane 2* coptisine, *lane 3* berberine, *lane 4* herb, sample preparation according to the DAB, *lane 5* herb, 20-min back flow, 70 °C, *lane 6* leaves, harvested in May, 20-min back flow, 70 °C, *lane 7* roots, harvested in September, 20-min back flow, 70 °C, *lane 8* milky juice from the rhizome, diluted with ethanol

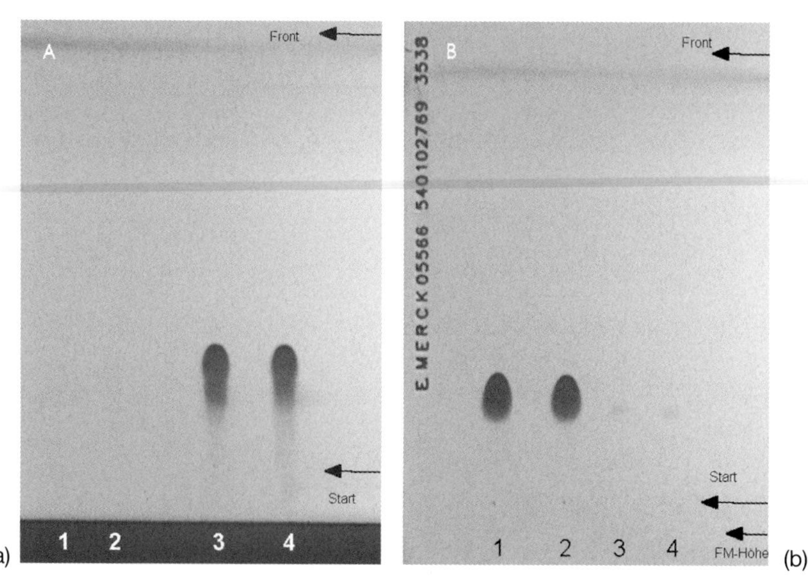

(a) (b)

Figure 25. Influence of the solvent level when the sample is located too near to the bottom edge of the plate
(a) Application zones were partially immersed in the solvent during development
(b) Solvent level was at a sufficient distance from the application zones

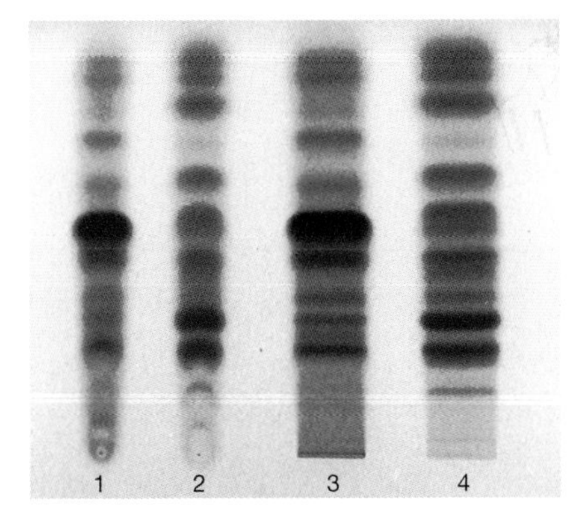

Figure 28. Influence of the method of application on the separation result
Lanes 1 and *2* 5 µl each pointwise, *lanes 3* and *4* 10 µl each bandwise, *lanes 1* and *3* dry extract of early golden rod herb, *lanes 2* and *4* dry extract of birch leaves

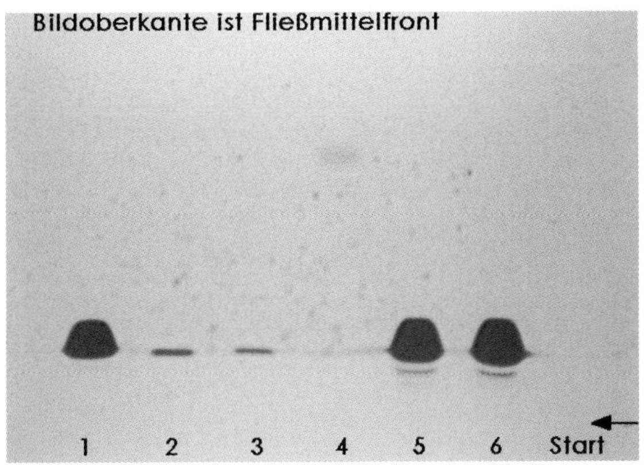

Figure 30. Application from too highly concentrated solutions resulting in uniform substance distribution on the TLC plate above the substance zone
The illustration shows the TLC plate after development. Substance particles (carbamazepine) that were spread over the layer below the substance zone have been moved by the solvent into the β front

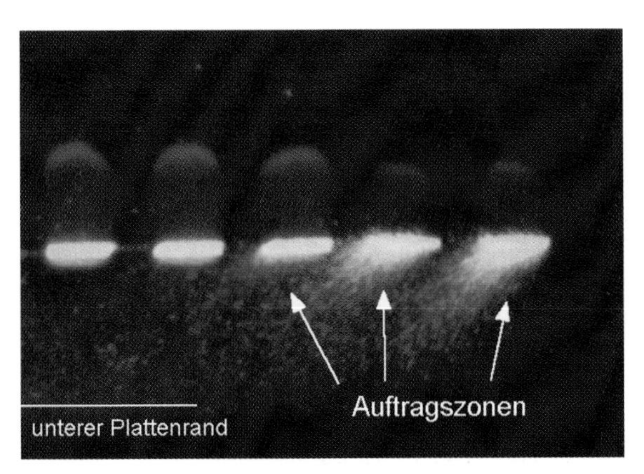

Figure 31. The right-hand part of this TLC plate shows starting zones in which the jets were not cleaned between each application
Here, the sample solution consisted of a concentrated preparation of a plant extract

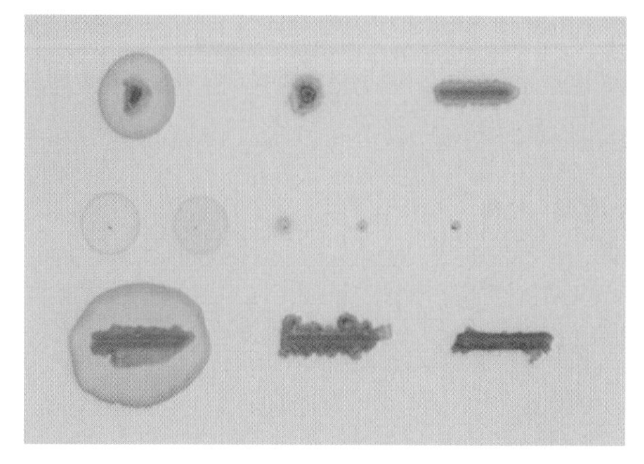

Figure 32. Application at different rates and using different solvents
Top line and bottom line concentrated plant extract solution applied using a Linomat IV
Top line 10-µl amounts (a) 4 s/µl pointwise, (b) 15 s/µl pointwise, (c) 4 s/µl bandwise (8 mm)
Bottom line 25-µl amounts bandwise (10 mm) (a) 4 s/µl, (b) 9 s/µl, (c) 15 s/µl
Middle line 5-µl amounts of chlorophyll-containing solutions: water, acetone, toluene, *n*-hexane, chloroform (hand application)

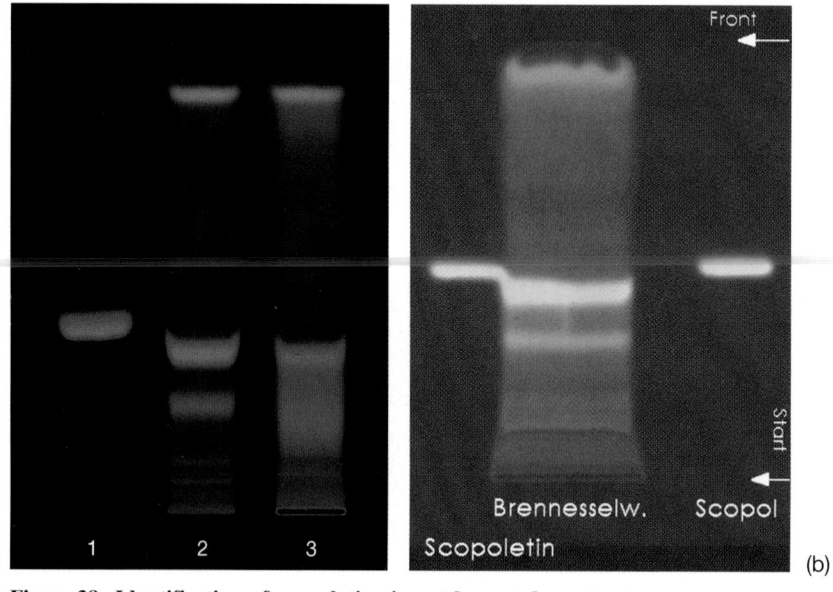

Figure 38. Identification of scopoletine in nettle root dry extract
(shots taken in 365-nm UV light)
(a) *Lane 1* 0.7 µg scopoletine, *lane 2* preparation of nettle root extract (35 mg extract/10-mm lane), *lane 3* preparation of the tablet (dosage form)
(b) *Middle lane* 125 mg extract/35-mm lane, *left lane* 3 µg scopoletine/30-mm lane, *right lane* 2 µg scopoletine/15-mm lane as comparison

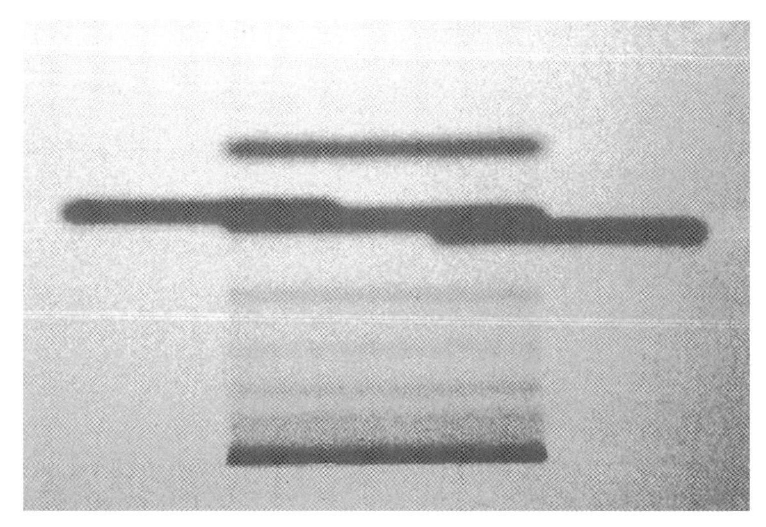

Figure 39. Overlapping application as an aid in questions of the identity of substances
Middle lane light-stressed sample of a pharmaceutically active substance
Left and *right lanes* known compounds formed in the synthesis of this active ingredient

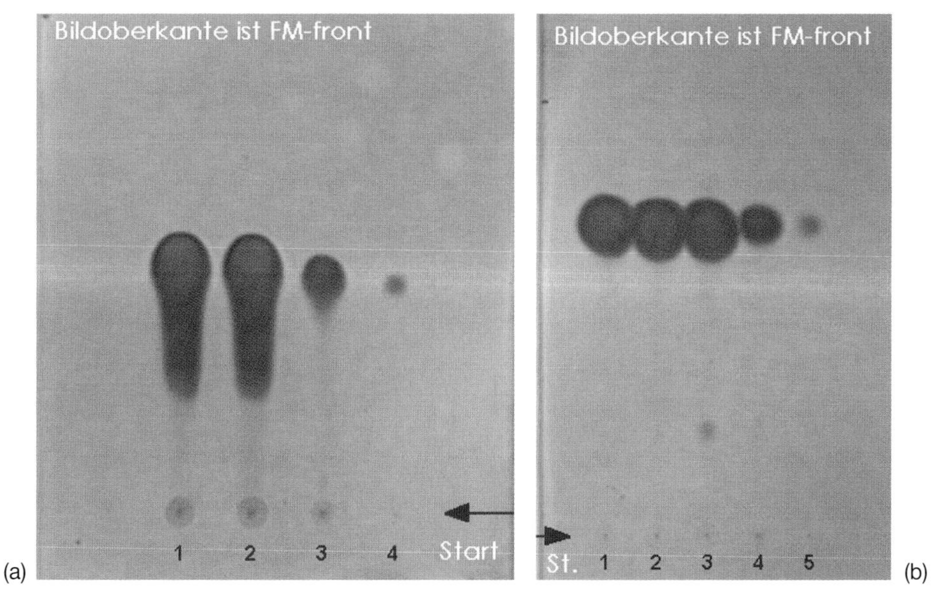

Figure 40. Influence of the positioning of samples and drying of the TLC plate before development
Purity test of a pharmaceutically active substance (top of the picture is the solvent front)
(a) 100 % amount of the reference substance is missing on the 1st application point
Lane 1 sample 1, *lane 2* sample 2, *lane 3* sample 3, *lane 4* 0.1 % ref. sub.
(b) *Lane 1* sample 1, *lane 2* sample 2, *lane 3* 100 % ref. sub., *lane 4* 0.5 % ref. sub., *lane 5* 0.1 % ref. sub.

(a)

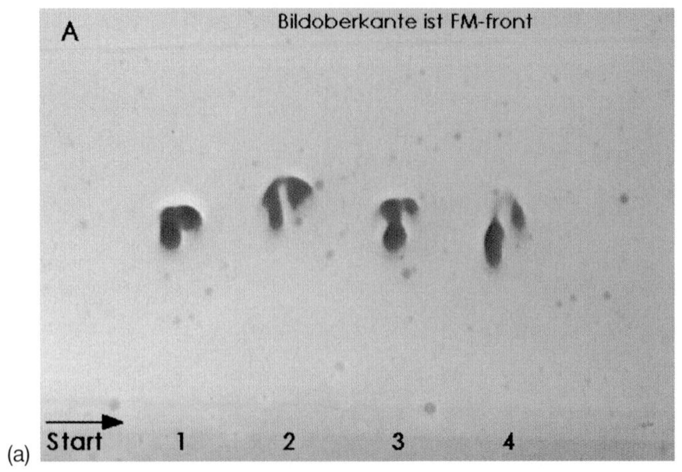

(b)

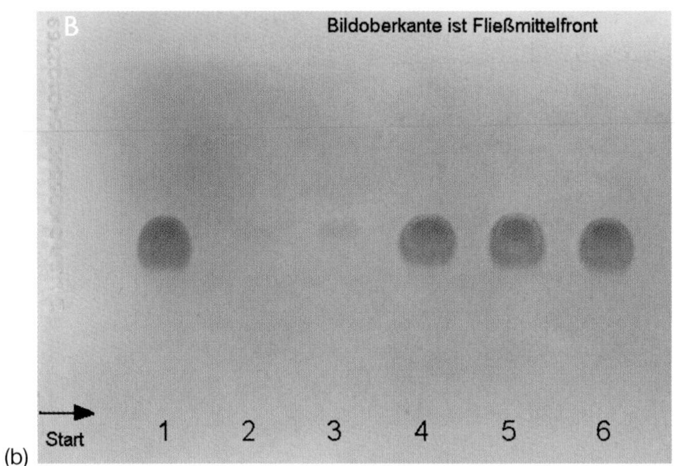

Figure 41. Influence of adequate drying before developing
Purity test of piracetam

(a) Inadequate drying before the development and use of a derivatization reagent that was too old
Spot 1 ref. sub., *spots 2–4* samples 1–3 (the limit test concentrations here were not visible because the reagent material was over 10 years old and "destroyed")

(b) Adequate drying before the developing and new reagent material
Lane 1: 1000 µg ≈ 100 % piracetam, *lane 2* 0,5 % piracetam, *lane 3* 1,0 % piracetam, *lanes 4–6*: samples 1–3
Derivatization: 2,4-dinitrophenylhydrazine reagent and 16 h in an iodine chamber

---

Figure 42. Influence of a consistent drying time of the TLC plate before development on the chromatographic result

(a) *Lane TE* sample preparation of marshmallow root dry extract, *lanes 1–4* prepared tea samples. Drying times: *lanes TE* and *1–3* 15 min, *lane 4* 5 min warm-air fan heater (grade I)

(b) Sample positions as (a). Drying time of the TLC-plate: 30 min warm-air fan heater (grade I)

(a)

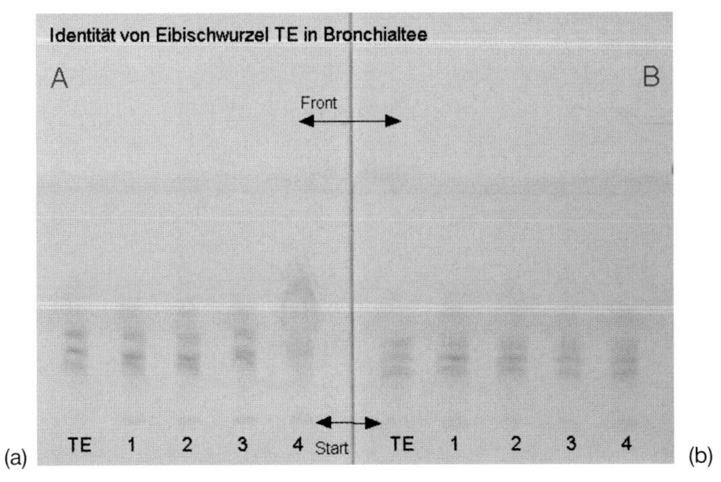

(b)

Figure 42 (legend see p. 282)

(a)

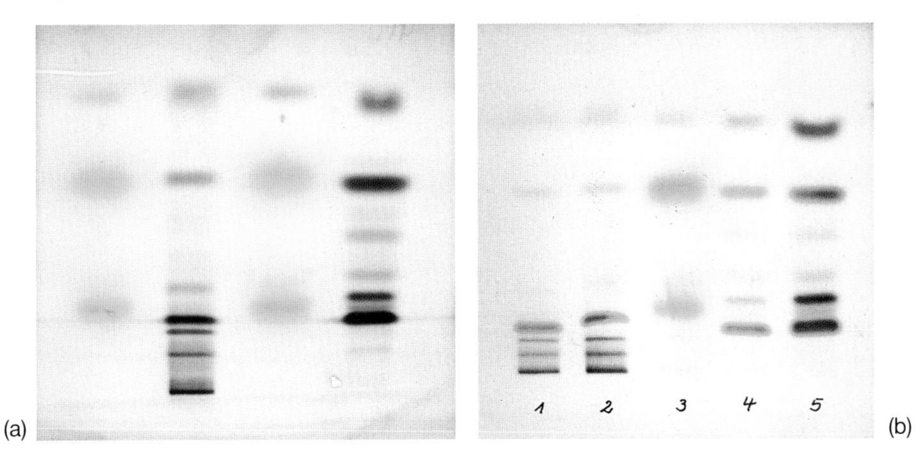

(b)

Figure 44. Influence of traces of water in the ethanol used to stabilize the solvent system chloroform

Sorbent:	HPTLC precoated plate silica gel 60 F_{254}, 5 × 5 cm
Solvent system:	Chloroform
Migration distance:	3.7 cm
Chamber:	Horizontal chamber without chamber saturation

(a) Chloroform stabilized with ethanol, used directly from the bottle:

Lanes 1 + 3	3 µl + 5 µl corresponding to *lane 3* in (b)
Lane 2	8 µl corresponding to *lanes 1 + 2* in (b)
Lane 4	8 µl corresponding to *lanes 4 + 5* in (b)

(b) Chloroform was dried over sodium sulfate immediately before use as a solvent system. Substances used were as follows:

Lanes 1 +2	3 µl + 5 µl dichloromethane extract of chamomile flowers (identity test according to the DAB)
Lane 3	With ascending hRf value: borneol, bornyl acetate, guaiazulene (3 µl each, reference solution according to the DAB)
Lanes 4 + 5	3 µl + 5 µl essential oil (0.2 ml oil/0.5 ml xylene) as for the assay in the DAB monograph "chamomile flowers"
Derivatization:	Anisaldehyde/H_2SO_4 (10 min at 120 °C)

(a) 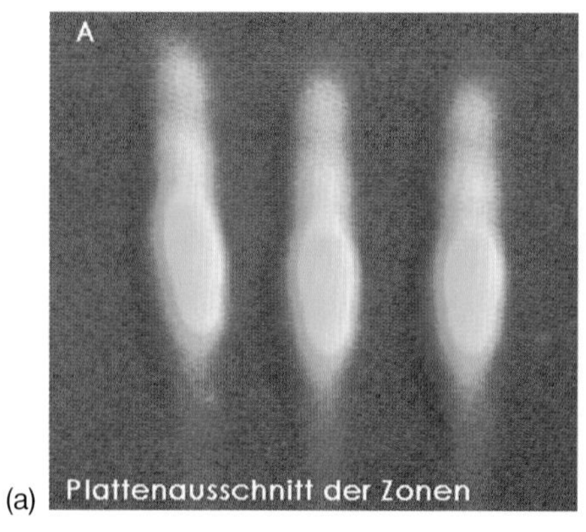

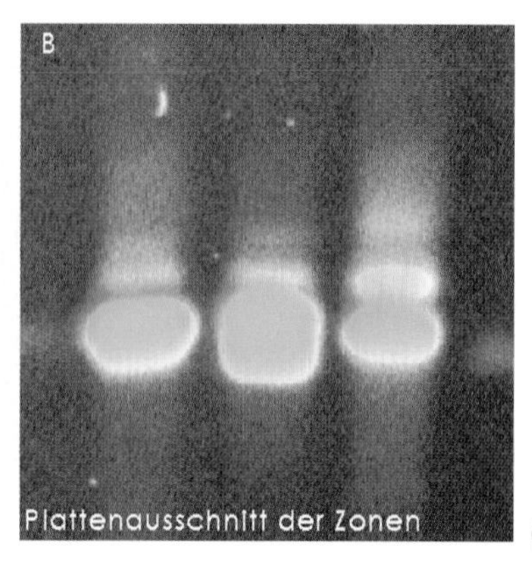 (b)

Figure 45. Influence of a "tailing reducer" on the chromatographic result (sections of the zones)
(a) Solvent system: Ethanol + water (9 + 1 v/v)
(b) Solvent system: Ethanol + water + glacial acetic acid (9 + 1 + 0.25 v/v)

Sorbent:	HPTLC precoated plate RP-18 WF_{254s} cut into 5 × 5 cm squares
Migration distance:	4 cm
Chamber:	Horizontal chamber without chamber saturation
Samples:	Various parts of the greater celandine plant
Detection:	365-nm UV light

(a) 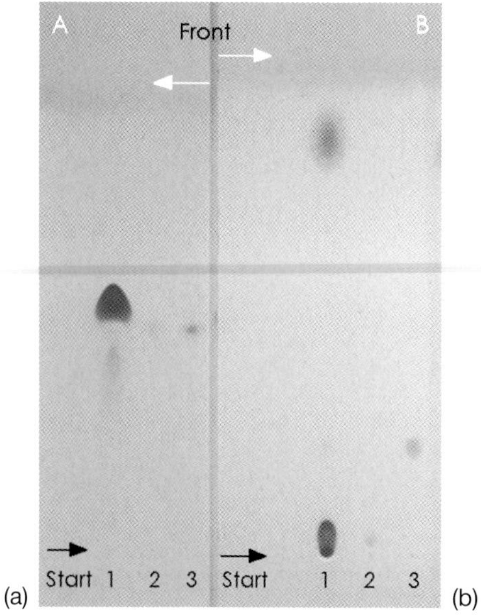 (b)

Figure 46. Effect of the presence of 1 % ammonia solution in the solvent system on the purity testing of dimenhydrinate according to the DAB (shots in 254-nm UV light)
(a) Solvent system: Dichloromethane + methanol (90 + 9 v/v)
(b) Solvent system: Dichloromethane + methanol + conc. ammonia solution (90 + 9 + 1 v/v)

Lane 1 200 μg dimenhydrinate (100 %), *lane 2* 0.1 μg dimenhydrinate (0.05 %),
lane 3 1 μg theophylline (0.5 %)

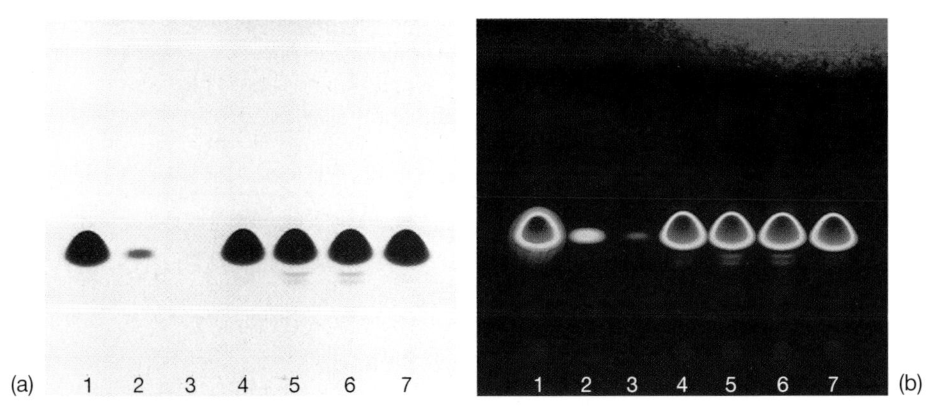

Figure 47. Testing of carbamazepine for related compounds according to the DAB
(a) 254-nm UV light before derivatization
(b) 254-nm UV light after derivatization with potassium dichromate-sulfuric acid

Lane 1 carbamazepine, working standard, *lane 2* 1 μg carbamazepine, working standard, *lane 3* 0.01 μg carbamazepine, working standard, *lanes 4–7* 100-μg amounts of carbamazepine from various producers

Figure 48 see page 286

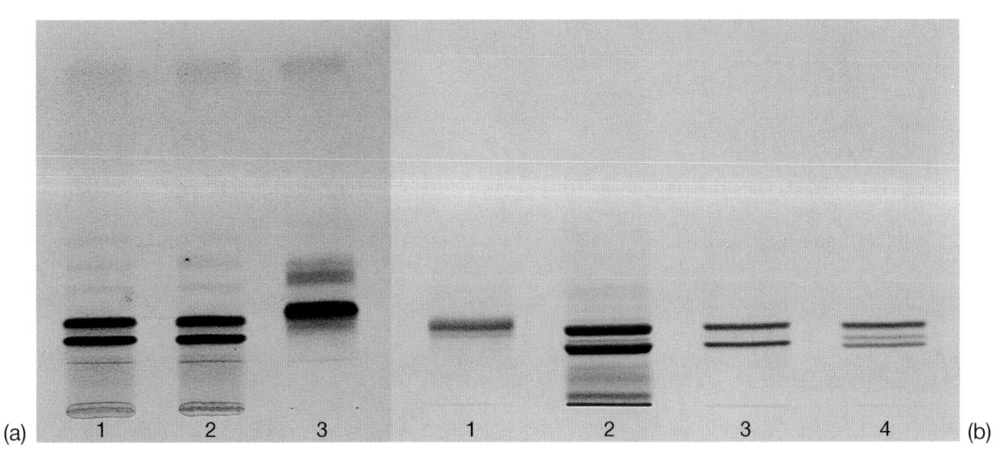

Figure 49. Identification of primrose root dry extract (DE) using two TLC systems after derivatization with vanillin-sulfuric acid reagent (shots taken in white light)

(a) According to the DAB
Lane 1 authentic primrose root DE, *lane 2* primrose root DE to be tested, *lane 3* aescin

(b) Alternative TLC system
Lane 1 aescin (only for comparison with the DAB system), *lane 2* primrose root DE for testing (as in *lane 2* above), *lane* 3 primrose root DE treated using an RP-18 cartridge (suitable for quantitative measurement), *lane 4* reference substance primula acid, sodium salt

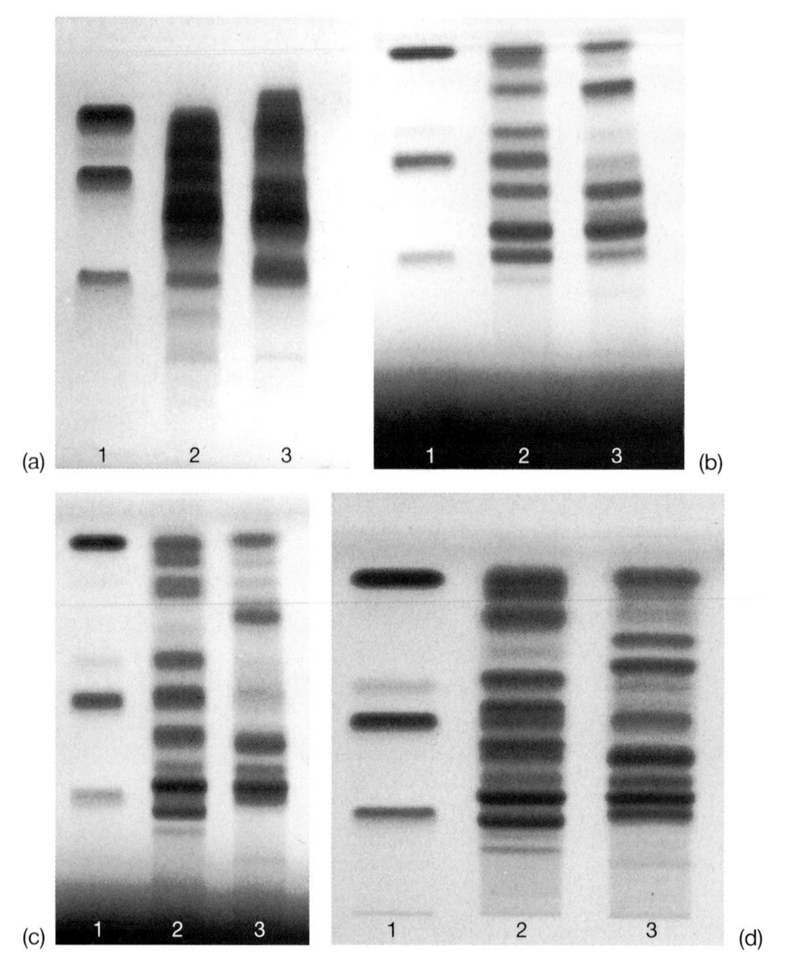

Figure 48. Identification of flavonoid-containing dry extracts (DE) using four TLC systems
(all shots in 254-nm UV light before the derivatization)
(a) According to the DAB
(b) According to Wagner et al [47]
(c) According to Pachaly (TLC Atlas)
(d) Alternative system

Lane 1 reference substances with ascending hRf values: chlorogenic acid, quercitrin, quercetin, *lane 2* 3.75 mg DE from birch leaves, *lane 3* 3.75 mg DE from hawthorn herb

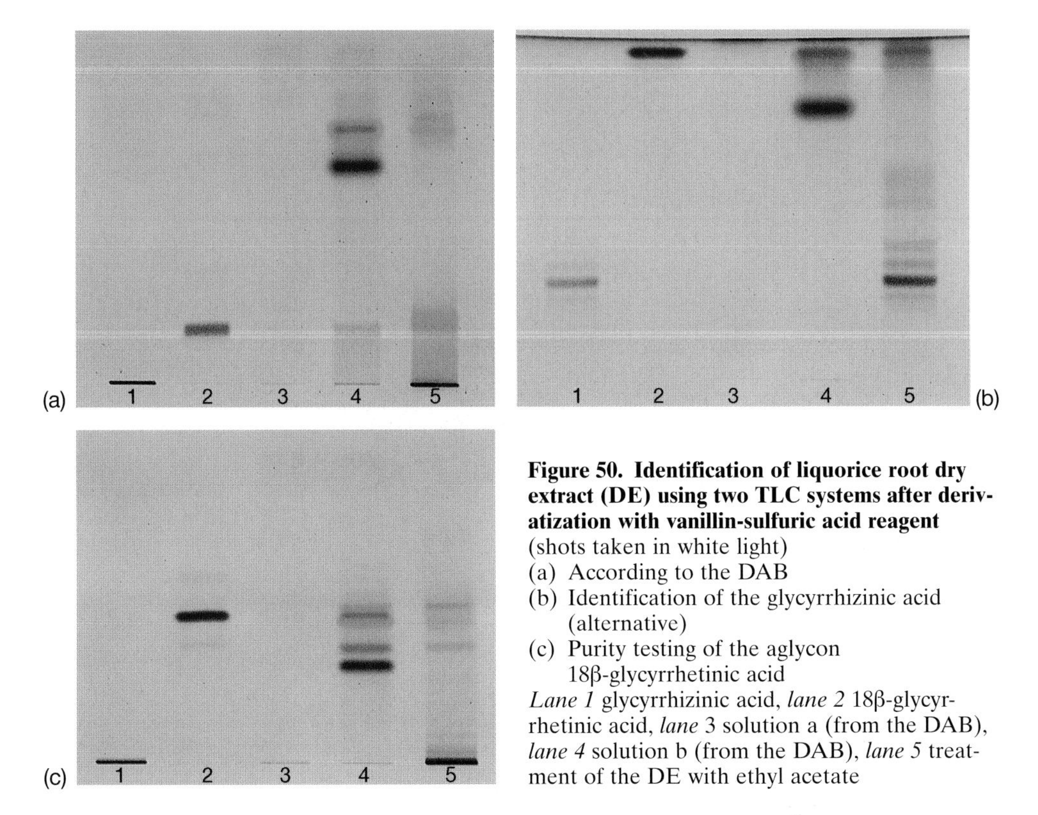

Figure 50. Identification of liquorice root dry extract (DE) using two TLC systems after derivatization with vanillin-sulfuric acid reagent
(shots taken in white light)
(a) According to the DAB
(b) Identification of the glycyrrhizinic acid (alternative)
(c) Purity testing of the aglycon 18β-glycyrrhetinic acid
Lane 1 glycyrrhizinic acid, *lane 2* 18β-glycyrrhetinic acid, *lane* 3 solution a (from the DAB), *lane 4* solution b (from the DAB), *lane 5* treatment of the DE with ethyl acetate

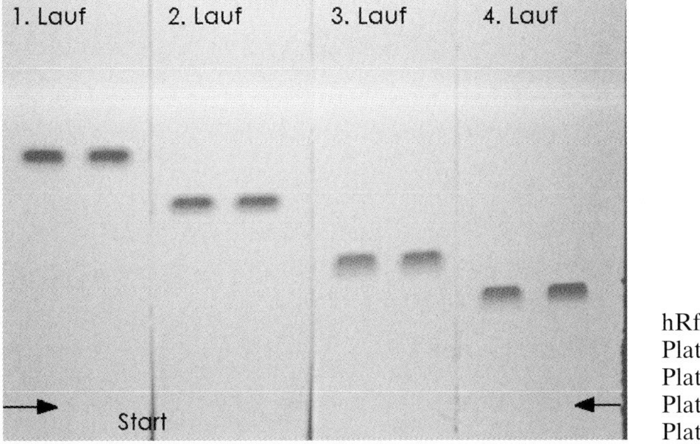

hRf values:
Plate 1: 54–60
Plate 2: 43–48
Plate 3: 27–32
Plate 4: 19–24

Figure 51. Influence of multiple use of solvent systems on the chromatographic result
(shots taken in 254-nm UV light)
Chromatography of 4 plates, each with 20 μg/10 mm metoclopramide HCl, developed one after the other in the solvent system chloroform + methanol + conc. ammonia solution (56 + 14 + 1 v/v) on TLC silica gel 60 F_{254} (Merck 1.05715)
Remark: "Lauf" means "run"

(a) 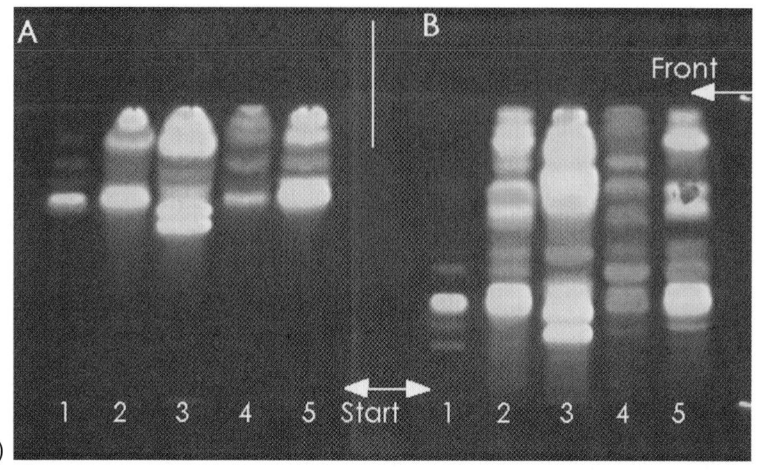 (b)

Figure 52. Difference between "matured" and fresh solvent system on the chromatographic result using the example of flavonoid-containing dry extracts (DE) (shots in 365-nm UV light after derivatization with the flavone reagent according to Neu)
(a) Solvent system 17 days old
(b) Solvent system freshly prepared

"Classical" solvent system for flavonoids according to Wagner [47]: ethyl acetate + formic acid + glacial acetic acid + water (100 + 11 + 11 + 27 v/v)

Lane 1 reference substances with ascending hRf values: rutoside, chlorogenic acid, hyperoside, *lane* 2 early golden rod herb DE, *lane 3* horsetail herb DE, *lane 4* birch leaves DE, *lane 5* hawthorn leaves and flowers DE

Figure 53. TLC of nettle root dry extract (DE) in various TLC systems
Test for phytosterols after derivatization with vanillin-phosphoric acid reagent (shots taken in white light):
(a_1) According to the DAB, development at 22 °C
(a_2) According to the DAB, development at 5 °C
(b) Alternative system
Lane 1 β-sitosterol, *lane 2* authentic DE from 1994, *lane 3* DE of 1995 to be tested

Test for scopoletine (365-nm UV light before derivatization):
(a_3) According to the DAB
Lane 1 1 µg scopoletine, *lanes 2 and 3* as in Fig. $50a_1$

(c) In the coumarin solvent system
Lane 1 0.5 µg scopoletine, *lane 2* authentic DE from 1992

(d) Direct optical evaluation of β-sitosterol and two batches of nettle root DE using the TLC scanner CD 60 (DESAGA) at 546 nm

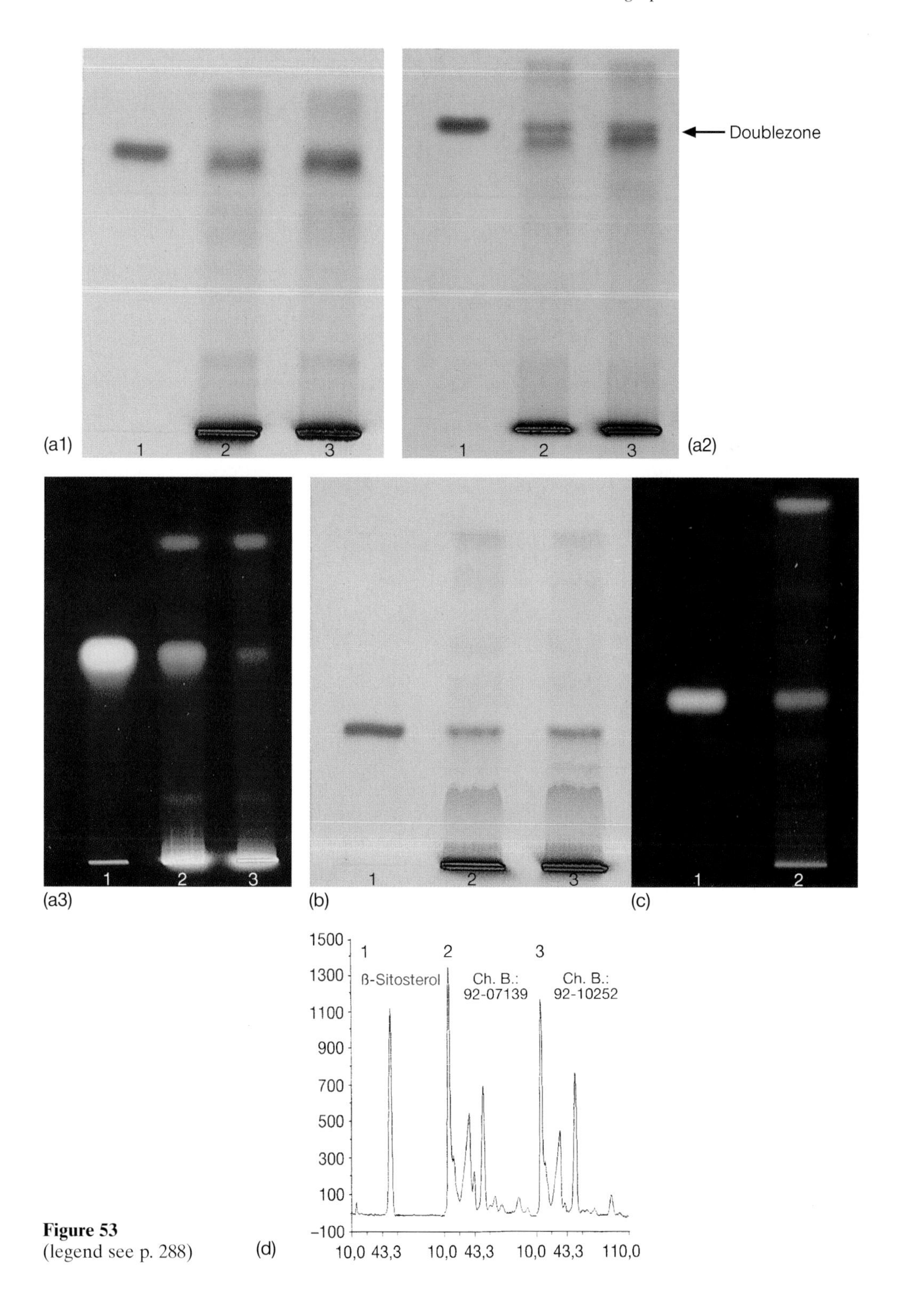

Figure 53
(legend see p. 288)

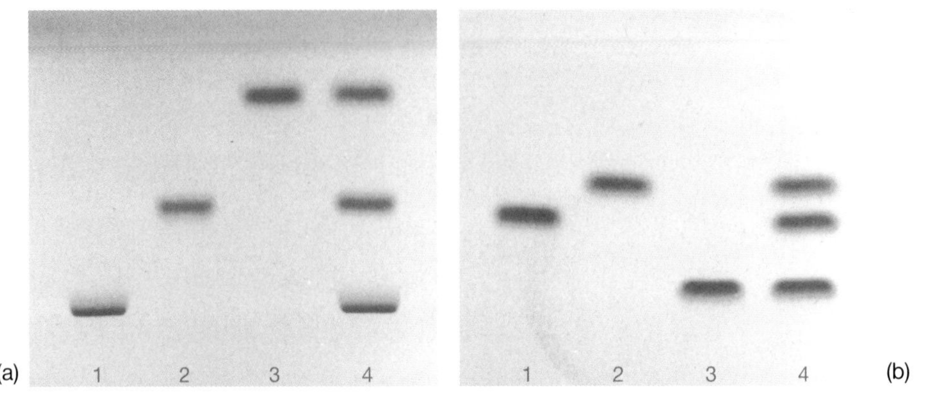

Figure 54. Identification of theophylline, theobromine and caffeine (shots taken in 254-nm UV light)
(a) Chromatography on TLC silica gel 60 F_{254} using CHC-containing SS according to the DAB
(b) Reversed-phase chromatography on RP-18 TLC foil with CHC-free SS

Lane 1 10 µg theophylline, *lane 2* 10 µg theobromine, *lane* 3 10 µg caffeine, *lane 4* 10 µg each of theophylline, theobromine and caffeine

---

Figure 58. Influence of the chamber atmosphere on the chromatographic results using the example of greater celandine samples (shots in 365-nm UV light)
(a) 10 × 10 cm plate, vertical development, unsaturated chamber
(b) 10 × 10 cm plate, vertical development, saturated chamber
(c) 10 × 10 cm plate, vertical development, S-chamber
(d) 5 × 5 cm plate, horizontal development, unsaturated chamber

Lane 1 Greater celandine, medicinal drug (dried), powdered, *lane 2*greater celandine leaves, *lane 3* greater celandine stem, *lane 4* greater celandine root, *lanes 2–4* one hour after harvesting the plants in July

The chromatographic conditions are given in Table 17

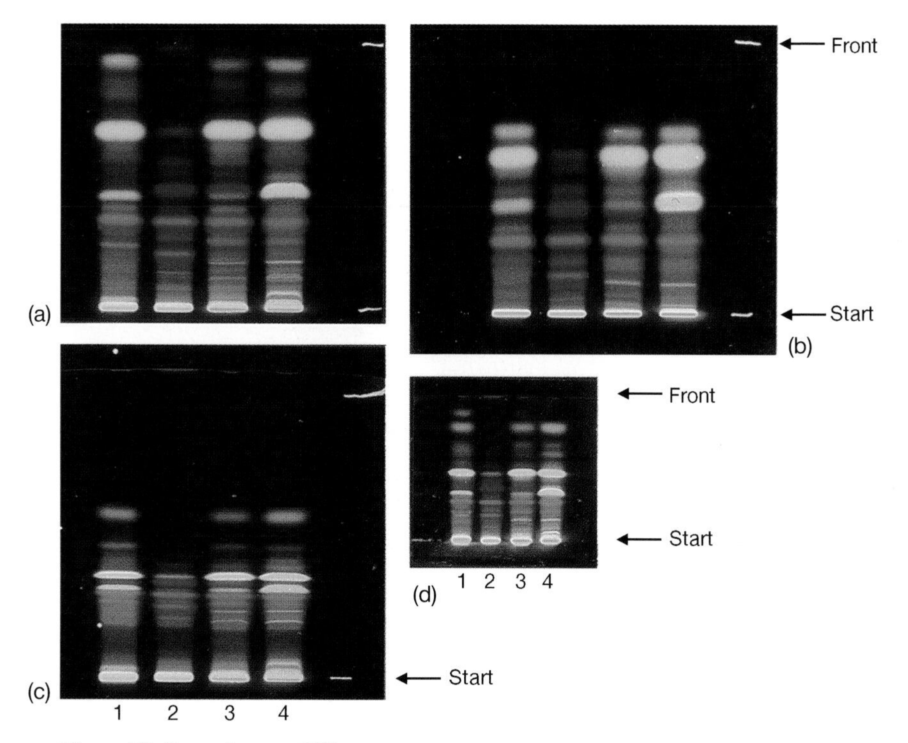

Figure 58 (legend see p. 290)

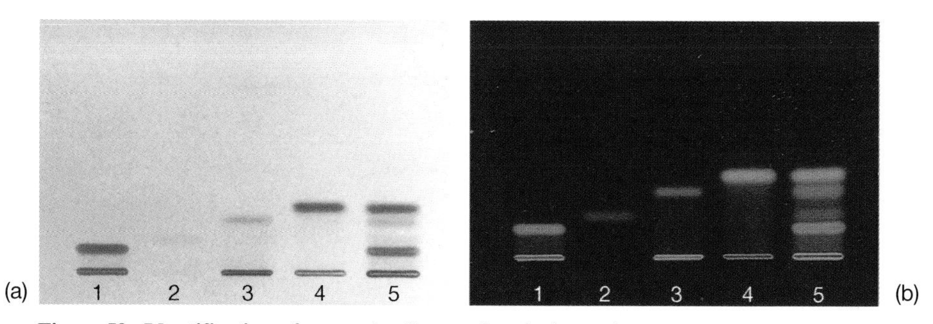

Figure 59. Identification of sugars by thermochemical reaction
Chromatography on HPTLC amino-modified silica gel 60 F_{254s}
(a) UV 254 nm after heating (15 min at 160 °C)
(b) UV 365 nm after heating

Lane 1 lactose, *lane 2* sucrose, *lane 3* glucose, *lane 4* fructose, *lane 5* mixture of all four sugars (top of pictures corresponds to ca. 67 % of the migration distance)

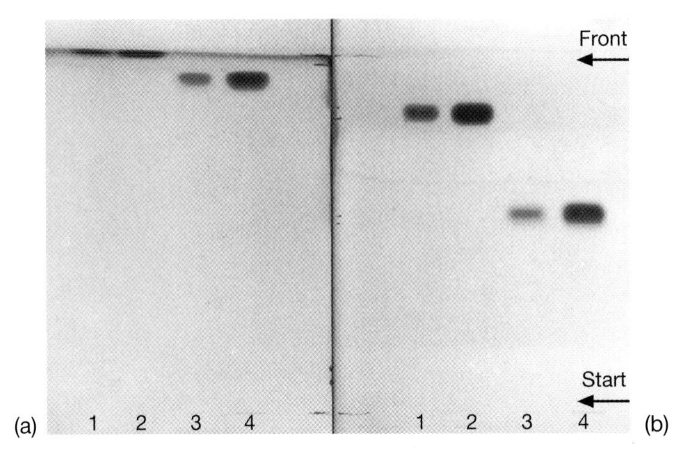

Figure 60. Influence of chamber saturation in the separation of spironolactone and furosemide (shots in 254-nm UV light)
(a) Without chamber saturation
(b) With chamber saturation

Lane 1 4 μg spironolactone, *lane 2* 20 μg spironolactone, *lane 3* 4 μg furosemide, *lane 4* 20 μg furosemide

The chromatographic conditions are given in Table 16

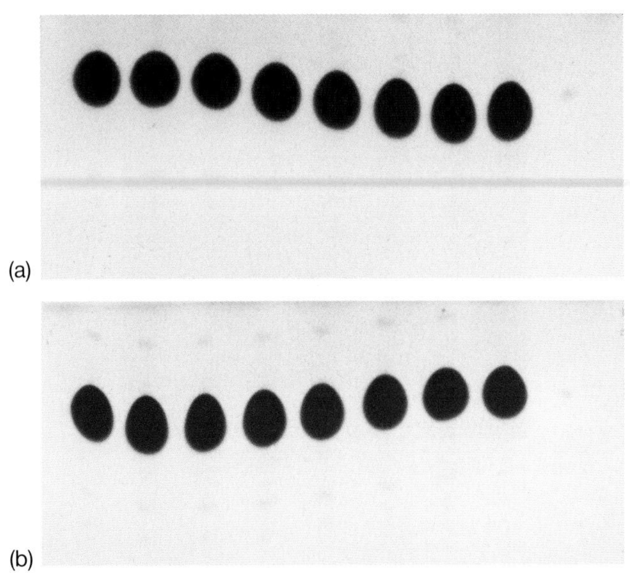

Figure 62. Influence of the presence of the continuous stream of air in a fume cupboard during development on the chromatographic result (shots in 254-nm UV light)
(a) and (b) Sections of two TLC plates, each with 8 × 500 μg of a pharmaceutically active substance per lane and the limit concentration of unknown impurities

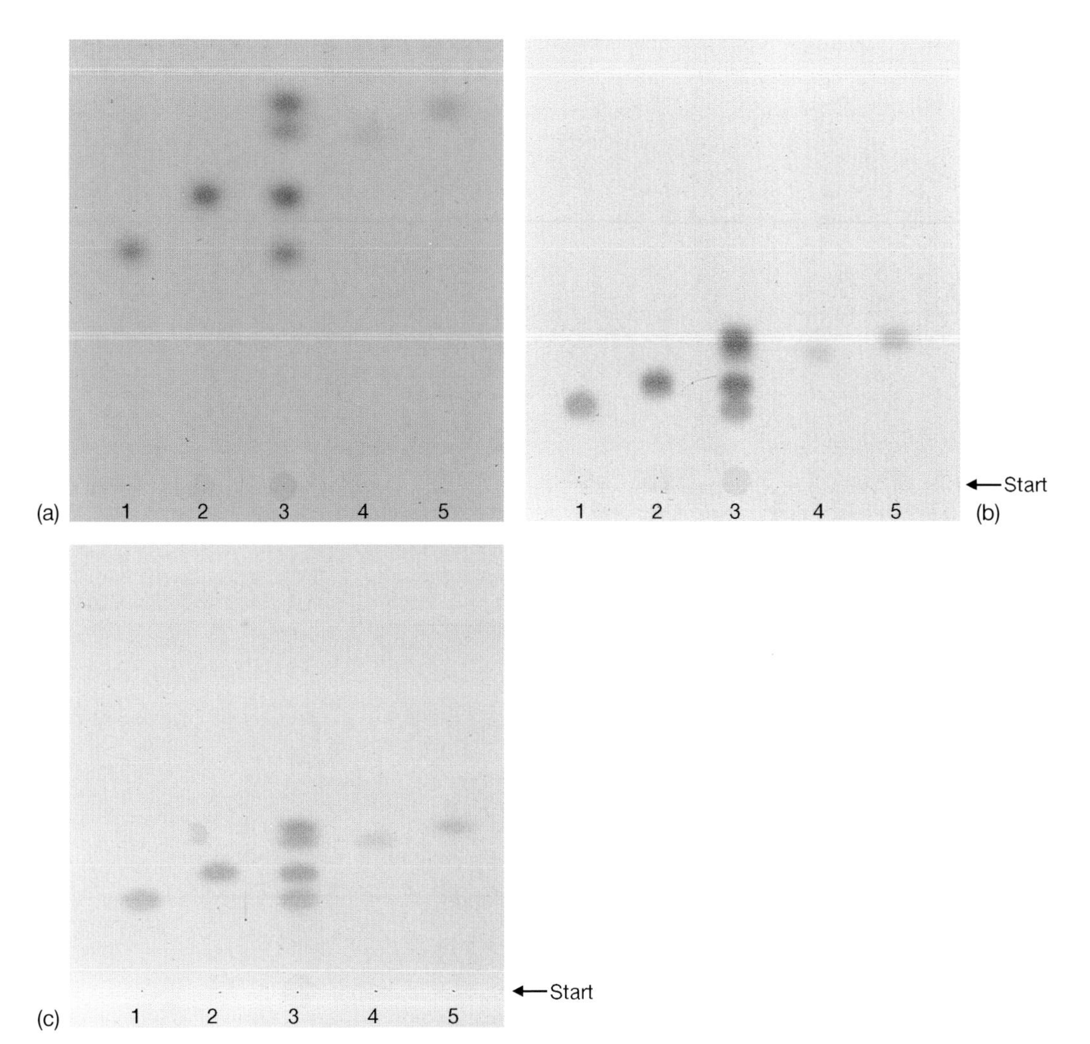

Figure 63. Identification of sugars after derivatization with the thymol-sulfuric acid reagent (shots in white light)

Chromatographed on TLC silica gel, 2×15 cm (a) and 1×10 cm (b), in the DAB solvent system (top of pictures = 50 % of the migration distance)
(c) Chromatographed on HPTLC silica gel 60, 2×7 cm, in the DAB solvent system (top of picture is the solvent front)

Lane 1 lactose, *lane 2* sucrose, *lane 3* mixture of the 4 sugars, *lane 4* glucose, *lane 5* fructose

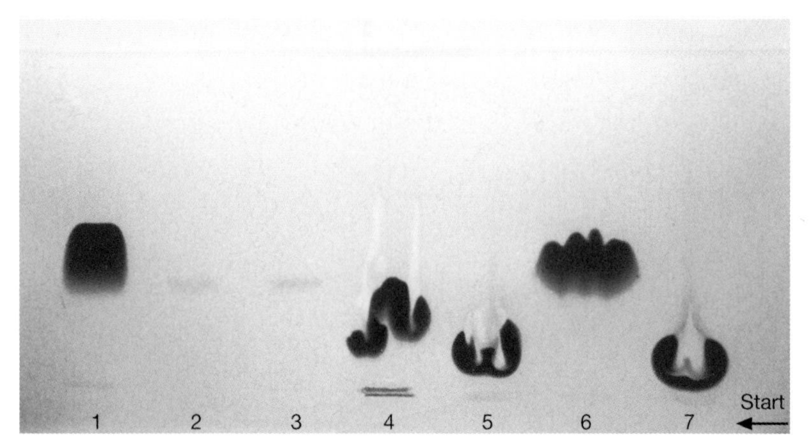

Figure 64. Influence of the tablet matrix substance on the chromatographic result after a single development in the purity test of a pharmaceutically active ingredient (shots in 254-nm UV light)

Lane 1 300 µg reference substance (RS) ≈ 100 %, *lane 2* 0.5 % RS, *lane 3* 1.0 % RS, *lanes 4–7* various storage conditions of the tablet samples used

(top of picture ≈ ca. 80 % of the migration distance).

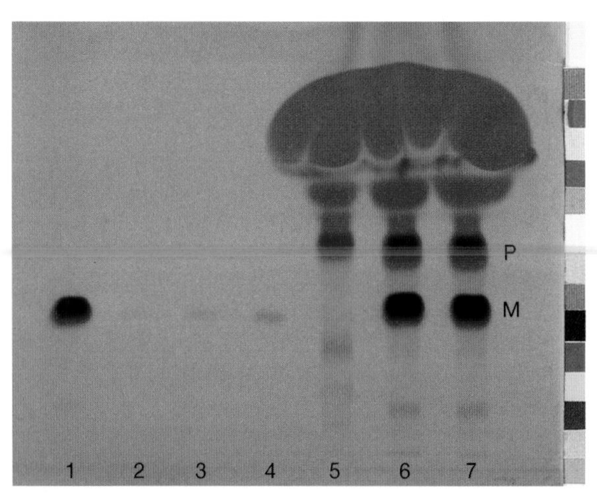

Figure 65. Repeated development with various solvent systems with decreasing migration distances (shots in white light)
Example: Metoclopramide base from Gastrosil® suppositories. The first solvent system, which is polar, displaces the suppository mass beyond the later solvent front of the main TLC run. A TLC plate after derivatization in the iodine chamber is shown.

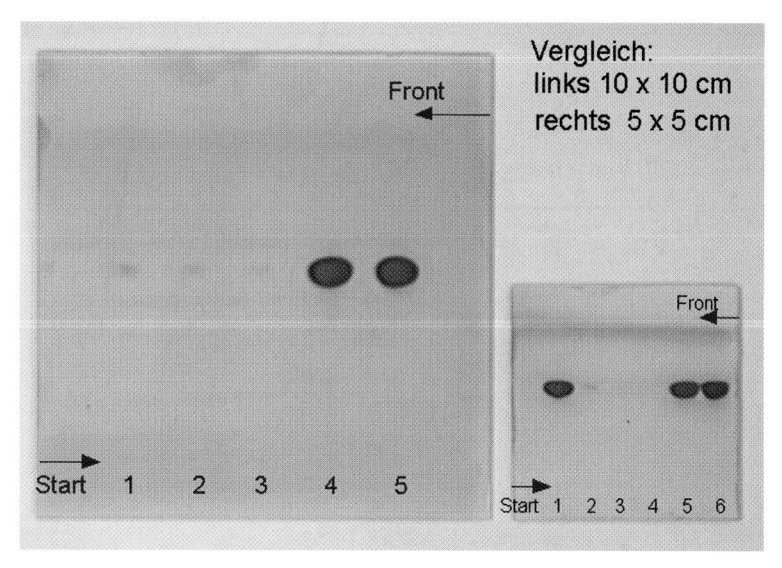

Figure 67. Purity test of cimetidine according to the DAB
(Shots taken in white light after derivatization)
The substance stress and the detection limit using a (vertical development) are compared with those using a 5 × 5 cm plate (horizontal development)

Sorbent: HPTLC precoated plate silica gel 60 F_{254}
Solvent system: Ethyl acetate + methanol + conc. ammonia solution (65 + 20 + 15 v/v)

10 × 10 cm plate:
Lane 1 0.2 µg ref. sub., *lane 2* 0.1 µg ref. sub., *lane 3* 0.05 µg ref. sub., *lane 4* 50 µg ref. sub., *lane 5* 50 µg sample substance

5 × 5 cm plate:
Lane 1 25 µg ref. sub., *lane 2* 0.1 µg ref. sub., *lane 3* 0.05 µg ref. sub., *lane 4* 0.025 µg ref. sub., *lanes 5 + 6* each 25 µg sample substance

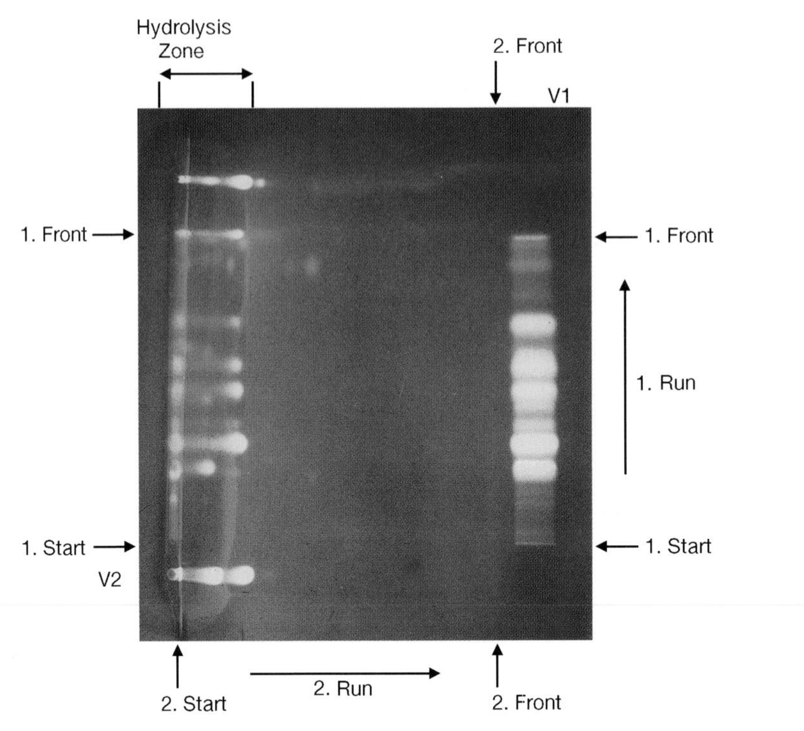

Figure 71. Two-dimensional TLC with intermediate substance hydrolysis using the example of birch leaves dry extract (DE) (shots taken in 365-nm UV light after derivatization with the flavone reagent according to Neu)
This TLC plate is discussed in Section 4.3.2 "Two-dimensional Thin-Layer Chromatography"

V 1 = Comparison 1 (birch leaves preparation, not hydrolyzed)
V 2 = Comparison 2 (aglyca, applied after hydrolysis of part of the plate)

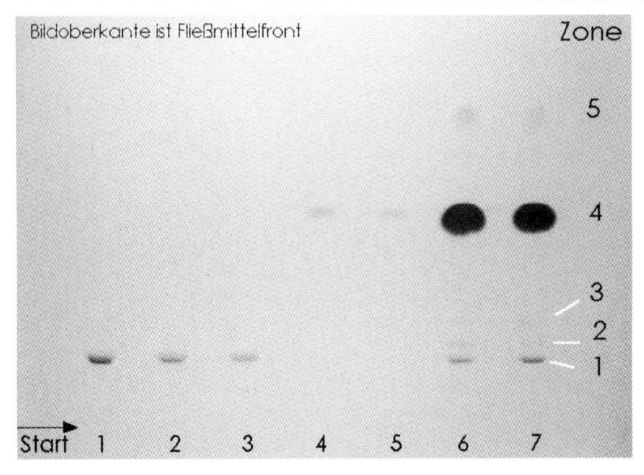

Figure 72. Purity test of chlortalidone according to the DAB (shot taken in 254-nm UV light)
This shot is also shown and described in Document 10 as a model chromatogram; the chromatographic parameters are also given in Document 11 (Section 9.4.1)

(a)

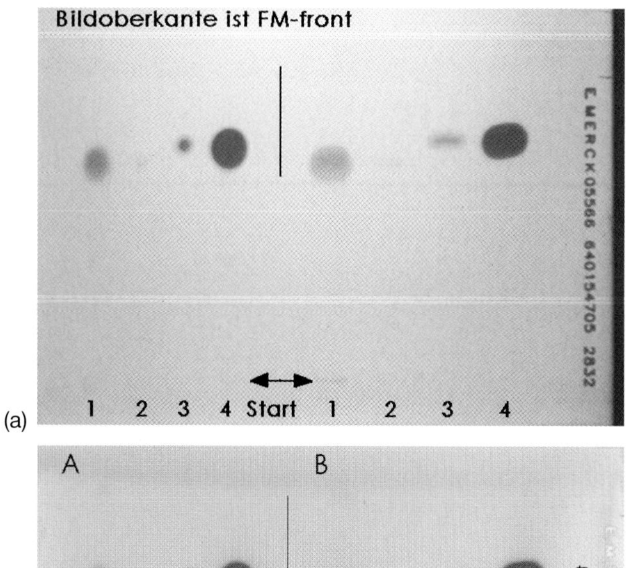

(b)

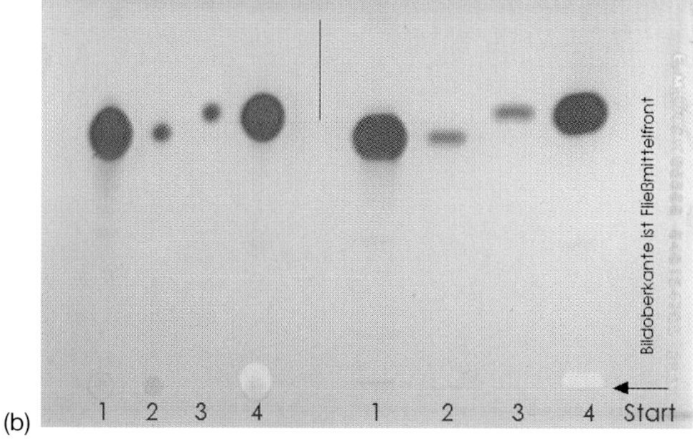

Figure 73. Evaluation of chromatograms of two active ingredients with different absorption properties in short-wave UV light

(a) Shot of the TLC plate in 254-nm UV light before the derivatization

(b) Shots in white light after treatment in the iodine vapor chamber for 16 h

Left-hand part of the plate: Pointwise applied sample solutions, 10 μl of each (2 × 5 μl)
Right-hand part of the plate: Bandwise (10 mm) applied sample solutions, 10 μl of each

Substance with weak fluorescence quenching: metoprolol tartrate
Substance with strong fluorescence quenching: metoclopramide HCl

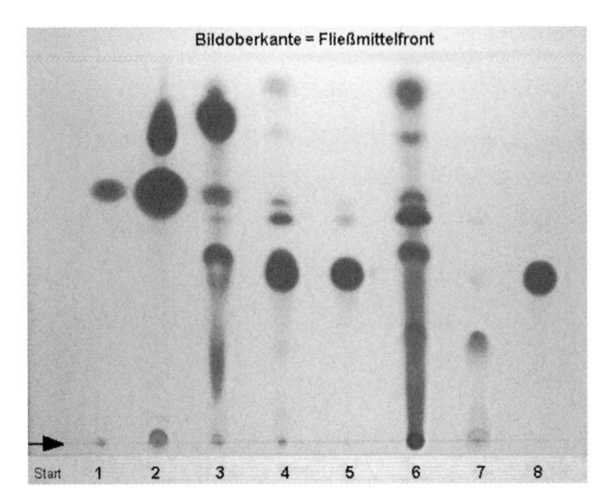

Figure 74. In-process control in the synthesis of a pharmaceutically active ingredient using TLC (shot in 254-nm UV light)
Lane 1 starting substance 1, *lane 2* mixture of starting substances 1 and 2, *lane 3* in-process control, *lane 4* end product, *lane 5* purified end product, *lane 6* mother liquor, *lane 7* wash water, *lane 8* reference substance (end product)

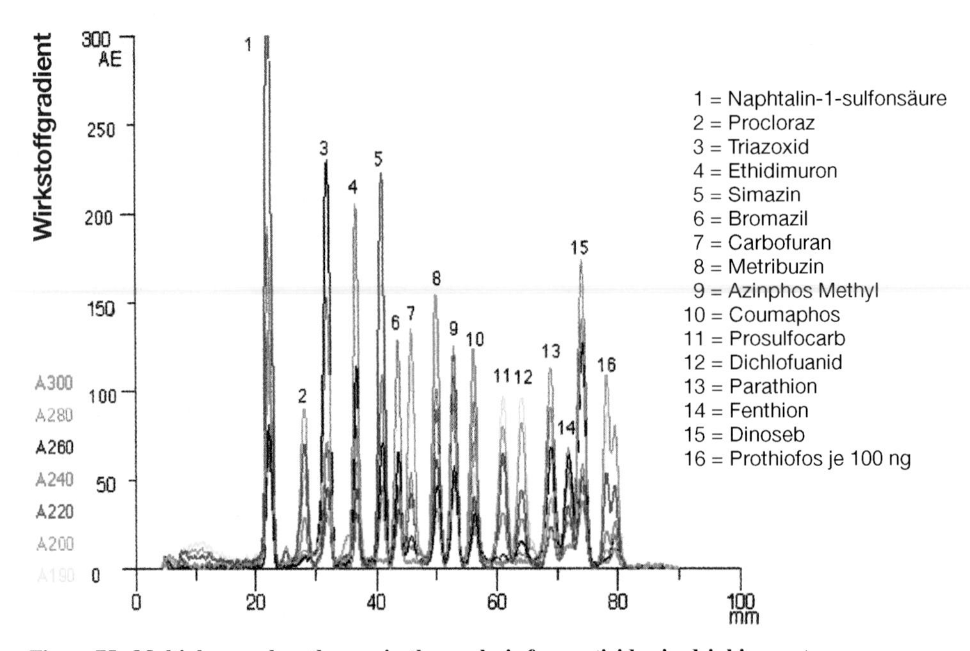

Figure 75. Multiple-wavelength scan in the analysis for pesticides in drinking water
Separation of the pesticides by AMD; measurement of the chromatograms successively at wavelengths 190 nm, 200 nm, 220 nm, 240 nm, 260 nm, 280 nm and 300 nm

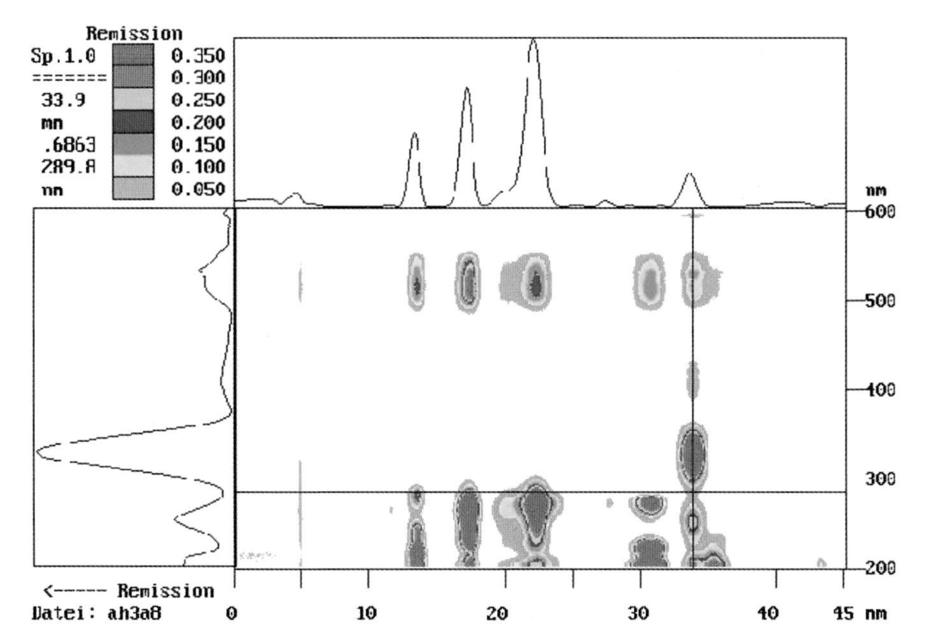

Figure 77. A *Kubelka-Munk* contour-plot, measured with a diode-array device. The sample-peak at 33.9 mm separation distance can be identified as flupirtine, a centrally acting non-opioid analgesic substance. Flupirtine shows absorptions between 200 and 350 nm and an additional fluorescence quenching signal between 500 and 530 nm. The weak signal at 400 nm is the flupirtine fluorescence spectrum which would be invisible if evaluated using equation (2). A transformation of the raw-data using equation (3) would show this signal only to reveal it as a fluorescence emission.

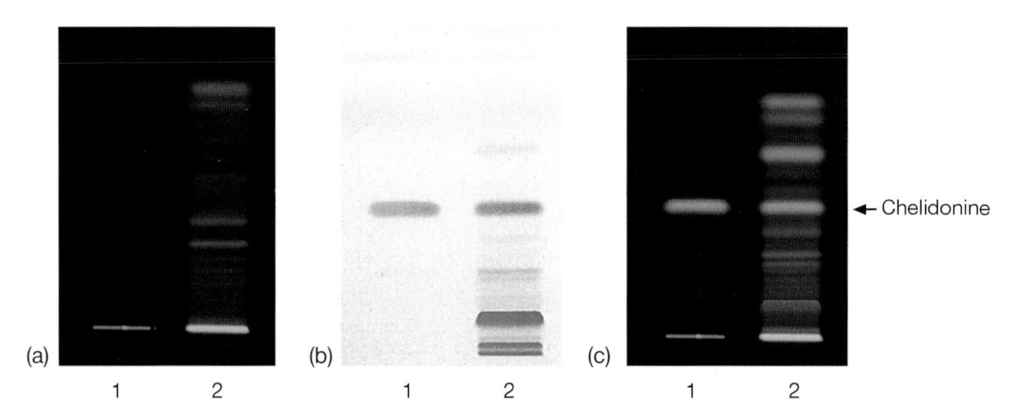

Figure 78. Reaction of chelidonine and other substances present in greater celandine DE with high-energy light
(a) Chromatograms after development in the dark before exposure to light
(b) Detection of UV-absorbent zones in short-wave UV light by fluorescence quenching
(c) Detection of photochemically activated chromatogram zones after irradiation of the plate with short-wave UV light

Lane 1 chelidonine, *lane 2* greater celandine DE

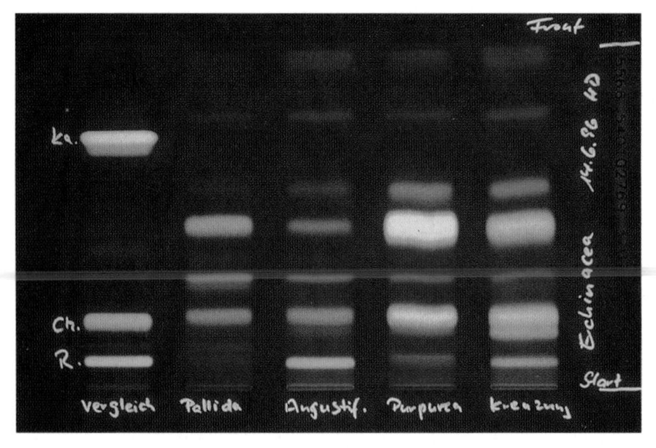

Figure 84. Identification of three species of coneflower (Echinacea) and detection of a hybrid of two of these after spraying with the flavone reagent according to Neu (shots in 365-nm UV light). All the plant material is from authentic cultivation in Middle Franconia (northern Bavaria), and the chromatography was performed 40 hours after harvesting (by the author) in June 1996; only above-ground parts of plants which were not in flower were investigated. All the parameters of the TLC are listed in Table 19.

Lane 1 reference substances with ascending hRf values: rutoside, chlorogenic acid, caffeic acid, *lane 2 E. pallida, lane 3 E. angustifolia, lane 4 E. purpurea, lane 5* hybrid between *E. angustifolia* and *E. purpurea.*

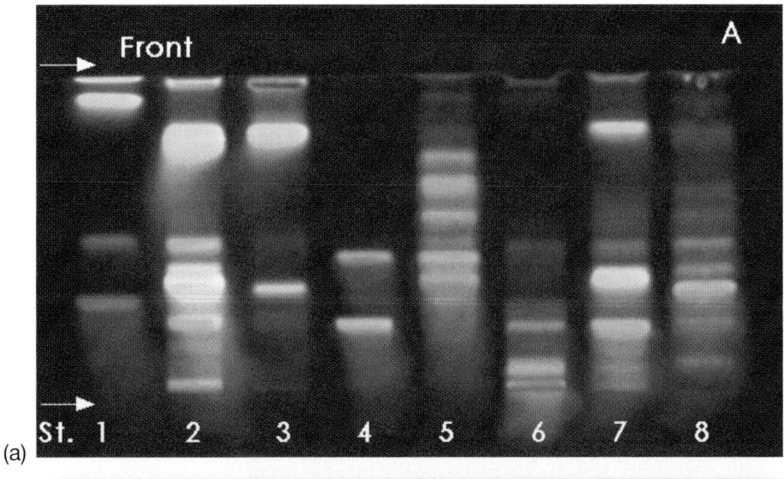

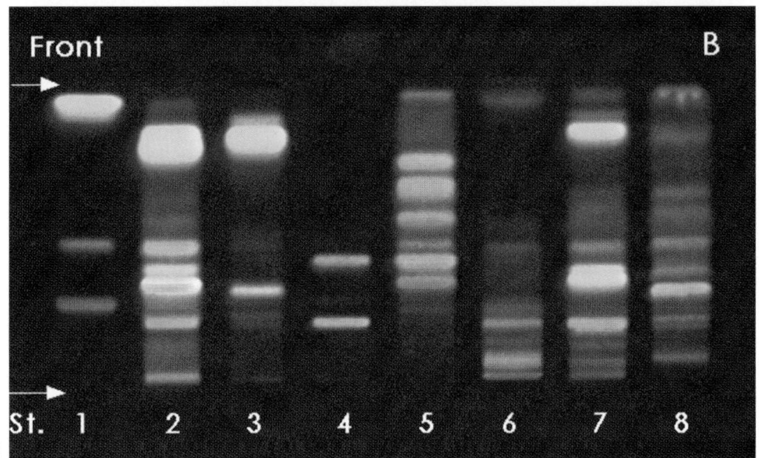

Figure 86. Comparison of a dipped plate with a sprayed plate (shots in 365-nm UV light after derivatization with the flavone reagent according to Neu)
(a) Plate derivatized by dipping
(b) Plate derivatized by spraying

Reference substances with ascending hRf values ($x - y$):
Lane 1 camphorol 3-rhamnosidoglucoside (23–27), luteolin 7-glucoside (43–46), caffeic acid (88–93), *lane 2* artichoke leaves, *lane 3* red coneflower (*E. purpurea*), *lane 4* rutoside, hyperoside, *lane 5* birch leaves (*Betulae folium*), *lane 6* primrose flower (*Primula flos*), *lane 7* elderflower (*Sambuci flos*), *lane 8* maidenhair tree leaves (*Ginkgo biloba*)

(a)

(b)

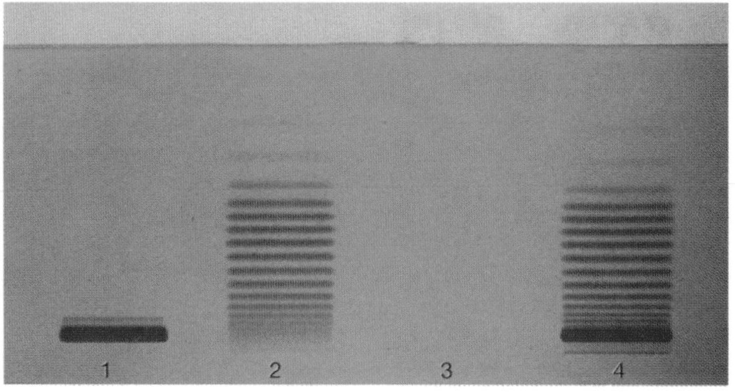

(c)

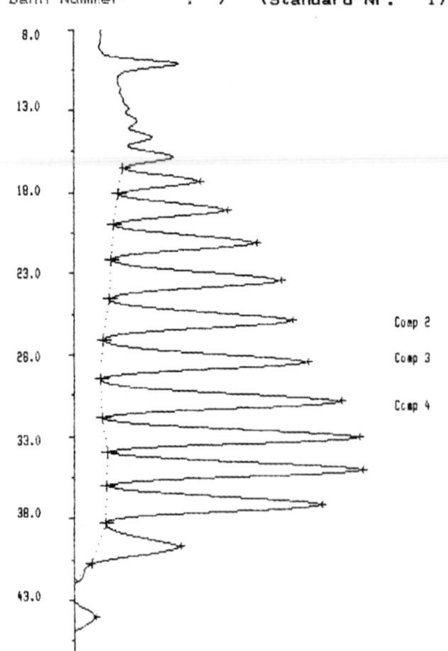

Figure 87. Identification of metoclopramide base and polidocanol in a suppository mass after dipping in a modified Dragendorff reagent
(shots taken in daylight at the window)

(a) Predevelopment in chloroform; development in SS 1: chloroform + methanol + ammonia solution (80 + 20 + 1.2 v/v)

(b) Development in SS 2: methyl ethyl ketone + water + acetone (40 ml MEK + 10 ml water are mixed together, and 30 ml of the upper phase is then mixed with 3 ml acetone)
Lane 1 metoclopramide base, *lane 2* polydocanol, *lane 3* witepsol (suppository base), *lane 4* suppository (dipping height is not the SS front)

(c) Direct optical evaluation of polidocanol after developing in SS 2 and derivatization with the modified Dragendorff reagent, measured with TLC Scanner CD 60 (DESAGA) at 510 nm

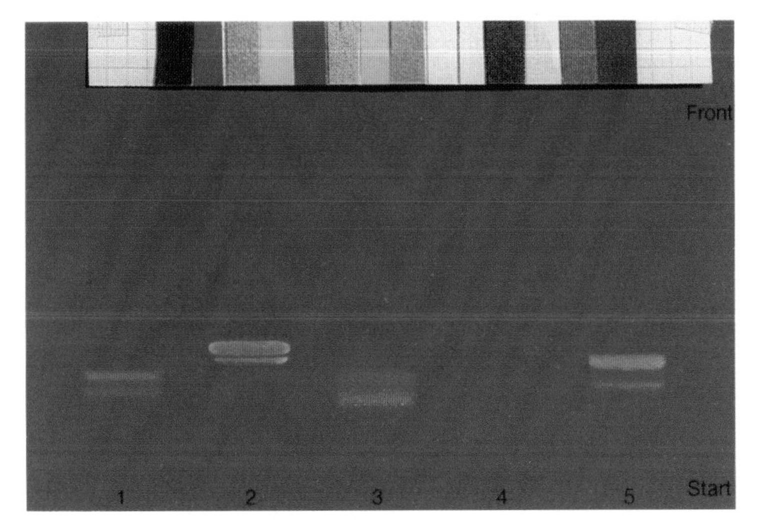

Figure 89. Identification of hemolyzing plant constituents after derivatization with the blood-gelatin reagent (shot taken in white light)

Lane 1 saponin (reference substance)
Lanes 2–5 are chromatograms of medicinal products prepared from the following plants:
Lane 2 early golden rod herb, *lane 3* soapwort herb, *lane 4* horsetail herb,
lane 5 primula root

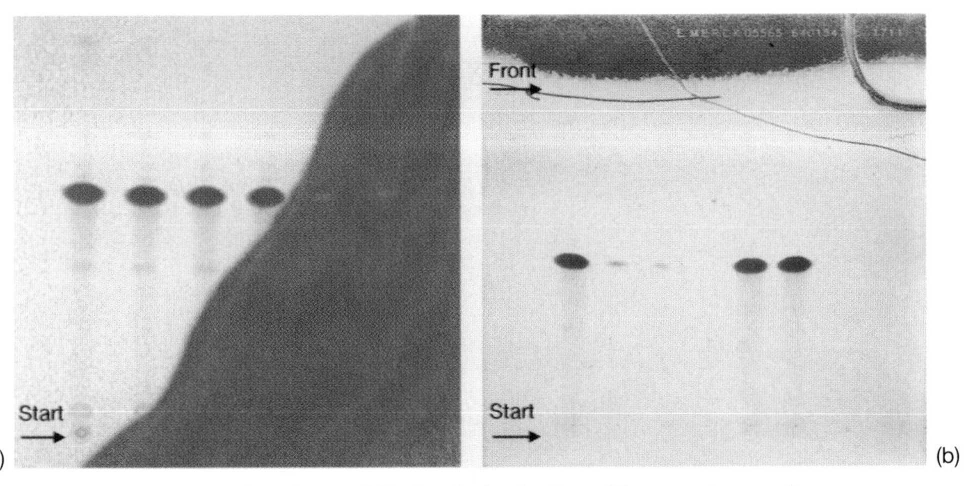

Figure 90. Purity test of ambroxol HCl after derivatization with successive reactions

(a) Plate with no line of sorbent scraped off to isolate a corner of the plate: hence immediate darkening (white light). Top of picture is SS front

(b) Plate with quarter circle of sorbent scraped off, after standing for a long time. The darkening has been stopped by scraping off another line of sorbent (flash photograph)

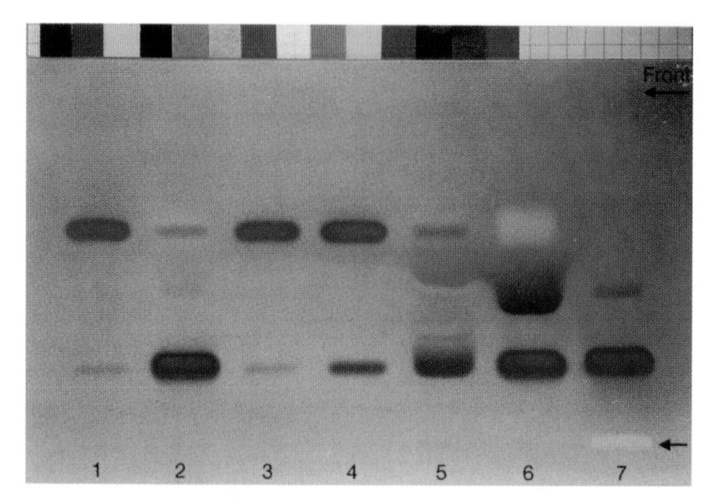

Figure 92. Derivatization of stressed dexpanthenol ointment samples with the 2,5-dimethoxy-tetrahydrofuran/4-(dimethylamino)-benzaldehyde reagent (shot taken in white light)
The completed raw data sheet "TLC III" for this chromatogram can be seen in Document 7 (Section 9.3)

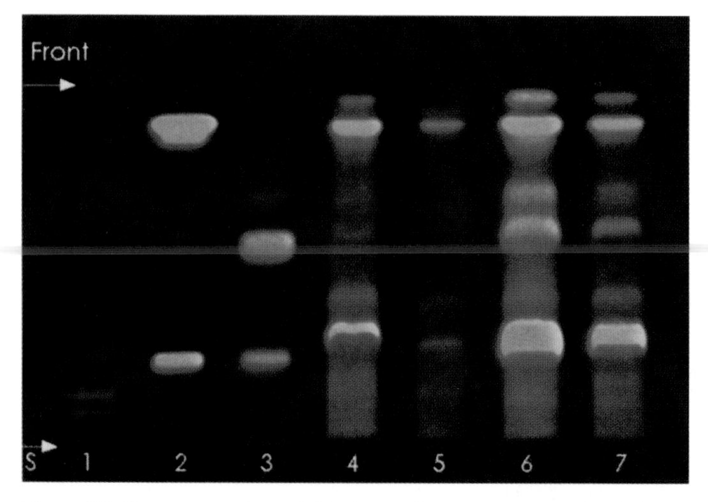

Figure 98. Documentation of thin-layer chromatograms with a video system (in contrast to a simple description of the chromatogram)
Investigation of juices expressed from *Echinaceae purpurea*
The parameters for the TLC are given in Table 19; the alternative system was used here

Reference substances with ascending hRf values: *lane 1* rutoside, camphorol 3-rhamnosido-glucoside, hyperoside, *lane 2* chlorogenic acid, caffeic acid, *lane 3* the crude drug *Echinaceae purpurea, lanes 4–7* various expressed juices of the same plant.

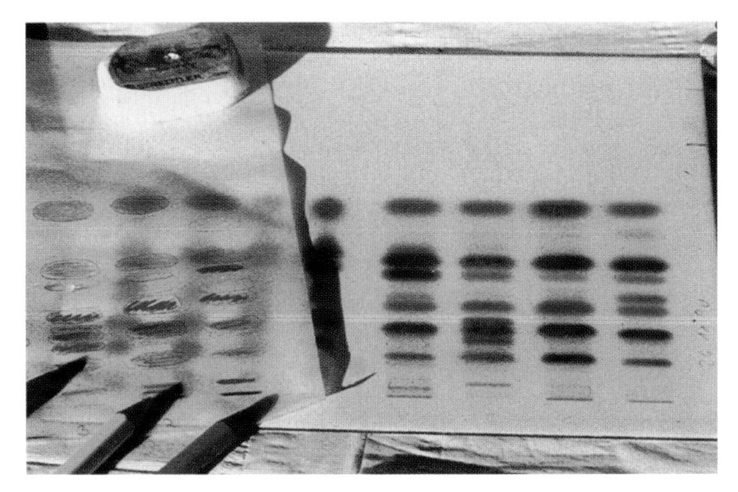

Figure 99. Documentation of thin-layer chromatograms using colored pencils on transparent paper

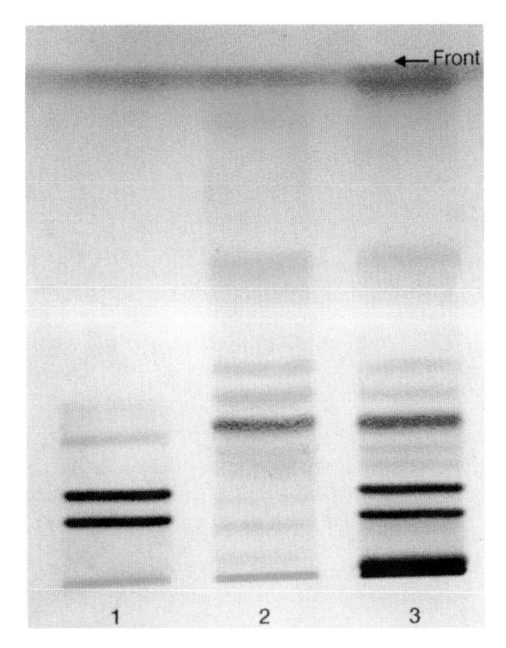

Figure 100. Identification of primula root DE and liquorice root DE in Heumann bronchial herbal tea Solubifix®
The method used to produce this photograph is described in Section 8.3.2.2
Lane 1 primula root DE, *lane 2* liquorice root DE, *lane 3* bronchial herbal tea

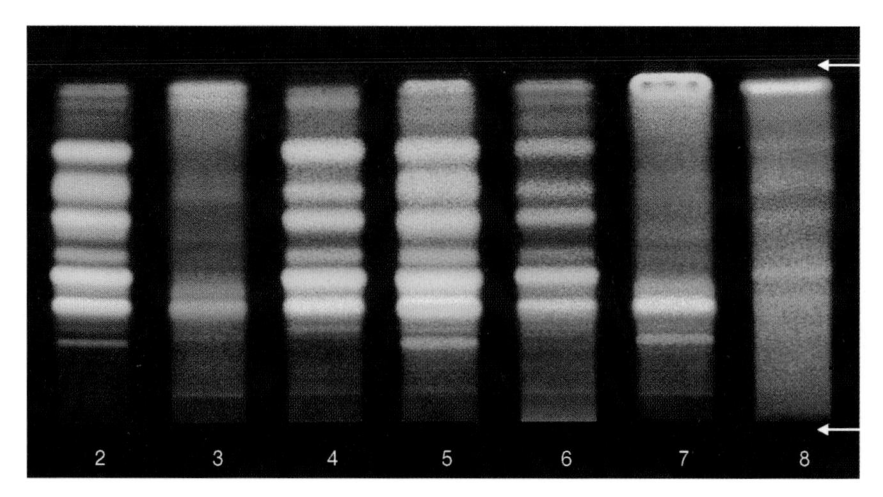

Figure 105. Controlled substance stress using birch leaves DE as the example (shots taken in 365-nm UV light after derivatization with the flavone reagent according to Neu)
The legend for this figure is given in Document 12, p. 6 (Section 9.4.2). *Lane 1*, which contains the reference substances, is not included for optical reasons

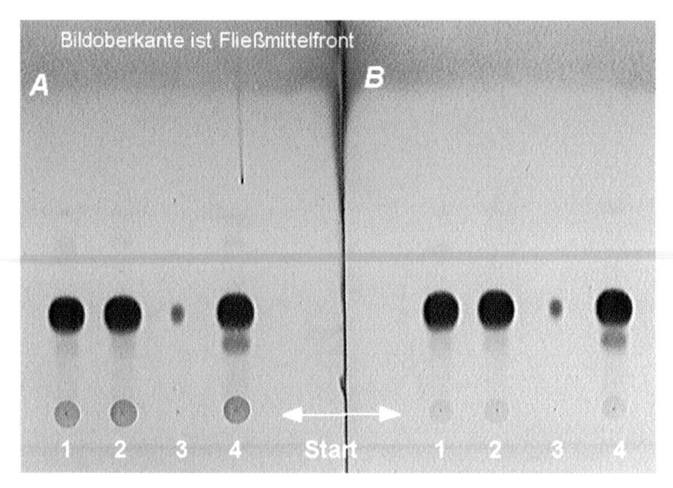

Figure 106. Influence of the action of light on a dissolved substance and on a substance already applied to a TLC plate (shot taken in white light after derivatization with the phosphomolybdate reagent)

Substance:	Ursodeoxycholic acid
Sorbent:	TLC silica gel 60
TLC system:	In accordance with the DAC
SS:	Chloroform + acetone + acetic acid (70 + 20 +4 v/v)
Chamber:	N-chamber with CS

Lane 1 100 µg from the solution stored overnight next to a window, *lane 2* 100 µg freshly prepared solution of a working standard, *lane 3* 0.5 µg as lane 2, *lane 4* 100 µg of a new batch for testing

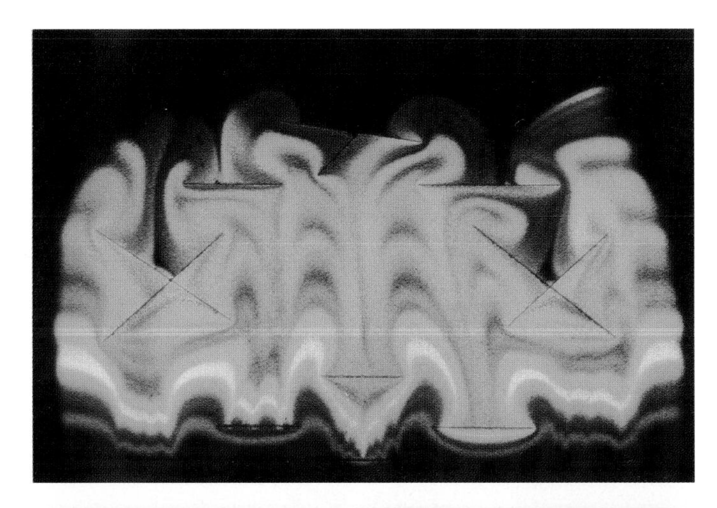

King of the Lions

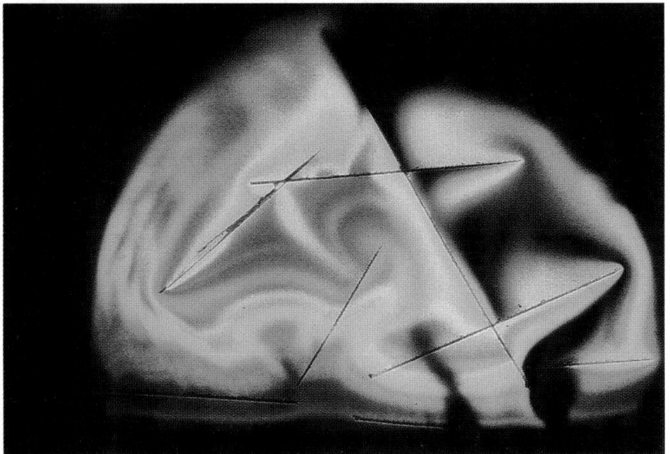

Destruction of the Earth

Long Legs

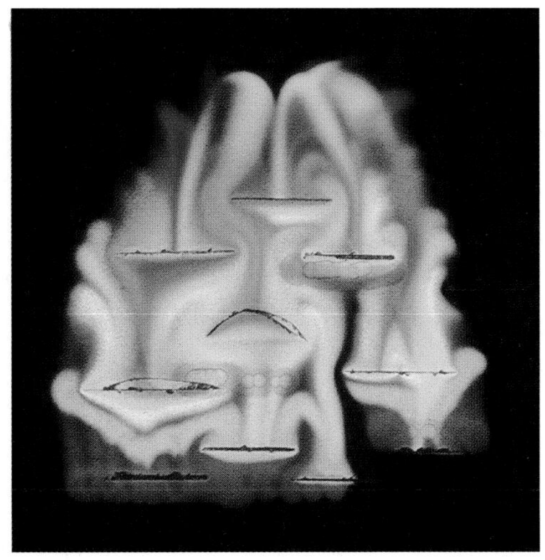

Fountain

I like TLC

Index

Applied Thin-Layer Chromatography: Best Practice and Avoidance of Mistakes, 2nd Edition
Edited by Elke Hahn-Deinstrop

ISBN: 978-3-527-31553-6